cyborgs &
CITADELS

Publication of the Advanced Seminar Series is made possible by generous support
from the Brown Foundation, Inc., of Houston, Texas.

SCHOOL OF AMERICAN RESEARCH ADVANCED SEMINAR SERIES

DOUGLAS W. SCHWARTZ, General Editor

CYBORGS & CITADELS / Contributors

GARY LEE DOWNEY / Center for Science and Technology Studies
Virginia Tech

JOSEPH DUMIT / Department of Social Medicine
Harvard Medical School

DONNA HARAWAY / History of Consciousness Program
University of California, Santa Cruz

DEBORAH HEATH / Department of Sociology and Anthropology
Lewis and Clark College

DAVID J. HESS / Department of Science and Technology Studies
Rensselaer Polytechnic Institute

FREDERICK KLEMMER / Department of Political Science
University of Hawaii

JUAN C. LUCENA / Science, Technology, and Globalization Program
Embry-Riddle University

EMILY MARTIN / Department of Anthropology
Princeton University

LAURY OAKS / Department of Anthropology
Johns Hopkins University

PAUL RABINOW / Department of Anthropology
University of California, Berkeley

RAYNA RAPP / Department of Anthropology
New School for Social Research

KAREN-SUE TAUSSIG / Department of Social Medicine
Harvard Medical School

SHARON TRAWEEK / Department of History
University of California, Los Angeles

ARIANE VAN DER STRATEN / Johns Hopkins University

SARAH WILLIAMS/ Women's Studies Program
Evergreen College

cyborgs &
CITADELS

Anthropological Interventions in Emerging Sciences and

Technologies / **Edited by Gary Lee Downey and Joseph Dumit**

School of American Research Press / *Santa Fe, New Mexico*

School of American Research Press
Post Office Box 2188
Santa Fe, New Mexico 87504-2188

Director of Publications: Joan K. O'Donnell
Editor: Jo Ann Baldinger
Art Director: Deborah Flynn Post
Indexer: Douglas J. Easton
Typesetter: G&S Typesetters
Printer: Thomson-Shore, Inc.

Distributed by the University of Washington Press

Library of Congress Cataloging-in-Publication Data
Cyborgs & citadels : anthropological interventions in emerging
sciences and technologies / Gary Lee Downey and Joseph Dumit., eds.
p.   cm. — (School of American Research advanced seminar series)
Includes bibliographical references and index.
ISBN 0-933452-96-9 (cloth). — ISBN 0-933452-97-7 (pbk.)
1. Science—Social aspects. 2. Technology—Social aspects.
3. Medicine—Social aspects. 4. Cyborgs. 5. Ethnology.
I. Downey, Gary Lee. II. Dumit, Joseph. III. Series.
Q175.5.C93   1997
306.4'5—dc21   97-20296
                                                        CIP

Cover illustration by Deborah Flynn Post.

CONTENTS

cyborgs &
CITADELS

123456789
1011
1011

PREFACE

T HIS VOLUME JOINS A COLLECTIVE EFFORT among anthropologists and
other researchers in science and technology studies to intervene in
emerging sciences and technologies through ethnographic approaches
that accept necessary participation while insisting on critical distance.
Ethnographic research can intervene in dominant images of science and technology by helping people recognize what established interpretations might overlook, ignore, or make less visible. Theoretical insights from different perspectives
can motivate people to reexamine their own interpretations of what they do
without necessarily demanding wholesale rejection or replacement of existing
conceptions. Ethnographic research on science and technology can thus make
a difference by calling attention to important but undervalued connections in
our worlds of technoscience, as well as by sometimes offering new relations.

The week-long seminar on "Cyborg Anthropology" held at the School of
American Research in October 1993 had its beginning in years of work among
a committed set of anthropologists who wanted to locate the roles science and
technology play in everyday lives, including their own. In particular, Emily
Martin's *The Woman in the Body* (1987) and Sharon Traweek's *Beamtimes and Lifetimes* (1988) remain basic texts in this nascent field. Sharon was a co-organizer
of this seminar, and her attention to the particulars of everyday interaction and
the cultures of academic worlds in conversation and writing has been a constant
inspiration. Her input was key to our structuring of the seminar itself as an intervention in the dominant practices of academic argument and discussion.

Alongside the contributors, other anthropologists to whose writings, organizational initiatives, and corridor work the seminar is indebted include Susan Ir-
win Anderson, Marietta Baba, Adele Clarke, Sam Coleman, Robbie Davis-Floyd,
Frank Dubinskas, Michael Fischer, Diana Forsythe, Sarah Franklin, Joan Fujimura, Shirley Gorenstein, Hugh Gusterson, David Hakken, Willett Kempton,

Barbara Koenig, Louise Krasniewicz, Jean Lave, Linda Layne, Henry Lundsgaarde, George Marcus, Linda May, Laura Nader, James Nyce, Julian Orr, Constance Perin, Bryan Pfaffenberger, Susan Leigh Star, Allucquere Rosanne Stone, Lucy Suchman, Christopher Toumey, Priscilla Weeks, and Stacia Zabusky. We offer special homage to Frank Dubinskas (1943–1993), whose work bridging the academic gaps between anthropology, science, and business remains visionary. Since the seminar, the list of anthropologists making significant contributions to the study of science and technology has grown ever larger. We thank each of you for your continued critical collegiality.

This book would not have made it to publication without the careful attention of many people. In particular, we thank George Marcus, whose perceptive review of the manuscript articulated the important goal of theorizing and practicing collaborations that do not become cloying. We hope the published version has taken some steps in this direction. We also thank the staff of the School of American Research, without whose conceptual and material contributions this volume would not exist. Thanks to Doug Schwartz, president, for inviting us to organize the seminar and for positioning SAR as a key facilitator of anthropological research; Duane Anderson, vice president, for supervising the seminar and helping us understand SAR history and contemporary practices; Cecile Stein, academic programs coordinator, for getting us all to Santa Fe; and Sarah Wimett, Jennifer McLaughlin, and Sarah Sandoval for preparing meals of a quality that was reason enough to spend a week at the SAR Seminar House. Thanks to Joan O'Donnell, SAR Press director, whose close reading and critical review of the manuscript became important gifts to every contributor; Jo Ann Baldinger, whose excellent copyediting was above and beyond the call of duty; and Baylor Chapman, marketing coordinator, who worked to make sure the book found its way into your hands.

We are also grateful for the constant support and encouragement of our families and friends, not least because these trusting relationships offer alternative models to the agonistic politics that still tend to dominate academic interactions. We dedicate this volume to these individuals and other siblings, biological and otherwise, with whom we share common challenges in the midst of our differences.

1234567891011

LOCATING AND INTERVENING / An Introduction

Gary Lee Downey and Joseph Dumit

> In dreams begins responsibility.
> —William Butler Yeats

W E ALL LIVE WITH A CONCRETE AWARENESS that we cannot say No to science, technology, and medicine. Even if we wanted to, we cannot say No to the medical complex that appropriates our bodies, defines our state of health, and positions us in a continuum of fitness from the temporarily abled to the permanently disabled. We cannot say No to the corporate/government information complex that wires our social security numbers, driver's licenses, bank accounts, credit ratings, tax returns, telephones, radios, televisions, electronic mail, and other technological vectors of identity. We cannot say No to the experience of science, technology, and medicine collectively as a disciplining center that polices other meanings and orders power relations in contemporary life.

But how can we go about understanding and taking account of these deep and abiding presences in our bodies, our persons, our selves? Furthermore, how are we to understand our often intense hunger to say Yes?

This volume contributes to a diverse and rapidly expanding set of anthropological projects that are seeking new ways of locating and intervening in emerging sciences, technologies, and medicines through cultural perspectives and ethnographic fieldwork. It is one product of a weeklong seminar held in October 1993 at the School of American Research in Santa Fe, New Mexico. Participants came together to map research questions, explore the extent to which we shared problems, practices, and objectives, and sort out some of the opportunities, limitations, and commitments in our work.[1] Because emergent relations in science, technology, and medicine often appear both haltingly strange and seductively familiar, every participant in the seminar arrived wanting help in exploring these elusive mechanisms of emergence. We still do. In this volume, we offer some ways of thinking through in cultural terms how science, technology, and medicine participate in everyday life. We also position our own career and research trajectories as ethnographic participants in the processes we study. At the same time, we continue to seek help in figuring out what we are doing, could be doing, and should be doing.

The main images in the title of this volume, "Cyborgs" and "Citadels," point to two related areas of questioning that concerned us throughout the seminar week. We devoted considerable time to unpacking what seminar participants came to call the Citadel Problem. The Citadel Problem is a problem of cultural boundaries: it calls attention to the centering effects of science, technology, and medicine within discourses of objectivity and practices of both legitimation and sovereignty. The word "citadel" denotes a small fortified city or a fortress at the center of a larger city that protects and oversees it. We use the term to highlight the ways in which prevailing modes of popular theorizing about science, technology, and medicine displace societal issues and concerns into expert and often expensive technical problems, thereby isolating participation and discussion while transforming the stakes involved.

One effect of the Citadel Problem is that science often appears as a culture of no culture (Traweek 1988:162). That is, what Bryan Pfaffenberger (1992) has called the "Basic Story" of science and technology regularly treats the two as developing according to autonomous logics apart from society. In this model, researchers are characterized as living in specialized technical communities whose deliberations are essentially opaque and presumably free of cultural content. This is also known as the diffusion model of knowledge in society (cf. Latour 1987; Martin 1987) in which knowledge, in the singular, is created by bright, well-trained people located inside the academy and then diffuses outside into the public arena through mechanisms of education, popularization, policy, and the impacts of new technologies. The tests of cultural significance for new knowledge occur "out there" in the public arena as it is used, abused, or ignored. The outward travel of knowledge preserves the autonomy of creation and separates creators from accountability for their products, even as these creators intervene, exist within, and make demands upon the public. In Johannes Fabian's terms (1983), we laypersons tend to understand Western science and medicine allochronically as existing in our future because they are the central source of new meanings, while we locate in the past those peoples who are far away in space, repositories of old meanings, and hence primitive (cf. Harding 1993).

Whether or not something is called a fact makes a great deal of difference to us. Statements that begin, "The fact of the matter is, . . ." lay claim to an important source of authority. Even when produced under the banner of "for our own good" (Ehrenreich and English 1973), one effect is to inscribe a boundary between those who achieve the authority to speak new truths and those who become card-carrying listeners (Gieryn 1983). Claims to knowledge that fall inside a citadel can gain status, privilege, access to resources and authoritative lines of descent, and the possibility of becoming seated as permanent facts. Claims that fall outside may have to struggle in a nether world of questionable legitimacy, marginal position, subsistence economy, and risk of punishment for acts of deviance.

For contributors to this volume, the Citadel Problem is not only about building and maintaining walls but also about flows of metaphors over, around, and through these walls, as well as connections between lives inside and lives outside. By unpacking the Citadel Problem in cultural terms, we hope to understand better how science gains and keeps the authority to direct truth practices and constitute power relations. We also note that the Citadel Problem remains visible and important even as the "hard" sciences and the dominant medicines increasingly come under fire themselves, as when physicists must cope with the cancellation of the superconducting supercollider and specialist physicians must struggle with the growing hegemony of managed care.

The image of "Cyborgs" is designed to call attention to ways in which science, technology, and medicine routinely contribute to the fashioning of selves. The cyborg concept originated in Cold War space research and science fiction to refer to symbiotic forms of life that involve both humans and machines. In "Manifesto for Cyborgs," now a citation classic, Donna Haraway (1985) claimed the cyborg as a feminist icon for identifying new opportunities for analysis and activism in an emerging blend of technoscience and multinational capitalism she calls in this volume the "New World Order, Inc." As hybrid creatures, Haraway pointed out, cyborgs refuse easy origin stories as well as discourses of purity and naturalism, insisting instead on more complicated accounts of the production and mixing of human and nonhuman agencies. Her challenge involved calling attention to dangers in the New World Order, Inc., while imagining how the future might be otherwise, an imagining that appeared less possible with simpler stories of bodily resistance to oppressive technology. That is, might it be possible to formulate new strategies for improving the conditions of humans that accepted mutual figurations of human and machine rather than necessarily premising authentic human existence upon a principled and permanent separation?

The SAR seminar took place less than a year after a double session at the 1992 meeting of the American Anthropological Association (AAA) titled "Cyborg Anthropology I: The Production of Humanity" and "Cyborg Anthropology II: The Empowerment of Technology," followed by an author-meets-critics session with Haraway.[2] One goal of these AAA panels, including the use of the title "Cyborg Anthropology," was to stimulate greater interest among anthropologists in studying emerging sciences, technologies, and medicines, for anthropology in the United States has been rather slow to embrace science studies. As recently as 1987 and 1988, for example, the AAA rejected sessions jointly proposed by Gary Downey and Sharon Traweek on the anthropology of science and technology on the grounds that such work did not fit under the AAA umbrella.[3] The 1992 panels were indeed successful, attracting standing-room-only audiences in a ballroom setting.

Along with Sarah Williams, a third seminar co-organizer, we speculated that one way to encourage expanded anthropological inquiry in this area might be to call attention to the human-centered foundations of anthropological discourse, extending poststructuralist and posthumanist critiques of the autonomous skin-bound individual to explore other sorts of human experiences with science and technology. That is, following ethnographically how people construct meaningful discourses about science and technology in everyday life could provide access both to emerging power relations, helping us to understand better how science and technology routinely constitute power relations without a great deal of overt discussion and deliberation, and to how science, technology, and medicine participate in everyday human experiences, helping us to understand better how we all, in effect, live as scientists.

We thus extended Haraway's concept of the cyborg from a label for specifically contemporary refigurations associated with the New World Order, Inc., to an adjective potentially marking a wide range of anthropological projects that explore how science, technology, and medicine contribute to the fashioning of selves, including the selves of ethnographers. Above all else, we wanted to encourage expanded attention to the concrete awareness that we cannot say No and often desire to say Yes to technoscience and biomedicine while recognizing that our research projects and our identities as researchers contribute to constituting and reproducing that awareness. The cyborg image helped by reminding us not to hide or overlook ambiguous or ambivalent human experiences of pleasure in, desire for, and anxiety over sciences, technologies, and medicines, whatever and wherever these might be.

While, for some, granting membership to the cyborg image as an anthropological concept legitimizes new strategies for excavating and making visible human experiences that blur cultural boundaries between humans and nonhumans, for others it conveys an MTV-like fascination with the technically superficial, a naive, anthropomorphic attachment to the unreal or virtual. It suggests a project dangerously gone native because it appears to accept stereotypic celebrations of new technologies that vest them with causal efficacy as a source—the main source—of human progress. Attending to pleasure becomes part of the danger, for getting caught up with following new developments in high technology threatens not only to reproduce a Euro-American centrism but also provides a skewed picture of what is emerging in Euro-American contexts. Far from a self-critical analytic for mapping and intervening in power relations and stories of origin, the cyborg risks becoming essentialized as a fad. This degeneration of ethnography is exacerbated if it comes across as an elitist activity that presumes to draw exclusive boundaries over what counts as proper fieldwork, correct writing style, or required citations.

When seminar participants themselves performed a version of this debate, the interaction demonstrated a shared desire to develop and maintain a welcoming stance that invites collaboration rather than inhibiting it. The passion was clear. This sort of hunger to work together may indeed be quite strong among scholars generally, even if hidden or rendered subordinate in an academy that emphasizes agonistic struggle among competing positions (cf. Downey and Rogers 1995). Might we find greater value in the theoretical differences that separate us and concentrate more on collaborating to make a difference if we made more visible the ways in which disciplines function in society as cultural projects, as intellectual activities that intervene in everyday theorizing?

Like all primates, we cling to the backs of others. Anthropologists have long explored the cultural positioning of forms of knowledge, practices of medicine, and engagements with the human body. They have long theorized relationships among humans and things, labeling those relationships with names such as tools, artifacts, fetishes, technology, built environment, medicine, and art. In the process, anthropologists have also carried out projects both to study worlds of human experience and to participate and make a difference within them.

Looking beyond important but underrecognized work in applied anthropology, a great deal of anthropological theorizing has provided valuable sources of insight for popular theorizing outside the academy, especially by challenging stereotypical images that elevate the West above the Rest. Might distinct theoretical perspectives already be engaging in de facto collaborations? For example, just as cultural anthropologies have worked to theorize diversities in human experiences that do not reproduce established hierarchies by race, gender, class, ethnic origin, and so on, so have self-described "scientific" anthropologies worked to theorize commonalities among human experiences for a similar end. Both have intervened in Western modes of theorizing superiority, shifting them from the status of nature to the status of cultural assumption or stereotype—still real, but located in time and place and implicit in human action. What sorts of collaborations may have been taking place here? What do they tell us about anthropological projects more generally? When are theoretical differences more or less helpful, valuable, or justifiable? The practice of collaboration is, as yet, undertheorized.

Turning to the question of what might be specific to these times, to the ways in which people today inhabit discourses of science, technology, and medicine, David Harvey (1989), Fredric Jameson (1984), Robert Reich (1983), and other political economists, historians, and culture critics have pointed out that during the 1960s and 1970s rich countries began to shift away from industrial, manufacturing-based economies into service and knowledge-based economies. Basic manufacturing has been moving to "develop" other countries, markets are becoming both global and highly diversified, and all kinds of labor and capital are moving more freely and "flexibly" around the world. This transition is sometimes likened to the one that took place at the beginning of the industrial revolution: that is, we might be participating in a worldwide social, political, economic, cultural, and intellectual transformation. Many anthropologists have been studying these local, regional, and global transformations from a variety of perspectives, exploring both changes and continuities (cf. Appadurai 1991; Escobar 1995; Harrison 1991; Ong 1987, 1991; Stacey 1990; Tsing 1993). In addition, research by academics and activists has called attention to the myriad ways in which Enlightenment connections between the production of knowledge and human emancipation have been undercut or unrealized, in some cases even producing greater inequalities and divisions (cf. ACT/UP New York Women 1990; Harding 1993; Lyotard 1984; Merchant 1980; Penley and Ross 1991; Shapin and Schaffer 1985; Sheehan and Sosna 1991; Yanagisako and Delaney 1995). But events of emancipation or hegemonic dominance only scratch the surface of human experiences with and participation within the citadels of science, technology, and medicine.

Seminar participants became caught up in the question of what might be *emergent* in the world today. Questioning emergence rather than positing a universal transformation from, say, modernity to postmodernity makes the new/old question especially relevant. What is really new here anyway? In the midst of apparent change, where do we locate ongoing forms of colonialism, racism, and sexism, as well as forms of liberation, equality, material abundance, and other continuities? At the same time, what new opportunities for resistance or change may be emerging in the midst of apparent continuities? In exploring emerging sciences, technologies, and medicines, might it prove help-ful *not* to presume we know what humanness is all about before going into the field to find out?

Starting out with emergence as a question is also valuable because, in ad-dition to asking what is new on the horizon, it suggests that contemporary prac-tices are unfinished, ongoing, continuously maintained, and something in which one's own practices can potentially intervene. As the seminar partici-pants shared stories about citadel boundaries and cyborg selves, we regularly found ourselves talking also about intervention. The "mapping" acts of location built into our ethnographic practices always seem to constitute interventions as well. What roles had we been playing as persons in and out of our field sites? The issue is only in part a question of writing. Although wanting to acknowl-edge that our work was always positioned, we still found ourselves both writing and speaking with declarative sentences. Even if we desired to avoid represen-tation, the act of "speaking for" others, "speaking as" their representative in the guise of disinterested objectivity, or "giving them a voice," our work can still be heard as joining and participating in contested fields and hence locating our-selves in relation to those fields (Spivak and Harasym 1990). Distancing oneself from totalizing representation does not free one from the problem of "speaking as." During the seminar, we found it significant that we all wanted to avoid the comforts of both progressivist enthusiasm and oppositional pessimism. Yet the question remains: How do we want to be heard and, perhaps more importantly, by whom?

Putting this in more general terms, we see a transition taking place in criti-cal intellectual work, from opposing or praising technoscientific practices to in-tervention, from necessary entrenchment to ongoing participation. We see a growing desire among scholars, whether located in colleges and universities or in other workplaces, to use the analytic tools they have inherited to both ana-lyze and participate in issues of contemporary science, technology, and medi-cine. In part, this change may be the product of individual interpellations into worlds normally cordoned off behind "experts only" signs. In part, it may mark a generational shift from forms of critical analysis to forms of critical participa-tion. And in part, it may indicate a fundamental change taking place in the academy itself. What positions inside, outside, around, and through the citadel walls might researchers, academics, and activists occupy at the end of the twen-tieth century?

Minimally, we find it important to locate activism away from old agencies that made all participation co-optation. What would constitute critical opposi-tion if one were positioned not in a clearly subordinate position outside but somewhere inside? If one were inside, oppositional politics could shift from something one accepts as a necessary part of critique to something one can choose or avoid, depending on the circumstances.

Beyond that, we want to better understand and theorize the connections be-tween the moments of location in our work and the moments of intervention, for we think the latter deserve as much attention as the former. As a first step, we editors have organized the contributions to this volume so as to highlight and map approaches to intervention. Although each anthropological project intervenes in more than one way, the sequence is designed to sort out some distinct pathways for intervening in emerging sciences and technologies through research on cultural boundaries, cyborg selves, and the cultural relocation of anthropologists. In providing an overview of the volume, we pay attention to links that methodological choices and theoretical dispositions tend to establish with different intervention pathways and try to identify some of the dangers and opportunities associated with each one. Although our interpretations draw directly from seminar discussions and readings of the papers, this account should be read as the editors' own summary statement rather than a series of self-reports by the fieldworker-authors. There is much room for continued discussion and debate.

Intervening through Cultural Boundaries

In the first contribution, Rayna Rapp tells a fascinating story of how the growing use of sonography in pregnancy is shifting fetal development into fast-forward, increasing the velocity at which the fetus becomes an independent entity separate from the mother and others gain a stake in a pregnancy. Granting authority to the technology as a diagnostic tool funnels everyone's consciousness into highly focused and routinized channels, downplaying the clues for which women act as gatekeepers and allowing physicians to bypass women in favor of a technological window to the fetus. A woman's generalized concerns about having a healthy baby become specific concerns about Down syndrome and other genetic disorders. Gaining access to "early baby pictures" frequently heightens male involvement. The doctor's role in the granting of personhood is magnified, from personifying the image on the screen to sexing the fetus. Narrowing the aperture and sharpening the focus on the fetus also increase the possibility that outsiders can speak on behalf of a fetus as a legal person, thereby contributing to polarization in the abortion debate. In going beyond the technologists and genetic counselors to interview pregnant women and their supporters, Rapp crossed important boundaries around medical knowledge and expertise.

Seminar participants were struck by the discovery of a shared impatience with bounded field sites. In conducting research, each of us tends to begin with a relatively defined group at a specific site. Then we notice there are leaks, flows of information, people, and resources into and from this place and time. We find ourselves moving to look at groups that were interconnected with our initial groups yet not always acknowledged by them. We are led from laboratory practices to classrooms, from activists to governments, from support groups to magazines and newspapers, from public meetings to laboratories. We follow connections into the past and back to the future.

Recognizing this shared restlessness was important because we realized we could learn as much from the methodological issues involved in trying to map the field as we did from interpreting the material we collected. Which information flows seem relatively easy and which more difficult? Who appears accessible or inaccessible? What different sorts of insights and commitments emerge from briefly encountering many people through interviews versus significant participant observation with a relatively small number of people? What insights become available, what commitments become reinforced, as we show up repeatedly at different sites? The point is that a methodological commitment to crossing boundaries through ethnographic fieldwork can be both an important step in mapping them and a potential source of intervention that troubles and remaps them.

One way of intervening through the concept and analysis of cultural boundaries is thus simply to make these boundaries visible, locating them in historical time and place. Rapp's ethnography demonstrates that a cultural boundary between medical expertise and women's experiences in pregnancy is moving into earlier and earlier stages, with potentially dramatic implications for other boundaries as well. Challenging the Basic Story of technological development as advancement earned through progressive impacts, Rapp provides a brief overview of the development of sonographic technologies, making it clear that the technology emerged from a specific history rather than as an immanent necessity of technological and human progress. Then she locates the cultural boundary around medical knowledge by tracing the technology's direct involvement in women's bodies and experiences. Images of impact are replaced by images of deep, often ambiguous, personal involvement, and altered boundaries—between women and physicians, women and fetuses, women and families, women and men, and so forth—do not necessarily follow stereotypic divisions by gender, race, or class. And these new boundaries, for better and for worse, change lives.

Making cultural boundaries visible can help people find out where they are positioned, understand how they got there, and perhaps establish the possibility of imagining how things might be otherwise. Rapp concludes by expressing hope that women seeking sonograms might better articulate and get what they want in the midst of enhanced medical and societal participation and surveillance. Medical hierarchy and authority would then be not removed but relocated from a fact of nature to a negotiated product of history, power, and desires. Participating in this shift in interpretation or theorizing can encourage people who desire change to work for it, and it can help those who are satisfied with the existing state of things to recognize that changing circumstances may sometimes justify change in this relationship.

Rapp builds plausibility for her claim that a key cultural boundary is shifting by quoting extensively many people who are located differently with respect to the technology and medical knowledge. She does not take us deep into the experiences of any individual in an intensive exploration of questions of selfhood. Rather, her work can be heard as speaking to a set of processes of cultural change. Such a stance is not an authoritative pronouncement from nowhere but is historically located within a network of subjects related through technological practices.

To avoid the danger of sounding all-knowing, Rapp traveled from the offices of geneticists and genetic counselors to the bedsides of pregnant women, cutting across class and religious lines, rather than staying in one place for an extended period. Leaving out the history or women's experiences would have limited the value of the work as intervention. Another danger in mapping boundaries lies in restricting oneself to the status of outside observer who notes the presence of a boundary but cannot legitimately intervene in its definition or participate in its direction of travel. Rapp wants to make sure that those participating in this change have both the understanding and resources to make informed choices, but she also preserves a continuing role for herself and her work by emphasizing that every position located with respect to sonography is unstable and likely subject to further change.

A second approach to intervening through cultural boundaries is to make visible types of theorizing that cultural hierarchies have rendered subordinate. For example, the citadel effects of science and technology render subordinate any theorizing that does not emanate from within a protected, neutral citadel of experts and diffuse outward into the realm of public use or abuse. Such subordinate theories might include the experiences of scientists and other experts who do not conform to mainstream science, as well as those of nonexperts. Although Rapp makes visible the experiences of pregnant women, this pathway is more centrally a focus of the next contribution.

Emily Martin and her co-workers Laury Oaks, Karen-Sue Taussig, and Ariane van der Straten show that nonexperts theorize too. Exploring how clients at an HIV/AIDS clinic theorize the origin of HIV/AIDS, the meaning of AIDS as a disease, the possibilities of a cure, and their faith in physicians and scientists, Martin et al. demonstrate that medical theorizing belongs not only to medical practitioners and that the substance of such theorizing is not random or purely individual, but likely varies with social position. They found that the former or current injection drug users they interviewed, nearly all lower-class African-American men in the inner city, position the disease, the research, and the institutions in the context of other class-based discrimination. Poor people who regularly experience police intrusions in their lives may find it quite reasonable to conclude that AIDS is a quasi-military attack on the body, that some governmental agency or other official organization played a role in the origin of the disease and might inhibit attempts to find a cure, that a possible governmental requirement for mandatory testing does not seem to be an especially new or egregious threat, and that one must take significant responsibility for one's own health because one surely cannot depend on others.

The ethnographer who makes visible a subordinate mode of theorizing may come to be regarded as a spokesperson for such theorizing. Why work so hard to make something visible, one might ask, if not to make sure it achieves visibility? Simply by exploring and articulating a subordinate perspective and then locating it with equal or comparable weight alongside a dominant perspective, one disturbs the hierarchical cultural boundary that made the dominant perspective dominant in the first place.

By joining the staff of the clinic, Martin, Oaks, Taussig, and van der Straten effectively became unpaid consultants, attached to the staff yet contesting certain features of citadel effects on behalf of injection drug users. As their contribution puts it, Martin et al. hope their work "gives conversational voice to people who have been silenced." The intervention lies in showing that if a perspective articulating discrimination as a feature of the meaning and power of HIV/AIDS exists out there, perhaps other perspectives exist as well. To what extent might official discourses that rely on citadel effects actually be missing opportunities to serve their patients?

13

The HIV/AIDS clinic that hosted this ethnography came into existence because its director had stepped out of the Basic Story in an extraordinary act of ethnographic intervention. Wanting to study the progression of HIV/AIDS in an inner-city community, he learned through his own interviews that a good way to attract participants in the study might be to offer them basic health care. Furthermore, Martin et al. helped staff members recognize the deep sense of gratitude felt by their clients by giving a presentation in which they quoted extensively from their interviews. To the extent that the director and staff members could have missed these insights, might narrowing the aperture to citadel effects actually constrain the experiences of medical practitioners in addition to those of consumers? Perhaps not even the experts themselves always benefit from having a sharp boundary drawn around them and their knowledge.

The choice of methodology in this project helped shape its pathway for intervention. Conducting on-site interviews at the clinic with over forty clients was a key strategy for plausibly establishing discrimination as a shared image among them. In other words, demonstrating the presence of shared meanings helps Martin et al. constitute injection drug users as a social group. This is a crucial step because the citadel effects in medicine tend to fragment sufferers into an array of unique, individual patients, each interacting with the whole of centralized medical science. For these ethnographers, establishing the presence of a group through the vehicle of shared perspective becomes a device for increasing the legitimacy of that group and that perspective, helping it to gain standing in public discussions and debates over the diagnosis and treatment of HIV/AIDS.

One danger in working to make visible a subordinate perspective lies in potentially establishing oneself as the de facto patron of the perspective and the people represented. If one helps a voice to be heard, then presumably one could help silence it as well, and assistance shades quickly into domination and denial of the Other. Martin et al. deal with this danger by presenting many long quotations, a writing technique that maximizes the extent to which informants "speak for themselves" in the text and allows multiple, personal, heterogeneous perspectives to potentially work against objectifying the group represented. Another danger in this approach to intervention is that one's work can be read as necessarily oppositional, as taking sides, even when one's goal might simply be to make one perspective visible without destroying another. Martin et al. manage this danger by reporting ambivalences in the experiences of drug users themselves. Even as they interpret HIV/AIDS through the lens of discrimination, the clients at this clinic also possess and reproduce a strong faith in scientific research. If it becomes difficult to construe their perspective as oppositional, then what might constitute opposition on the part of the ethnographers becomes more complicated as well.

One final issue raised by this pathway to intervention involves the choice of topic itself. Anthropology's participation in dominant institutions of colonialism, multinational capitalism, foreign policy, domestic policy, and various arenas of political economy has been the subject of much investigation (Escobar 1995; Fox 1991; Harrison 1991; Marcus and Fischer 1986; Said 1989). A set of questions that haunted the seminar concerned dealing with our own tendency to value projects according to contemporary hierarchies of capitalized sciences, technologies, and medicines. Why have we granted disproportionate interest to the high-technology, high-profile areas of biotechnology, genetics, physics, information technologies, and specialized medicine over such less visible areas as routine health care, water supply, agriculture, electrical power, and engineering? Echoing Said (1989), to what extent are our choices to study these topics accepting direction from capital, through the availability of money to study some problems and not others, as well as from our own particularly American nostalgia for the new? Even as we critique this predilection, are we not also participating in it by identifying prestigious topics as those most worthy of study? Does intervening in concentrated centers of cultural authority provide opportunities to contribute to novel shifts of power, or might we be fulfilling a function of public but ineffective critique? What are we choosing to hide even as we work to make alternate modes of theorizing more visible?

Deborah Heath's contribution to this volume emphasizes a third pathway to intervening through cultural boundaries: the mediation of relationships across such boundaries. This ethnography draws directly on extended periods of participant observation among scientists, laboratory workers, clinicians, and activists involved with a disease called Marfan syndrome. Through fieldwork as both a DNA sequencing technician and a cell culture technician, Heath describes how bench workers and principal investigators enact the hierarchical cultural boundary separating mind from body. Novices are given access only to body activities at first, and promotion involves movement into activities that necessitate ever greater engagement of the mind, with the mental activities of principal investigator located at the top.

Ethnographically working through this hierarchy, Heath focuses on the importance of "good hands" and on the value that lab workers place on developing a "mindful body." Although forms of body knowledge might not fit well with the image of science as directed by creative *minds*, they do show up routinely in the daily practices of lab workers as one moves from science to science, lab to lab. Through her extensive fieldwork in the worlds of both bench workers and principal investigators, Heath gained the experience and authority to represent each perspective in the midst of the other. Exploring features of body knowledge calls attention to the knowledge contributions of bench workers, and highlighting the struggles of one principal investigator alerts bench workers to the extent to which the investigator values them and treats them with respect. In other words, Heath is able to relocate scientists for bench workers and bench workers for scientists beyond the terms of science as authoritative knowledge and in ways that reduce differences between the two.

Heath also works to mediate relationships across the boundaries that separate scientists who do research on Marfan syndrome, clinicians who treat the disease, and activists who build solidarity among patients and seek greater recognition for their problems. Not only did she participate directly by organizing a meeting at a national conference that brought together representatives of all three perspectives, but her text indicates a routine strategy in both written work and conversations of confronting stereotypic expectations with experiences that belie them. In particular, she helps one scientist confront and challenge her own desires to keep the concerns of clinicians and activists out of her lab.

Participant observation is a crucial methodological choice for Heath in this contribution because it helps her establish credibility on all sides of the boundaries she examines. As a fieldworker physically moving and communicating across standard flows of knowledge—technician and principal investigator, activist and scientist—Heath creates new forms of partner theorizing (see below, Downey and Lucena). This fieldwork strategy demands an investment of time sufficient for one to be heard by each side as (at least potentially) an authoritative member of the other. Interested in the scientific and medical goals of all of the groups in her expanded field, Heath works to allow these goals to crosscut each other as all being relevant to the production of science, technology, and medicine. In short, earning the right to mediate demanded an enormous investment of self on Heath's part.

The main danger in mediation lies exactly in the question of membership, for with membership come commitments that can last. One can gain the opportunity to participate comfortably in a consulting role, offering valuable advice that helps each perspective take account of others. But to what extent does achieving membership make it more difficult to distance one's work and one's self? To what extent does one limit one's role to a consultant politics, stuck in the job of helping others concoct strategies to fulfill their objectives? Heath makes it clear that she became friends with the scientist who hosted her fieldwork. She handles the danger of friendship and consulting by making sure she never stops moving back and forth across the boundaries that separate this scientist from clinicians and activists. In other words, living constantly on the boundary, however lonely that might be, can preserve the status of insider and outsider simultaneously, keeping one in a position of power as a representative of other groups in the midst of each.

Another issue that inflects the strategy of mediation through cultural boundaries concerns how cultural boundaries are conceptualized theoretically. If one treats a boundary, as Heath does, as a feature of a dominant mode of theorizing with which everyone has to deal, then mediation can consist of blurring the boundary by making visible all those experiences on both sides that both enact and contest the dominant mode of theorizing. In formulations such as this one, the word "culture" tends to designate the simultaneous identification of meaning and power, and the main problem of analysis is to establish the extent to which sharedness and, hence, groupness exists. However, if one treats the boundary as a border between distinct cultures, each of which is internally structured and coherent, then mediation is more a matter of getting each side to recognize and accept the legitimacy of the other than blurring the differences that separate them. In formulations such as this one, the word "culture" tends to designate the shared meanings that constitute each side as a group, and power is located in the relationships between the groups that hold such shared meanings.

In sum, anthropological projects that intervene in emerging sciences, technologies, and medicines through cultural boundaries distinguish between relocating the authorities of science and wishing or dissolving those authorities away. Challenging the citadel effects of science and locating scientific practices within cultural narratives need not be the same as practicing a popular theory of antiscience. The "antiscience" label serves as a rhetorical political tool for devaluing that which cannot be labeled "proscience" or is otherwise not wanted. The point is not to question science per se, but to characterize the roles of sciences, technologies, and medicine in our lives and imagine ways in which our lives might be better.

Intervening through Cyborg Selves

With Joseph Dumit's contribution, the volume shifts theoretical emphasis from identifying and following traffic over, around, and through cultural boundaries to exploring the participation of science, technology, and medicine in the fashioning of selves. Dumit is interested in how facts become incorporated into how people understand themselves. He understands facts as always "facts-in-the-world" to call attention to the specific stories, explanations, and experiences through which we learn facts or, alternatively, through which facts find us, without our ability to pass independent judgment about their truth. Dumit illustrates the role facts play in the formation of persons and categories of personhood by examining the history and everyday uses of a brain-imaging technique called positron emission tomography, or PET scanning. PET scanning provides images of a living brain in action as it thinks and experiences emotions.

Used with increasing frequency to diagnose forms of mental illness, especially schizophrenia, PET scans are understood to provide solid biological facts about otherwise contested behavior. In a society where stereotypic popular theorizing locates all agency in the intentional will of individual human decision makers, the presence of new biological facts can shift or rearrange rather dramatically the identities of schizophrenics and those close to them. For example, trial lawyers sometimes rely on PET scans in the sentencing hearings of convicted murderers to portray their clients as not fully responsible for their actions even though not certifiably insane. Also, locating schizophrenia as a fact of nature rather than a product of nurture can provide patients and their families, especially the oft-accused mothers, with the comfort of knowing that the illness was not their fault. Such changes are examples of what Dumit calls "objective self-fashioning"—the fashioning of selves through facts.

The main conceptual move in exploring the fashioning of selves is to construe experiences of self as the product of connections and relationships involving science, technology, and medicine rather than as their essential precondition or core substance. Human experiences and personhood at any given time and place are, accordingly, understood through analysis rather than asserted or assumed at the outset. Development of a stable, coherent self over a period of time despite new encounters and interactions thus becomes an achievement rather than an assumption. What people come to attribute to distinctively human or nonhuman agency depends upon how and where selves are located in fields of meaning and power. For example, just as the pregnant women interviewed by Rapp found themselves worrying about Down syndrome after admitting facts from reproductive technologies into their bodies and selves, so might the mother of a child diagnosed with schizophrenia by a PET scan find herself transformed back into a good parent after having accepted the facts of the matter. In each case, the transformation of personhood involved people attributing the agencies of personhood to nonhuman sources—self as cyborg. Dumit concludes by expressing the hope that learning about and following these circuits of fact distribution might help both laypersons and experts play a greater, even critical, role in their own understandings of themselves.

The methodological strategy of traveling across cultural boundaries is important here. Dumit first works to build a convincing account by linking together seemingly unrelated cases, such as the struggles of anthropologist Victor Turner to incorporate facts from neuroscience and medicine into social theory, and stories from the nonfiction best-seller *Listening to Prozac* about how this antidepressant alters people's behavior and experiences. He also provides a more extended ethnographic tour through organizations and people involved in PET scan development, mapping internal differences between work sponsored by the National Institutes of Health and work sponsored by bank loans in order to preempt interpretations of this story as monolithic technological progress. Thus, in ways that parallel Rapp's methodological travels across cultural boundaries, Dumit's work can be heard as speaking to self-making and meaning-making in participation with technological and medical facts.

The main danger in this approach to intervening through cyborg selves lies in casting the anthropologist as a virtuoso observer and interpreter of human experiences. How can an outside observer gain access to emotional experiences of body and self? Is it not presumptuous of someone who is simply observing people's behavior to claim to get inside their heads and experiences? This danger is not only a risk to intervention; it is also a significant methodological entanglement. By locating selfhood theoretically as associated with cultural position and identity, one makes a methodological commitment not to draw sharp distinctions in advance between mind and body, thoughts and emotions, inside and outside, and so on, including in one's own fieldwork and writing.

This pathway to anthropological intervention thus relies wholly upon the ethnographic interpretation of meanings and power relations encountered in fieldwork rather than on analytically separating emotional moments of empathy and sympathy from cognitive moments of observation. The ethnographic challenge is to identify, describe, and present such meanings, including the cultural attribution of emotions and thoughts, in ways that readers who live with a cultural distinction between emotions and thoughts would find plausible and convincing. In this contribution, using a best-selling book and overt expressions of comfort and relief as evidence for emotional reactions helps Dumit achieve plausibility because they suggest a sharedness that is widespread. Such a strategy, it is hoped, reduces the possibility that readers will judge the work as arrogant virtuosity rather than solid ethnographic analysis.

A second approach to intervening through cyborg selves is to concentrate on a specific category of scientist self-fashioning over time and across cultures, challenging a specific citadel effect: the belief that scientists are born and not made. Sharon Traweek has followed the lives and selves of physicists in the United States, Japan, and other countries for over twenty years, examining everyday practices to identify what she calls "themes" or "patterns" as well as "faultlines" among them. In her contribution to this volume, Traweek explores how images regularly displayed on the walls of physics laboratories, classrooms, and corridors, such as charts, maps, timelines, and photographs, actively serve as indicators and expressions of selfhood both to physicists and to outsiders. For example, one poster of Einstein draws on a common iconographic motif of Roman Catholic art, using backlighting to suggest a radiated divine grace, while another of him awkwardly riding a bicycle is one of several images that juxtapose intellectual subtlety and simplicity with childlike pleasures and the flaunting of social conventions. Also, a timeline documenting progress in scientific discoveries with a gap between AD 530 and AD 1453, the so-called "Dark Ages," demonstrates that knowledge is perishable if society acts to inhibit its development.

These images and their associations are especially important in the context of widespread debate over the superconducting supercollider because they locate physicists as individual geniuses whose curiosity should be encouraged (i.e., funded) to make discoveries and facilitate human progress. In similar fashion, contrasting layouts of laboratories in the United States and Japan indicate a significant difference in cultural patterns between a "dominating gaze" and a "glance" in the organization of physics knowledge, while the increasing appearance of simulations in place of log books suggests a generational shift in the aesthetics of physicists and physics knowledge away from taxonomies, classifications, and stabilities and toward complexity, variations, and instabilities.

Such concentrated attention upon one category of person or self defines a pathway to intervention that can involve helping people understand and assess the different ways they position themselves, even if the meanings involved are contradictory. The main images physicists display for themselves and others tend to locate physics securely within a citadel, placing physicists at the core of autonomous knowledge development that diffuses outward to the rest of us. The superconducting supercollider, however, was not approved. Might acknowledging and examining how physicists fashion themselves as intellectually subtle but childlike people who live outside of social conventions improve their abilities to reformulate and adapt their funding strategies to changing national agendas? Might acknowledging and paying more attention to faultlines of ethnicity, gender, age, and so forth within and across the boundaries of national physics communities improve the ability of physicists to work together, both in collaborative theoretical or experimental projects and in mechanisms of professional development? By serving to mediate one category defining someone's personhood in relation to another category, the anthropologist following this pathway might begin to resemble a group counselor or management consultant.

Traweek's choice of methodology, two decades of sustained participant observation, is important to this intervention pathway. Presenting the selves of physicists to physicists confronts her with the dual problem of constituting physicists as a social group and convincing its members to locate her amidst them. Just as Martin et al. demonstrate above, the anthropological finding of shared meanings serves to constitute a social group as well as to represent it. Without long-term participant observation, Traweek might have greater difficulty establishing her claims of sharedness. The issue of membership is trickier. Undergoing advanced training as a physicist herself would have been one possible way for Traweek to help herself become located among physicists. Sustained participant observation is another, for as physicists have come and gone over the years, Sharon Traweek has been there.

A main danger in this pathway to intervention involves losing the delicate balance between the identities of insider and outsider. Is Traweek an apologist for physicists, a critic of physicists, a patron of physicists, an outsider observer of physicists, or what? The answer is, roughly, yes. One way Traweek has maintained this ambitious, ambiguous status has been to concentrate her analytic attentions on everything but the mathematics of physics knowledge. She has thus avoided being positioned as a physicist-wannabe while becoming authoritative on much that is embedded, and often hidden, in physicists' bodies.

A third pathway to intervening through cyborg selves involves direct participation in self-fashioning, a practice that Gary Downey and Juan Lucena refer to in their contribution as "hiring in." Downey and Lucena explore how undergraduate engineering students experience engineering education as an outside challenge to personhood, a test of one's ability to integrate the practices of engineering problem solving into one's body and self. Downey and Lucena describe, for example, how solving an engineering problem involves drawing a sharp boundary around the problem, abstracting it out to solve in mathematical terms, and then plugging the mathematical solution back into the original problem. Engineers learn to view this method as rigorous and invariant; they come to believe that interfering with it in any way by allowing personal interests, desires, or concerns to creep in constitutes a serious violation of sound engineering practice. In contrast with, say, physics problem solving, in which the main challenge is to learn to "think like a physicist" (White 1996) so one can bring that unique genius to bear in a process of discovery, integrating engineering problem solving into one's body involves sharply separating "self" from "work." Downey and Lucena seek not only to make visible and help students understand better the diverse strategies through which they meet or reject this challenge, but also to participate directly in the education of engineers and the ongoing formation of curricular policies for engineering education.

As a metaphor of employment, "hiring in" indicates a willingness to allow one's ethnographic work to be assessed and evaluated in the theoretical terms current in the field of intervention, to become employees in a sense, paid or unpaid. "Hiring in" acknowledges that theorizing within established power relations captures one within those relations (cf. Rapp on abortion and Hess on capturing, this volume). Downey and Lucena conduct their research in the context of significant debate among engineers about engineering education as well as substantial national policy changes in engineering curricula. Whether these fieldworker-authors desire it or not, their written work will become located somewhere in the midst of these debates and changes, unless it is simply ignored as irrelevant. In particular, their work relates to ongoing concerns about the underrepresentation of women and minorities in engineering in the United States. The problem Downey and Lucena face concerns how to have their work received by engineers as participating significantly in the problem of underrepresentation without having to force their data about curricular self-fashioning into artificial, predefined groups of "women" and "minorities." These groups are interesting to Downey and Lucena as cultural categories that people apply to themselves rather than as distinct types designed for analytic purposes.

The strategy Downey and Lucena adopt for hiring into this contested field of education involves experimenting with what Downey and Rogers (1995) call "partner theorizing," which envisions all acts of theorizing as undertaken with interlocutors in collective, but temporary, negotiations of knowledge production. A practice that Downey and Rogers recommend for academic theorizing in general, partner theorizing involves looking for ways of factoring into one's own thinking the views of those one seeks to convince, without necessarily seeking consensus. Applied in this case, partner theorizing entails going beyond showing that the strategies engineering students use in accepting or rejecting curricular self-fashioning do not divide up neatly according to gender and race. Yet Downey and Lucena try to account for how underrepresentation does occur once people are demographically divided by gender and race, as in the studies that engineers and policy makers use. Furthermore, partner theorizing involves accepting limits on possibilities for change. Downey and Lucena seek new policies for engineering education, from designing a new course to recommending changes in everyday pedagogy, that take account of the current structure of engineering education and do not demand the resources that would be necessary to redesign curricula from scratch.

As with the previous approach to intervention, methodological strategies in this case necessitate convincing people to locate oneself among them; however, where one gets located shapes where one can legitimately contribute. In addition to long-term fieldwork, such strategies could include actually accepting employment, as many anthropologists of science, technology, and medicine have done. Fortunately for them, Downey and Lucena can expect some measure of credibility for their work among engineers by virtue of their undergraduate degrees in engineering. They can also cite active involvement in teaching undergraduate engineering students as well as research support from the National Science Foundation, which has been a major player in reformulating curricular policy in engineering.

The main risk in hiring in is co-optation—allowing one's work to be subsumed completely by the categories, the goals, and hence the power relations that define the field of intervention. Downey and Lucena work to reduce this risk by focusing their attention on ways in which engineering curricula contribute to the fashioning of selves, which tend to promote a student-centered perspective on engineering education rather than reinforcing a citadel model of education as knowledge transmission or diffusion. A second, equally dangerous, risk is social engineering, the arrogant presumption that one's expert knowledge grants one the authority to legislate new mechanisms for fashioning the selves of others. Downey and Lucena rely on partner theorizing to avoid this outcome, trying to formulate recommendations in terms that actually fit current debates over engineering rather than emanating from an elitist position.

In sum, these anthropological strategies of intervening in emerging sciences, technologies, and medicines begin with very local notions of how selves are fashioned in relations with technologies of education and mentorship, with ongoing medical redefinitions of normality and disease, and with scientific disciplinary divisions. By ethnographically attending to the lives of researchers and managers alongside the lives of students, subordinates, sufferers, and activists, these fieldworkers work to produce better accounts of the contingent co-production of selves and better practices of self-making.

Intervening by Relocating Anthropologists

David Hess's contribution intervenes in emerging sciences and technologies by relocating the position of anthropologist in science and technology studies (STS), the interdisciplinary study of science, technology, and society. Cautioning anthropologists who might be moving into this neighborhood that other researchers already live there, Hess explores how researchers in one branch of STS, the sociology of scientific knowledge (SSK), have regularly used anthropology as a resource in their work, sometimes in ways that cultural anthropologists trained in the United States might not recognize. For example, appropriations of the term "relativism" might blur cultural relativism into epistemological relativism, anthropology might be taken to be synonymous with the practice of ethnography, and ethnography itself might be thought to depend upon maintaining a clear sense of distance from the practices one studies.

Hess describes opportunities in STS for incoming anthropologists by presenting a "counternarrative" of STS development that makes visible a diversely organized "wing" he calls "critical STS" and outlining five "interrelated strands" that together make up a "distinctive anthropological/cultural studies contribution to STS." He concludes by locating anthropologists provocatively in the midst of STS by identifying ways in which they might use SSK concepts such as "impartiality," "enrollment," and "obligatory passage points" as resources in building anthropological work that is at once political, cultural, evaluative, and intervening. By thus relocating anthropologists and STS researchers simultaneously, Hess hopes to encourage development of an engaged anthropology of science and technology that "not only theorizes but also does more about exclusion, marginalization, hierarchy, and difference."

Relocating anthropologists intervenes by rearranging geometries of relationships both inside and outside of the academy, that is, among researchers and between researchers and nonresearchers. Intervening directly in the practices of one's own colleagues is one way of exploring and changing how they live and intervene in the worlds they study. Intervening ethnographically in the practices of anthropologists involves grappling with cultural boundaries and cyborg selves at the same time, for the process of redrawing the cultural boundaries that define anthropological work also refashions the selves of anthropologists. At the same time, participating critically in one's own mechanisms of professional development and practice offers distinct methodological challenges and poses unique dangers to the ethnographer.

Hess's choice of pathway for relocating anthropologists is to reformulate the genealogies that locate them in the present, that is, to redraw the boundaries around their work in order to make visible what has heretofore been hidden. This approach can help anthropologists not only to recognize that they are working alongside others but also to accept their own desires to make a difference through their work. In other words, cultural anthropologists turning to study science, technology, and medicine do not have to play by what appear to be the established rules if such rules hide important opportunities to make a difference.

Hess builds plausibility for his genealogical vision through the methodological strategy of a literature review. Mapping published literatures can be a key ethnographic strategy for identifying boundaries for people who locate themselves professionally through publications, so Hess travels across cultural boundaries through reading and lets people speak for themselves through citations. Although he also reports personal conversations and informal interviews, Hess provides no texts of such encounters, since these would likely appear as idiosyncratic opinions rather than disciplined interpretations. His explicit theoretical commitment to a culture and power perspective (Hess 1995) also helps locate his methodological priorities, leading him to focus on contrasts among distinctive cultural communities in order to sort out the power relations between them and place less emphasis on the contrasts he finds within each community. Identifying shared meanings is a strategy for constituting each collection of researchers as a social group.

The stakes in reformulating an academic genealogy become quite high when one's professional identity as a scholar figures in the analysis. Where is one located as an ethnographer? Is one an outsider to one particular group and an insider in another? If so, on what grounds can one claim to map anything but one's own space, and is one's work necessarily opposed to work in other categories? Furthermore, what if those one locates in one's own group do not see themselves as members? Having worked for several years as an anthropologist in an academic STS department, Hess manages these risks by locating himself as both an insider and a stranger to STS. His account offers detailed reflections that indicate many years of patient, systematic observation and interpretation in both anthropology and STS. Hess thus distances himself from the critiques that different schools of STS scholars make of one another and seeks ways for anthropologists to collaborate with researchers in other groups instead of rejecting them.

A second pathway for intervening by relocating anthropologists is to make visible writing and conceptual practices that might otherwise be hidden. In her contribution to this volume, Sarah Williams examines the presence and power of "fetish objects" among anthropologists, including participants in this seminar. Like the Arunta's sacred Churinga described by anthropologist Michael Taussig, the fetishes of anthropological researchers are "unrepresentable" objects whose presence can be "strenuously noted yet not reflexively recognized." The key example in this case is the prominent anthropological concept of cultural diversity, which not only reifies isolable cultures as objects of empirical knowledge and elevates anthropological interpreters into authoritative, expert knowers but also inhibits anthropologists from acknowledging "the complicities of knowledge and power that cannot be spoken yet empower the force of research itself." In other words, reflexivity in theory does not translate easily into reflexivity in practice. A New Zealand archaeologist of Maori ancestry, for example, finds himself unable to reconcile being Maori with treating Maori culture as an object of study.

Similarly, seminar participants found it difficult to acknowledge and discuss feelings of vulnerability in the field, the ethical complexities involved in taking money, and nagging pressures not to do fieldwork "the wrong way." They also had difficulty recognizing ways in which their concepts establish new fetish objects, and, most revealingly, trusting the presence of an ethnographer in their midst. In other words, seminar participants may be doing a better job of theorizing a new game than living it themselves.

By confronting one contribution to anthropological selfhood with another, making their practices more visible can help anthropologists to understand and assess the ways they position themselves, even if the meanings involved are contradictory. For example, to the extent that anthropologists find themselves struggling to move beyond the concept and politics of cultural diversity just as it has gained currency outside the discipline, perhaps understanding the ways in which this concept still shapes their academic practices might help anthropologists reformulate those practices and relocate their discipline. The question is, How might anthropologists be able to live without setting themselves up as the experts of Otherness? Achieving such change will have to involve more than theoretically sophisticated meta-anthropology; the *practices* of anthropology will have to be meta-anthropological as well.

It is important to this pathway that Williams's main ethnographic strategy is participant observation, revealing the practices that literature review alone would hide. Her account of the New Zealand anthropologist draws on her experiences as a colleague and a taped interview, while her account of SAR seminar participants draws on her experiences as a participant and a taped, on-the-record session in which she served as interviewer. In addition, structuring her article explicitly in terms of a traditional scientific paper allows Williams to adopt an ironic stance vis-à-vis the fetishes of academic research and finesse the problem of moving reflexivity from theory to practice. This saves her from having to elaborate in significant detail how her work itself produces or avoids producing fetish objects.

A key danger in trying to make anthropological practices visible lies in the tension between insider and outsider. How can one live powerfully on the margins of a group with whose members one competes for employment and funding? What are the implications of offering either affirmation or critique? That Williams manages these issues in several ways illustrates the extraordinary risks one must assume in undertaking an anthropology of anthropology. She begins with an account of her fieldwork experiences in Africa, thereby establishing her credentials as an anthropologist while describing how her interests shifted from the Turkana to the anthropologists studying them.

Williams's contribution displays an understanding of orthodox genealogies in anthropology even as it draws theoretical inspiration from the work of Michael Taussig and Homi Bhabha, who have positioned their work around the margins of the discipline. Relying extensively on direct quotes allows Williams to decrease the extent to which anthropologists might read her as a presumptuous outsider, even as she offers interpretations with which her informants might not agree. Using a formal research protocol in both experiments (New Zealand and the SAR seminar) also maximized the extent to which her work would be interpreted as legitimate research rather than muckraking journalism. Finally, Williams acknowledges that the authorship of her text itself is ambiguous, and that she lives with the risks of membership and/or estrangement.

Paul Rabinow's contribution illustrates a third pathway to relocating anthropologists—reformulating anthropological practices themselves. Although Rabinow seeks in part to "reinvent" some anthropological practices by making these "more visible" and hence "more available," this work goes beyond excavating the daily practices of anthropologists to reformulate key notions of practicing science. Its main objective is to retheorize practices in the human sciences, including anthropology, by articulating and exploring possibilities in their "ethical" dimensions.

Rabinow distinguishes two ideal types of ethical scientist, locating them in two different "sites." The first type, the "vigilant virtuoso," is the archetypical citadel scientist who keeps himself [sic] out of his work. Pierre Bourdieu serves as a key sentinel for this approach to mastery through knowledge, and the academic conference serves as its main site. The second type is the "attentive amateur," whose main site is the relationship among friends and whose features Rabinow articulates through Michel Foucault's "framework for analyzing ethics." For Foucault, ethics is "the kind of relationship you should have with yourself," and ethical self-constitution has four distinct aspects. Rabinow uses the first, "ethical substance," to call attention to reflective curiosity in human science, which he thinks is both valuable and underrecognized in recent science studies. The "mode of subjectification" in this ethical type involves serving as something of a philosophic observer who problematizes the world rather than mystifying it. The "ethical work" involves the challenges of participant observation rather than participant objectification, and the "telos" involves accepting the limitations of attentive engagement rather than seeking mastery.

This pathway to intervention lies in identifying new, theoretically possible patterns of conduct and then working to convince others of their value. Going beyond participant observation to make alternate practices more visible, it involves the refashioning of scientific selves through retheorizing their contents. Similar to Downey and Lucena's approach to participating in engineering education, this pathway indicates a willingness to allow one's work to be assessed and evaluated in the theoretical terms current in the field of intervention. The difference here is that the field of intervention is one's own professional home.

Rabinow's ethnographic methodology combines participant observation with philosophical exegesis. Attendance at a professional conference becomes fieldwork to identify the vigilant virtuoso's dominant mood of indifference, and systematic fieldwork in a biotechnology corporation identifies the site of friendship for the attentive amateur. At the same time, Rabinow rereads the classics, especially Aristotle, to relocate "virtue" as an epistemological practice. Thus the alternative type of human scientist Rabinow identifies not only lives in the present but also embodies a tradition every bit as pervasive and long-lived as the tradition of scientific mastery.

A major hazard of reformulating anthropological practices lies in defining and maintaining the ambiguous position of leadership so as to avoid both pedestrian short-sightedness and elitist self-centeredness. Having already earned senior status in disciplinary anthropology, Rabinow can feel secure that his reformulation will be read and cited—if he writes it, they will come. Rabinow signals this status implicitly by locating himself with Foucault in a debate with Bourdieu, a relationship that intrigues and interests anthropologists. Someone with less-established credentials would likely not be able to rely on first-person accounts but would need additional fieldwork strategies to attribute patterns to the community of human science as a whole. Rabinow avoids the dangers of elitism through ethnography, shifting the spotlight from him to us. Perhaps he is reinventing us, but only by showing us what was there all along.

A final pathway to intervening in emerging sciences and technologies by re-locating anthropologists is by setting an example oneself, that is, by locating one's work and, hence, one's self, as something for readers to assess and (if all goes well) to emulate. Every anthropological study adopts this pathway to the extent that it seeks to be cited and used in subsequent work. The pathway is much trickier when one tries to convince members of another discipline to read and find value in one's work, including its dreams. This is the task Donna Haraway takes up in the final contribution to this volume. If pedagogy can be understood as a practice of leading people somewhere, then Haraway has much to teach.

Characterizing herself as applying for "a visa for an extended stay in the permeable territories of anthropology," Haraway challenges anthropologists moving to study science, technology, and medicine to examine and reconsider their fundamental assumptions about who they are and what they are doing as researchers. She locates anthropologists in the midst of the set of emergent relations she calls the New World Order, Inc., by adopting the position of anthropologist herself. At first glance, this anthropologist is located not in a human body but in the bodies of laboratory mice, whose "mutated murine eyes give me my ethnographic point of view." After a while, however, the separation between human and animal dissolves away as we learn that their genealogies are the same, together experiencing the "force of implosion" through technoscience that brings together the "technical, textual, organic, historical, formal, mythic, economic, and political dimensions of entities, actions, and worlds."

Haraway outlines an interpretative framework that calls attention to figures and stories, examines mechanisms of "materialized refiguration," explores science as both "practical culture and cultural practice," and analyzes the "tangle of sticky threads" in nuclear and genetic worlds. In the process, she challenges anthropologists to find theoretical insight in science and technology studies, to find symbolic significance in the messy details of contemporary corporate life, and to recognize *how* their own work is always located. Perhaps by recognizing their participation in the New World Order, Inc., anthropologists might be more motivated to explore and contest what counts as "rational," "natural," and "technical," accepting full engagement in the contemporary worlds of techno-science.

Haraway's main methodological strategy is to conduct anthropology for anthropologists, demonstrating a thorough understanding of the cultural position of anthropologist by performing and playfully parodying it at the same time. After a sense of familiarity has been established, Haraway disrupts it by introducing the foreign, the strange, challenging the assumptions that locate the position of anthropologist. Anthropologists are not just students of culture; they also contribute to the emerging New World Order, Inc. Haraway is able to stand for the New World Order in the midst of anthropologists because she was trained as a biologist, became an accomplished historian of biology, and is now a renowned culture critic. Nevertheless, if the "anthropologist" as a cultural identity can be separated successfully from the human substrate in which it resides, perhaps readers who call themselves anthropologists might be more likely to redefine what that means.

The main danger in this pathway is marginalization, the act of locating oneself irretrievably on the margins of the field of intervention. Such a position might be risky for a less established scholar but, in addition to having long demonstrated a willingness to stake her career in the pursuit of her dreams, Donna Haraway is a public intellectual who is at risk only if everyone marginalizes her from their work—something that seems unlikely to happen. Haraway also manages this danger herself by avoiding direct critique of or oppositional confrontation with anthropology, which might have made it easier for some anthropologists to reject her message without listening.

Making Intervention Visible

Taken together, the contributions in this volume challenge readers to ask, What if researchers devoted half of their research time to theorizing and practicing intervention? While desires and concerns about intervention are likely present in every step of a research project, from sorting out the right questions to pursue to making sure that a written product sounds right, the Basic Story that researchers tend to tell themselves has often hidden these desires and concerns or devalued them as the "applied" implications of good work. Might we be able to share, discuss, and debate more openly the sorts of differences we hope to make through our work and how we go about achieving those differences? Might sorting out research projects according to how they intervene make it easier for each of us to accept the value of other perspectives and to conceptualize and practice collaboration?

Although they are an idiosyncratic array of anthropological pathways to intervening in emerging sciences and technologies, the papers in this volume do suggest, regardless of the area of study, that theoretical dispositions, methodological strategies, and the identities of researchers as persons together scope out fields of intervention and available pathways for participating critically in those fields. Theory matters, for it locates one in relation to the forms of theorizing prevalent in the field of intervention. The opportunities to intervene that contributors identified depended not only on how they conceptualized culture or self within the study but also on the relationship between these academic formulations and the concept(s) of culture or self encountered in the field. Methodology matters, for it establishes the steps through which one becomes located in a field of intervention. Just as extended participant observation establishes different steps than a series of taped interviews or the analysis of documents, so might entirely different methodological choices or configurations of choices establish still different pathways. The analysis of quantitative data, for instance, locates one especially well in fields of intervention that call themselves "populations," such as the polling of electorates.

Finally, one's identity as a researcher matters, both shaping one's initial location with respect to a field of intervention and establishing what might be necessary methodologically. For example, being able to claim prior membership in the field can open many doors, but not without also adding special burdens. In this case, the question of positioning sometimes shifts from figuring out how to get in to figuring out how to get out.

In this volume, each anthropological project seeks to make visible lives and practices hidden by features of the Basic Story of knowledge creation, diffusion, and utilization. This commonality derives from a shared commitment to cultural perspectives and ethnographic fieldwork. Take away either one of these, and the pathways to intervention change.

An important responsibility in recognizing our participation in that which we study involves working on limitations in our own fields of vision. Seeing through our work how well communities can inhibit or prevent self-reflection, we want to take care to recognize what our perspectives ignore, silence, or make invisible. Seminar participants repeatedly expressed interest in investigating and critiquing the desires, values, and assumptions built into our projects. We want to work on our own Euro-American centrisms by making an effort to notice and name the vehicles through which they live in our work.

Owing to the limits of the organizers' egocentric networks, an explicit desire to focus on connections between analysis and intervention, and biased attractions to the lives and practices of big sciences, technologies, and medicines, this volume does not venture into questions of environmental justice, public health, popular epidemiology, third- and fourth-world issues of technological equity and survival, and a range of other arenas that would make Euro-American centrism a more central and sustained focus of discussion. As privileged first worlders studying privileged first-world science, we must each devise ways to question and, it is hoped, trouble such practices.

The seminar week was intense and instructive. It did not answer all the questions or satisfy all the desires participants brought to the exchange. We were acutely aware of the people not present and the variety of perspectives not represented in our small group. At the same time, experiencing several days of sustained collaboration awakened and nurtured in each of us a profound sense of the challenges and potential importance of locating and intervening in emerging sciences, technologies, and medicines through cultural perspectives and ethnographic fieldwork. We hope here to share that sense of challenge and opportunity and to ask for your help. In work begins responsibility.

Notes

1. Seminar participants included Gary Downey, Joseph Dumit, Donna Haraway, Deborah Heath, David Hess, Emily Martin, Paul Rabinow, Rayna Rapp, Sharon Traweek, and Sarah Williams. Coauthors who were not present include Juan Lucena, Laury Oaks, Karen-Sue Taussig, Ariane van der Straten, and Frederick Klemmer.

2.   Participants in these three sessions included Gary Downey, Joseph Dumit, Michael Fischer, Deborah Gordon, Donna Haraway, Deborah Heath, David Hess, Emily Martin, Constance Penley, Paul Rabinow, Rayna Rapp, Allucquere Rosanne Stone, Lucien Taylor, Sunera Thobani, Sharon Traweek, Sherry Turkle, and Sarah Williams.

3.   There was evidence of growing interest prior to 1992. The 1991 AAA meeting, for example, included an invited session on "Cultural Perspectives on Information Systems Development," organized by David Hakken and Linda May; a panel on "The Ethnography of Scientific Practice," organized by Alan Stockdale; a panel organized by Allen Batteau and Elizabeth Brody on "Anthropology and Engineering"; and another invited session on "Nation, Culture, and Power in Science and Technology," organized by Gary Downey. Also in 1991, Joseph Dumit presented the paper, "Cyborg Anthropology: Brain-Mind Machines and Technological Nationalism."

REAL-TIME FETUS / The Role of the Sonogram in the Age of Monitored Reproduction / **Rayna Rapp**

> With her own eyes, she could now pretend to see reality in the cloudy image derived from her insides. And in this luminescence, her exposed innards throw a shadow over the future. She takes a further step—a giant leap—toward becoming a participant in her own skinning, in the dissolution of the historical frontier between inside and outside.
> —Barbara Duden

ULTRASOUND FETAL IMAGING has rapidly gained wide diffusion as a screening device during pregnancy. In the United States it is estimated that more than 50 percent of pregnant women currently undergo at least one sonogram examination during their pregnancies, and in urban areas like New York City, where my research was conducted, informal estimates place the population under surveillance at 90 percent.[1] Some medical authorities have suggested that the technology is being dramatically overused, given the difficulties of assessing its long-term safety effects. The rapid diffusion of ultrasound in pregnancy has raised epidemiological questions on two fronts: consensus studies suggest that the technology is safe in the short run, but only the analysis of much greater longitudinal data will reveal whether or not it has low-level biological effects on fetal auditory or neurological systems. Important questions concerning its cost-effectiveness have also been raised (LeFevre et al. 1993; NIH 1984; Stratmeyer and Christman 1983). Yet most obstetrical clinicians consider ultrasound to be the best invention since sliced bread and use its instant visualizations to measure, date, position, and intervene in pregnancies, while "reassuring" their patients that their fetuses are developing in a normal manner.

Wherever the evolving debate on over-routinization may lead, some uses of the technology seem unambiguously acceptable to epidemiologists, obstetrical and radiological clinicians, and pregnant women and their supporters. Most notably, second-trimester sonography is routinely used prior to amniocentesis. This procedure has clear medical utility, enhancing the safety of the amniotic tap and screening for some physical fetal anomalies as well. This essay is drawn from my experience in observing and interviewing women (and sometimes men) about the use of ultrasound accompanying amniocentesis. I want to stress that such amnio-related sonograms constitute a very small percentage of all the sonograms performed in any obstetrical service.

The problems and possibilities of sonography in pregnancy are legion. As many feminists have pointed out, the epidemiological debate on cost-benefit analysis and safety systematically overlooks other mundane but widespread effects of the technology. These are usually labeled its "psychosocial impact" in the medical literature, referring to the effects of sonograms on the individual women whose fetuses are being screened (Black 1993; Kolker and Burke 1994; McDonough 1990; Villeneuve et al. 1988). As I hope to illustrate, this label itself is highly reductive; technologized bodies are eminently social, public, and contested, as well as individual.

Much of the feminist concern with ultrasound points toward the enduring critique of the medicalization of pregnancy. The widespread dissemination of "first baby pictures" suggests that there is an evolving politics of representation affecting many if not most pregnancies in the United States, in which the fetus has both a domesticated and a public presence. Compelling as virtually all parents-to-be may find them, "blurry baby pictures" provide yet another instance of the extension of medical control over pregnant women, constructing what a fetus is and how it should be treated. Sonograms may also be used to further the separation of the fetus as an independent entity, a potential patient, and occasionally a contested ward in the legal system. The powerful visual representations of fetuses provided by sonograms make a contribution to notions of fetal personhood, simultaneously effacing the "maternal vessel" within which its gestational life is lived and calling into question her "selfishness" in the way she conducts (or ends) her pregnancy (Cartwright 1992; Duden 1993; Hartouni 1991, 1992; Stabile 1992). Fetal images thus cast an aura well beyond the obstetrical suite.

I am quite sympathetic toward this critique, having penned versions of it myself with a fair degree of frequency. I particularly wish to honor the collective work of feminist analysts, which suggests that the stakes involved in advances in reproductive medicine affect far more than individual women and their pregnancy outcomes. Such effects are highly social, indeed, political, for they produce both a newly independent fetus and a medically reinscribed mother under the powerful sign of normative maternal-child health. Feminist researchers have been calling for an examination of the social and nonmedical consequences of reproductive technologies, including sonography, for some time (e.g., Pechesky 1987; Rothman 1989). Representations of these contradictory actors and factors also affect relations of gender, kinship, and political community in ways we have just begun to explore.

As an anthropologist interested in finding out how a diverse range of women exposed to reproductive technologies interpret the benefits and burdens of these interventions, I also have competing methodological commitments. We know too little about how pregnant women of different racial-ethnic, class, national, and religious backgrounds experience and frame their own interpretations of these newly visualized "facts of life." The recent history of Western feminism, and of feminist anthropological scholarship, beckons us to diversify and complexify the terms of our social and political analysis. Some of the original and quite passionate feminist polemic against what have loosely come to be called "the new reproductive technologies" was premature in my judgment because of its partial and problematic political epistemology. How can "we" speak to women's experiences of this technology when so much of our evidence is anecdotal and disproportionately represents the constructions of women from the same privileged communities, broadly speaking, as their enthusiastic, Doppler-trigger-happy obstetricians and radiologists?

To put it another way, the defense of a unified and embodied "maternal experience" against an encroaching "male" medicalization is too often framed in a discourse that exhibits an oppositional rhetoric supported by a social similarity. It is an enduring irony of contemporary feminist thought that "we" who are most likely to benefit from the technologies of modernity often count ourselves among their strongest critics. Our discourse too often mirrors a familiar conundrum: Either we march in lockstep with obstetric-technologists with whom we share a worldview of interventionist cultural control over nature (Davis-Floyd 1992), or we are aligned with an antitechnological romanticism to which we often attach a feminine label (cf. Corea 1985; Rothman 1989). Not only do such oppositional discursive alliances endlessly replicate one another; they also reproduce the centrality of the middle-class worldview in which both privileged access to technologies of pregnancy and the development of its critique are located. All other experiences and their interpretations then remain marginal. When we accept a polarized discourse on technology, we run the substantial risk of ignoring, silencing, or misinterpreting what a heterogeneous population of women might tell one another about their criticisms of and aspirations for their reproductive health care. Fetal images may cast a powerful light on the political space in which shifting notions of maternal responsibility, abortion rights, and disability rights are evolving in contemporary American culture. How are women's hopes and conflicts about this space articulated in all their diversity? Our collective ability to think politically in fast-forward about the emerging technopolitics of maternal-fetal (dis)embodiment implied in ultrasound imaging is impoverished without this richer base of knowledge.

In this essay I want to suggest two possible empirical strategies to unsettle this theoretical logjam. One involves learning how women of diverse backgrounds interpret the impact of ultrasound examinations, so that our feminist concerns are not so prematurely narrowed. In the second, I join with many feminist theorists who are chipping away at the medical framework within which the ultrasound debate has been conducted, pointing toward the public nature of fetal images and their uneven seepage into the interstices of social life in contemporary America. Before turning to the social location and cultural interpretation of sonograms, I offer a brief sketch of the history of prenatal diagnostic technologies.

Imaging Technological History

In 1958 Jerome Lejeune, a French geneticist working at the Hôpital Saint-Louis in Paris, peered through an aged microscope at a sample of smooth muscle tissue taken from three patients with Down syndrome. His cardiologist-colleague Marthe Gauthier had used the then-innovative techniques of tissue culture to treat the sample. The full complement of human chromosomes had only recently been confirmed as forty-six in 1955–56. Lejeune was trying to assess whether his patients with Down syndrome lacked one human chromosome, as some abnormal fruit flies lacked one fruit fly chromosome. Instead, he (or, perhaps more accurately, he and Gauthier) discovered that they had a surfeit of chromosomes: then, and in subsequent studies, tissue samples taken from people with Down syndrome yielded a chromosome count of forty-seven. With great hesitation, Lejeune published his results in 1959. At the same time, a research group at the University of Edinburgh independently arrived at the same findings, confirming the Paris research. As a provost at University College wrote to pioneering English geneticist Lionel Penrose, "It must be one of the most important things that have happened in genetical studies for a long time" (Kevles 1985:248).

A few years later, in 1967, American researchers reported the first detection of a fetal chromosome problem in a sample of amniotic fluid drawn from the womb of a pregnant woman. Amniocentesis—the technique of extracting amniotic fluid transabdominally through a catheter—was first performed and described in Germany in 1882 by a doctor attempting to relieve harmful pressure on the fetus of a pregnant woman. It became an experimental treatment for polyhydramnios (excess fetal fluid that threatened fetal development) but was not widely used until the 1950s, when researchers in Great Britain and the United States discovered they could deploy the same technique to test for maternal-fetal blood group incompatibility and to assess the severity of Rh disease. In critical cases, amniocentesis led to intrauterine transfusion. It also permitted the assessment of fetal lung maturity, so that fetuses with serious disease could be delivered as early, and as safely, as possible.

Once it became possible to isolate and grow fetal cells from amniotic fluid, the examination of their chromosomes quickly followed. Some of the earliest "fishing expeditions" inside women's wombs were undertaken on mothers who had hemophilic sons and wanted to know the sex of their present pregnancies. If the fetus was determined to be chromosomally male, it ran a 50 percent risk of having the disease. By the time researchers in Sweden, Japan, Great Britain, and the US began experimenting with amniotic prenatal chromosome diagnosis, the technology to invade the uterus and extract its liquids was well known (Cowan 1992, 1993; Davis 1993; NIH 1979). In the same decades, abortion reform throughout much of the West made the termination of problem pregnancies more easily available; even today, in the midst of concerted efforts in both legislatures and courts to dismantle abortion rights in the United States, surveys report strong public support for legal abortion in the case of serious "fetal deformity." The success of prenatal diagnosis must thus be situated in its social and legal, as well as its medical, contexts.

The evolution of amniocentesis was accompanied by other, closely related technological developments. In the same year that Lejeune observed the karyotype (chromosome picture) of his Down syndrome patients, midwifery professor Ian Donald and his colleagues in Glasgow published an article titled "Investigation of Abdominal Masses by Pulsed Ultrasound," in which they described the adaptation of sonar naval technology to observe fetuses inside their mothers' wombs (Oakley 1984). While the idea of sonar (sound navigation and ranging) had been patented in England directly after the Titanic disaster for the detection of icebergs at sea, it was the French who developed it to detect enemy submarines during World War I. In the decades after the Great War, commercial engineers and physicists in the Soviet Union and later the United States investigated the technology's potential for revealing metallurgical flaws. Initial medical experimental uses focused on the energy generated by sound waves as a rehabilitative therapy. It took decades of clinical experimentation before the image-producing capacities of the technology were fully recognized and methods to harness it developed (Yoxen 1989:281–96).

Diagnosis of pregnancy was thus neither the first nor the most obvious use claimed for sonography, which was initially (and wrongly) thought to be beneficial in scanning brain tumors and, later, kidney masses. While the sonogram has certainly proved to have many medical uses (especially, in cardiology), its most routinized successes developed in obstetrics. Pulse-echo sonography (ultrasound) works by bouncing sound waves against the fetus, creating an image as the waves return to a cathode. After decades of perfecting transduction (the ability to attach the machine to the exterior of a patient's body so as to image interior soft tissue) and sectoral scanning (the capacity to render a three-dimensional object in two dimensions in regularized segments), it became clear to Donald and his colleagues that sonograms offered the possibility of normalizing representations of the fetus throughout its gestation. After their reports, sonography's diffusion in obstetrics was rapid and dramatic, and physicians hailed it as "totally safe" long before any actual safety studies were conducted (Oakley 1993).

As many feminists have pointed out, the technology intervened in the doctor-patient relationship dramatically, allowing the physician to bypass pregnant women's self-reports in favor of a "window" on the developing fetus (Mitchell 1993a, 1993b, 1994; Oakley 1984, 1993; Pechesky 1987). Additionally, radiologists and obstetricians working together could use sonograms in the developing technology of amniocentesis. Sonography enabled them to picture where the fetus wasn't and the fluid was, rather than groping blindly for a pocket of liquid into which to insert the amnio-bound catheter. As "real-time" sonography became available, doctors were able to observe a moving image of the fetus while sampling its environment. When used in conjunction with sonography, experimental invasive techniques of the womb became safer, and the miscarriage rates attributable to these procedures dropped dramatically.

The technology of prenatal diagnosis continues to evolve at a rapid pace. During the decade in which I have been investigating its social and cultural geography, another intrauterine technology has waxed and waned. Initially hailed as a revolutionary replacement for amniocentesis, CVS (chorionic villus sampling) works on preplacental tissue, allowing diagnoses to be completed within the first trimester. Despite the attraction of ever earlier diagnosis, serious safety objections have been raised (Firth 1991). More recently, experiments in early (twelve-week) amniocentesis are moving from the anecdotal to the clinical, although controlled national trials have yet to be completed. Additionally, inexpensive MSAFP (maternal serum alfa fetoprotein) screening has become routinized throughout the US. When elevated values for this biochemical marker produced by the fetus are found, they may indicate a neural tube problem. But the test itself produces a large number of false positives, leading to increased use of amniocentesis as a backup diagnostic technology.

California has become the first state to mandate offering MSAFP screening to all pregnant women, with psychosocial results that are fundamentally unknown (Browner and Press 1995; Press and Browner 1993). MSAFP is also being used in conjunction with multiple biochemical markers (acetylcholinesterase, HGC) to develop new, non–age-related screens for Down syndrome. These biochemical triangulations are most effective when used in conjunction with level II (high-resolution) ultrasonography. Currently, experiments in maternal-fetal blood centrifuges and sorters, especially PCR (polymerase chain reaction)-assisted FISH (fluorescent in-situ hybridization), point toward future technologies which will invade the maternal body less deeply, while tracking the fetus through more intersecting biochemical values and organ systems. Ultrasound is likely to be implicated in all of these developments.

It is hard to evaluate what drives the proliferation of prenatal diagnostic technology. Consumer demands certainly enter in, as genetic counselors serving high-income populations are quick to point out; every time an experimental technology appears in the Tuesday *Science* section of the *New York Times*, they get inquiries from anxious parents-to-be. But the real technological action must also be traced through political and economic pathways. Ultrasound machinery has had a huge commercial success, appealing to obstetricians and their patients alike. "Clearly, what is beginning to appear is a narrative that constructs new needs for women as it constructs new markets for imaging" (Treichler and Cartwright 1992:8). Moreover, prenatal diagnostic technologies have moved out of academic settings into commercial labs with great speed over the last decade, and the most experimental technologies are often available, for a price, long before their accuracy is fully tested (Milunsky 1993; Natowicz 1994).

Additionally, the impact of molecular biology's focus on the Human Genome Initiative, for which prenatal diagnosis serves as one obvious public rationale, should not be underestimated (see Rabinow and Heath, this volume). We live in a technological "history of the present" in which each innovation speaks from the profit margins of biotechnology and obstetricians, radiologists, geneticists, and genetic counselors increasingly are cast as academic, consultant, and service-providing handmaidens, despite the considerable uneasiness many providers have expressed both publicly and privately.

The Role of the Sonogram in the Age of Monitored Reproduction

Above all else, a close encounter with prenatal testing increases women's worries about the specific health status of the fetuses they carry. Generic pregnancy fears might once have crystallized around the desire to carry a "healthy" baby. For women having amniocentesis, there is now a focus on precise conditions; chromosomes and AFP levels index a panoply of anxieties with newly medicalized names. The specificity and reality of childhood disability becomes exquisitely focused through prenatal testing, engaging a complex mix of science and superstition as pregnant women and their supporters encounter potential diagnoses:

Down syndrome, I knew about Down syndrome. What I didn't know about was all that other stuff. There's more to worry about than just Down syndrome, now I know there's other heredity problems. And this spine business [spina bifida], I wasn't exactly acquainted with that. Something more to worry about. (Lacey Smythe, African American secretary, 38)

I remember thinking, "Oh, my God, it's like a message direct from inside." In the old days, our mothers certainly never knew this, the picture of the inside of their wombs, the small swimming thing. But we do. We're the first ones to follow pregnancy in books, day by day, with photos. We know exactly when the arms bud off, when the little eyes sew shut. And if something goes wrong, we know when that happens, too. They called it "an error in cell division." It feels like the cells could have a car crash, and produce this wreckage, and that's the extra chromosome, that's Down's. (Pat Gordon, white college professor, 37)

Suddenly, I'm starting to see all these kids with Down syndrome on the street. Who knows if they're really Down's kids, or if I'm imagining it. And now you're asking all these questions, and I'm trying not to think about spina bifida. I never even knew spina bifida was a problem. But after counseling, I do. (Enid Zimmerman, white municipal service planner, 41)

The power of the sonographic imaging which accompanies the test has complex effects, funneling the pregnant woman's consciousness of her fetus into highly focused and routinized channels (Mitchell 1993a; Oakley 1993; Pechesky 1987). But how are these channels constructed through imaging? The gray-and-white blobs of the sonogram do not speak for themselves but must be interpreted. As many sociologists and historians of science and technology have pointed out, the objects of scientific and medical scrutiny must be rendered; they are rarely perceived or manipulated in their "natural" state. It is their marking, scaling, and fixity as measurable, graphable images that enable them to be used for diagnosis, experimentation, or intervention (Fyfe and Law 1988; Lynch 1985; Lynch and Woolgar 1990). The power of scientific images may, in large measure, be attributed to their mobile status: they condense and represent an argument about causality that can be moved around and deployed to normalize individual cases and theoretical points of view (Latour 1986, 1990b). Viewed on a television screen or snapped with Polaroid-like cameras, sonograms may appear to pregnant women and their supporters as "babies." But the particularity of the object women view is deeply embedded in the practices of its scientific representation.

The partial rotation of the beam and the electronic recording of the echoes as spots of light thus "renders" . . . the internal two-dimensional structure of an organ or a limb or a test object in a given plane. The resulting image is certainly not artifactual. It registers features, like the fat-muscle interface, that really exist. Yet it picks out only those features that reflect ultrasound. (Yoxen 1989:292)

But surely pregnant women and their supporters are not thinking about the embedded, reductive, and normalizing aspects of imaging technology as they "meet" their baby on a television monitor for the first time. Such uterine "baby pictures" are resources for intense parental speculation and pleasure, for they make the pregnancy "real" from the inside, weeks before kicks and bulges protrude into the outside world. The real-time fetus is a social fetus, available for public viewing and commentary at a much earlier stage than the moment of quickening, which used to stand for its entry into the world beyond the mother's belly.

Perhaps sonograms also enable fathers and mothers to "share" what was formerly an entirely female experience of early pregnancy, increasing and hastening men's kinship claims (Taylor 1993, 1994, 1996).[2] And surely they increase the speed with which fetal development is recognized as a process independent of the mother's embodied consciousness. As one white college teacher commented to me, "It put my pregnancy into fast-forward." She thus neatly aligned sonography with videotapes, that other near-ubiquitous forum for home-viewing. One couple, who disparagingly referred to themselves as "yuppies," brought their own video camera and tape recorder to the sonogram examination because they wanted to capture and domesticate the fetal heart beat. The acceleration of a subjective connection to the pregnancy thus passes through and is augmented by technologies external to the pregnant woman herself.

Of course, modern imaging technologies provide powerful framings for the health and meaning of a pregnancy that appear radically new and individualistic, but they do not hold exclusive rights to the air space in which the image of pregnancy is interpreted. Public commentary on pregnancy has ebbed and flowed with the development of religious discourse, the representational arts, and the history of science and medicine. Current biomedical interpretations pass through other "images that possessed power within their own time and to which other images and ideas clung" (Stafford 1991:xvii). In the process, pregnancy is constantly relocated as an object of speculation, investigation, and intervention.

Contemporary feminists have alerted us to the changing relationship among a pregnant woman, her fetus, and the social world indexed in reproductive medicine; they have also provided ample evidence for older representational politics. Sonograms reinscribe prior debates and interpretations about the meaning of pregnancy which have deep roots in Western history; residues of those discourses shape what we take to be modern notions of sex and its biological embodiment. Pregnancy, for example, figures in the tensions and agreements between medieval theology and natural philosophy. In that period, women's "fleshiness" was associated with the body in body/soul dualities, and her fetus was a cause for speculation about maternal and paternal contributions to God's purpose and perfection. Women's reproductive capacities invoked reflection on divine regulation and causes of oddity. Medieval texts and artifacts evinced enormous curiosity about pregnancy, monstrous births, and the relation of blood to milk, couched as problems of permeability and stability (Bynum 1991). Because biological sex was thought, for example, to be extremely labile, "the nature and cause of hermaphroditism, as of other embryological anomalies, was much discussed" (Bynum 1989:187).

In a later period, the eighteenth century, "The activity of visibilizing, or incarnating, the invisible became endowed with a special urgency in early modern art and medical experimentation" (Stafford 1991:17). A fascination with conception—with eggs, fetuses, grotesques, and biological monsters—provided the ground for artistic and scientific debates about how matter was formed—all at once, as the preformationists asserted, or in developmental stages, as the epigenicists insisted. Embryology and pregnancy were two fertile fields for the representation of theories of causality (Stafford 1991; cf. Duden 1993). Beginning in the same period and continuing throughout the nineteenth century, the study of anatomy focused on sexual difference and its representations, using models, cadavers, and works of art (Jordanova 1989; Schiebinger 1989).

By the turn of the twentieth century, representations of the body (especially the female body) were once again relocated, this time at the flourishing intersection of photography, early cinema, and medical research, where "the body emerged as a (visual) apparatus . . . an embodiment of . . . the techniques of the observer" (Bruno 1992:249; Cartwright 1992, 1995). The imbrication of theology, natural philosophy, artistic and media conventions and emergent medicine thus provided verbal, plastic, and artistic representations of women and fetuses long before the advent of the new reproductive technologies.

Women and their fetuses were embedded and represented in social and power-laden discourses long before the present moment; this much is not new. But in prior times, individual fetuses made their presence public only slowly, over a period of months, and that presence was attached to signs—whether miraculous, mundane, or scientific—that passed through the woman's codification. A woman's physical and emotional state might reveal internal signs of pregnancy in hormonally induced swollen breasts, skin changes, energy loss, dizziness, or nausea, all of which were experienced kinesically and holistically. Later a midwife or physician might pick up a fetal heartbeat through a wooden trumpet, a stethoscope, or, more recently, a Doppler machine. But the passage from internal to external signs was slow, and almost all of the cues depended on the pregnant woman's reportage.

Now, however, sonography bypasses women's multifaceted embodiment and consciousness, providing independent medical knowledge of the fetus. Moreover, the technological framework reduces the range of relevant clues for whose interpretation women act as gatekeepers. A technology of exclusively visual signs that renders "a collection of echoes" into a representation of a baby substitutes for prior, embodied states. This reduction also sharpens the focus from a diffuse knowledge of women's embodied experiences to a finely tuned image of the fetus as a separate entity or "patient." This visual representation can then be described by radiologists, obstetricians, and technologists in terms that grant it physical, moral, and subjective personhood (Mitchell 1993a).

Indeed, one ethnographic study of sonographers and their pregnant patients powerfully described the code-switching that medical professionals perform. Among medical peers, sonograms are described in the neutral language of science. But when speaking to pregnant women, sonographers attribute motives to fetal activity and presence: a fetus that is hard to visualize is "hiding" or "shy"; an active fetus is described as "swimming," "playing," or even "partying" (Mitchell 1993b). "Showing the baby" drives its personification (Mitchell 1993a, 1993b). Thus the routinization of a new reproductive technology (or, more properly, a technology whose routinization is most powerfully occurring in the prenatal context; sonograms are also used to visualize the human heart, and abdominal masses, but I doubt whether these uses are personified) provides medical professionals with a "toy" through which they can simultaneously provide a compelling service and stake their claim to authority. The need to both monopolize a new professional turf and popularize its value here contributes to radiologists' and technologists' perhaps unconscious desire to personify the fetus (cf. Brown 1986).

Perhaps the most powerful aspect of that personification process is the sexing of the fetus. The technology often (although not always) allows radiologists to visualize fetal sex organs at the mid-trimester examination that precedes amniocentesis. And whether or not the radiologist "can tell," the chromosome analysis always reveals fetal sex. As Barbara Katz Rothman's study pointed out a decade ago, knowledge of fetal sex increases the velocity of a pregnancy; in our culture a sexed fetus is no longer a developmental imaginary but a "little slugger in a Mets uniform or a ballerina in a pink tutu" (Rothman 1986). Lost in the rush to fetal sexing is the slower process by which even a newborn may remain relatively unsexed, or at least episodically sexed, in the experiences of new parents.

Not everyone wants to know the sex of the fetus. In the interviews I conducted, genetic counselors reported that about half their clients would rather retain the mystery. But in my interviews with pregnant women, less than a quarter didn't want to know, and they were almost always those bearing a second or subsequent child. For first-time parents, knowing the sex is a powerful lure. From personal experience during a second pregnancy in which I didn't want to know the sex of the fetus, I know it is difficult for obstetricians to keep their mouths shut once fetal sex has been entered into the charts of their pregnant patients. Genetic counselors often caution those who would rather not know to announce their preferences firmly when they enter the radiology suite. Otherwise, a loquacious medical staffer is likely to point out the sex.

Some of the lure of sexing is based on control of knowledge. To the question "Why do/don't you want to know the sex of your fetus?" many people (and virtually every Jewish person in the sample![3]) answered, "Because if the doctor [or technologist, geneticist, or clinic secretary] knows, then I should know, too."

> I didn't like the idea that someone knew something about my baby that I didn't know . . . I don't care whether it's a boy or a girl, it really isn't that, it's merely that information exists, and other people have it. (Laura Forman, white theater producer, 35)

> As long as it's known, I feel the parents should know, you know. I mean, we shouldn't be the last to know, it's that kind of a feeling. (Carola Mirsky, white school teacher, 39)

For such respondents, once technology exists to provide the information, ignoring it constitutes deprivation. Such a structure of sentiment surely drives the proliferation of knowledge generation and consumption. For others, the need to know is cosmological:

> Just like that, because it's a miracle of science to know what God provides for you, that's why I want to know. (Felicia Bautista, Dominican factory worker, 37)

> They tell me it's a boy. After three girls! I still don't believe it. I'll believe it when I see it. I heard from a neighbor they sometimes make mistakes. I'll believe it when I see it. But knowing, that's a gift. (Cynthia Baker, African American homemaker, 40)

For some, fetal sex knowledge genderizes in conventional ways:

> Because if it's a girl, you got to be more careful with girls. You can't just let anyone take care of them. (Rafael Trujillo, Puerto Rican unemployed worker, 43)

> I want a girl, but my name is dying. If it's a girl, well, we'll just have to plan for a second. (John Freeman, African American computer technologist, 32)

> Let's face it, knowing the sex made it go from a fetus to a child. I can't tell you how, but now I feel more protective, it's more real. And because it's a girl, I feel more connected to it, to my mother, to my sisters. Jeremy asked me which sister I want to name it for. I don't know if I want to do that. But the possibility made her more of a baby, a full kid, a living child. (Marise Blanc, white college professor, 35)

Several women from working-class families claimed they wanted sex information for practical reasons:

> I figured at this point, financially, instead of buying all those different kinds of clothes, you just buy one specific set. (Angela Carponi, white homemaker, 33)

During the course of my research, I was invited to a baby shower for a pregnant genetic counselor with whom I worked closely. Her colleagues (who had analyzed her fetal chromosomes) had purchased appropriately pink items, but she refused to take them home, saying:

> It's gonna cause a war in my family. My mother wants a girl. His mother wants a boy. They'll both be happy at the birth. But if they find out now, they'll tear each other apart.

In some cases the prospective parents may have a difference of opinion. The decision to know or not to know must then be negotiated:

> I want to know, but Frank doesn't want to know, he absolutely doesn't want to know, he doesn't want some doc, you know, telling him before he has the real experience, finding it out together, in life, not as information. (Marcia Lang, white psychologist, 37)

Like amniocentesis itself, which feeds on age-old pregnancy anxieties, the curiosity and mystery of fetal sexing is now reified and revealed through technology. Old cultural preoccupations with genderizing "who the baby will be" are thus put through the sieve of new technologies of knowledge.

Many women are delighted to claim this new knowledge as their own, using it for old purposes:

> I wanted Frank to get more involved. I didn't say, "Come with me to the first sonogram because it will make you more involved." I think I just said, "I'd really like it if you came with me to this sonogram thing." He's not that affective, he isn't really that connected to the baby yet. But he is connected to me. So I knew the sonogram would get him more connected, through me . . . And I think it was true because he seemed moved, emotionally moved, and for the first time, after the sonogram, he started talking to people about the baby. (Marcia Lang, white psychologist, 37)

> I was frustrated that so many women came alone. I brought my husband, it's important, very important. For good and for bad, it's women's burden, and her husband should know about it, and share it. If there's a miscarriage, he should share the pain. It's his creature, too, and if he sees it on the television, he will know it as his own, love it from the very beginning. He will see God's work, marvel, and share. The family should be united for the test, so that men share the power of God's work and the joy of a new baby. (Juana Martes, Dominican home care attendant, 38)

> I took my husband and my son to see it. I thought he was, you know . . . [RR: Not committed?] Yeah, uncommitted. I thought it would touch home. That's why I did it. I had a feeling. Use a little psychology. My husband, he was amazed by what he saw. (Diana Mendosa, Puerto Rican nurse, 35)

In such stories, a new technology supports an old female strategy of attempting to heighten male involvement with pregnancy.

Many women are also happy to align their descriptions to what technicians and physicians orchestrate:

> It was wonderful, I said, "It's great, can I leave now?" I mean, I didn't want the amnio, I just wanted to see my baby. I saw the spine, the bladder, the orbs for the eyes, the penis, everything, I saw all of it, I loved it. That was very satisfying. Maybe they do it as bribery, so you won't jump off the table. I feel like there's not much discrepancy between the sonogram and what it feels like inside me. (Alicia Williams, African-American public relations executive, 36)

> It's a creature from the moment of conception. On the TV screen I saw it all, a little head, a beating heart, even fingers and a back bone. It looked like a baby but indistinct, blurry . . . As it grows, it will get bigger, and more distinct, almost like tuning in the television. It corresponds to what's inside me now. (Juana Martes, Dominican home care attendant, 38)

It looked very alien, like a little space creature. It was clasping and un-
clasping its hands, and it had its fist under its chin . . . it was moving
around so I could see the arms and fingers, which was nice, then it kind
of got up on its legs, kind of pushed itself up, and you could see the
whole spinal column, and the heart and the eye sockets and the shape
of its skull. It's like a halfway baby now, yes, it's a halfway baby, and it's
an inside-out feeling. (Marge Steinberg, white social worker, 39)

Like the pregnant woman who used the video analogy, Marge Steinberg was
drawing on the fetus-as-voyager imagery that moviegoers and television watchers recognize from films like *2001: A Space Odyssey* and the Right to Life's *Silent
Scream*, and, more recently, the Volvo advertisement selling safety to pregnant
couples and their "passengers" (Taylor 1992). The ad, which presents an ultrasound fetal image accompanied by the message, "Is Something Inside Telling
You to Buy a Volvo?," was withdrawn after public protest over what was widely
perceived as capitalizing on a sacred terrain. Sonographic fetal images perform
practical and aesthetic service in the world at large, where women get to know
them long before they arrive in the obstetrics suite.

Many women also recognize that their viewing is orchestrated and their internal state has been interpreted:

It was nothing, really, it looked like nothing. Then they showed it to
me, and made it something. (Ileana Mendez, Ecuadorian-born baby-
sitter, 37)

To tell you the truth, it didn't really look like a baby, I couldn't really tell
what it was, they had to tell me. (Letty Sharp, white hospital clerk, 36)

You could see at certain points. Towards the end, I couldn't really tell
what was what, and then there was the feet. I saw the legs crossed, and
then it looked like a little baby, cute. After they told me what to look for,
then I knew I was really pregnant. (Sandra MacAlister, African American administrative secretary, 41)

That baby was so active, jumping around, I swear it was mugging for
the camera. But I couldn't tell what was what, I know somebody's in
there, but I don't know who. (Lauren Smith, Anglo-American lawyer, 39)

The "mysterious voyager" image provided by sonograms is compelling,
ubiquitous, and unavoidable. When I asked women to describe their internal
images of their pregnancies, most of them depended on the stereotypes of fetal space creatures: "Like *2001*," "Just like in *A Child Is Born*, you know, kind of
pinkish-creamy," "Floating," and "A little traveler inside me" were common answers. Only a few could imagine other descriptive referents, and these women
had luxuriant animals and vegetables blooming in their bellies:

I could just imagine it like a little fish, you know, the one that jumps a
lot, like a sardine, no, not a sardine, it goes uphill. [RR: Do you mean a
salmon?] Yeah, that's right, a salmon, that's what I feel, this child goes
so low sometimes it jumps like it's going to go through my vagina, that's
how it jumps, all alive. (Angela Carponi, white homemaker, 33)

43

It's got lumps and bumps, and they're growing, organic, you know, sort of like a cauliflower. (Marcia Lang, white psychologist, 37)

For most women, internal images of their pregnancies had been refocused through the lens of sonography, eclipsing any alternative, less standardized, embodied notions of what a fetus felt like. Their internal states were now technologically redescribed. This process of reinscription has a visual history: in 1965, *LIFE* Magazine published Lennart Nilsson's photographs of autopsied embryos, presented in "living technicolor," as the "Drama of Life Before Birth." Greeted with far more awe and credibility in the US than in his native Sweden, Nilsson's imagery was once again presented in the rejuvenated *LIFE* of 1990, where endoscopic photography enabled him to track sperm and egg uniting, as well as fetal development. In the quarter-century separating the two photographic essays, the mother had become not only transparent but also a potentially hostile environment for the sperm, eggs, and fetuses she carries from "The First Days of Creation" (Stabile 1992). We cannot argue that women (and their supporters) were re-educated to see themselves and their fetuses as separate entities exclusively by *LIFE* Magazine's enormously popular articles and Nilsson's (1977) best-selling book, *A Child Is Born*. But these images, which came to permeate sex education, right-to-life literature, and obstetricians' examining rooms, surely contributed to the narrowing of aperture and sharpening of focus on the fetus rather than on pregnant women.

While media presentations of technologically assisted viewing are presented as sources of visual pleasure, actual sonograms may provoke anxiety as well. If the fetus has become "real" through its imaging, as mysterious as an underwater dream and as intimate as a videotape, it has also become vulnerable:

I saw the sonogram of the twins, and I was thrilled. But I really couldn't read it, I didn't know what it meant. They had to interpret it for you, to say, "Here's a heart, these are arms." Afterwards, it made me queasy— they made the babies real for me by telling me what was there. If they hadn't interpreted, it would have just been gray blobs, and now I'm more frightened to get the results of the amnio back. (Daphne McCarle, white college professor, 41)

Because as soon as you see the sonogram, it's very real. They focused on the heart and it was beating, and then you could see the head . . . and the doctor was really terrific, like, there was all this excitement in the room, and she gave me a picture and they're all very positive . . . but you're trying to contain yourself from feeling that way because you know the only reason you're having this test is because you're more likely than the average person to have a problem. So I walked out of there pretty high . . . but I really have been trying to hold back the feeling pending results. (Laura Forman, white theater producer, 35)

With sonography and amniocentesis, one *can* be "just a little pregnant." Laura Forman's comments on self-containment surely echo Rothman's (1986) analysis of the effects of having a "tentative pregnancy." A woman's growing awareness of the fetus she is carrying is here reshaped by her need to maintain a distance from it, "just in case" something wrong should be discovered and she should be confronted with the choice to end or continue the pregnancy. Even as the sonogram personifies the fetus, the amniocentesis puts its situation in question. Simultaneously distanced and substantiated, the pregnancy is suspended in time and status, awaiting a medical judgment of quality control. Personal and social constructions—of maternal sentiment on the one hand and disability beliefs on the other—activate and agitate the fetal imaginary powerfully pictured through these technologies.

Choosing to Choose

The intertwined technologies of sonography and amniocentesis underline the liminality of pregnancy, etching the burdens as well as the benefits of "choice" into the heart of the experience. Occasionally, viewing the sonogram enables male partners of pregnant women to articulate their own engendered anxieties:

> It's definitely a woman's choice, but it's heavy. I think guys should be there for the sonogram and the amnio. It's all very heavy. You really see something moving, it makes it into a person for you. If something goes wrong and she has to have an abortion, after that the guy should know what she's going through, take responsibility for that. (Steve Schwartz, white lawyer, 36)

After encountering his fetus on the screen, one father asked anxiously,

> Where are all the other fathers? Why aren't they here to see this? (John Freeman, African American computer technologist, 34)

But most of the technological augmentation of anxiety is expressed by women, not only because pregnancies happen inside of women's bodies, but because most (perhaps all) cultural constituencies in contemporary America assign the benefits and burdens of making and raising babies to women (Collier and Delaney 1992). The idea that women are responsible for the outcome of pregnancies, including the production of anomalous or ambiguous births, has a long history in Western theology, natural philosophy, and medicine (cf. Bynum 1991; Jordanova 1986; Stafford 1991). In its present incarnation, women's responsibility for children takes place inside an imaginary and bounded female-centered domain in which women are drafted as our contemporary philosophers of the "private":

> So I went off to have the sonogram, and I had these two guys, lab technicians, I mean, we're all in a dark room, semidarkness, and they begin to refer to the fetus as he, it's like there's a real ba-. . . I mean, they were joking, but I was traumatized. It became a real baby. I didn't realize what a sonogram really was, what they show you up on that screen. All of a sudden the baby, the fetus turned its face toward me. And, Rayna, there was a real face. Almost twenty weeks of face. And the technician said, "See it," and I thought for a moment, "He's looking right at me." He looked like that image from *2001*, I mean there was a person there, inside my body, looking out at me. It was too strange. And too traumatic

to have an abortion after that. That's what the sonogram did. (Carol Seeger, white college professor, 43)

I was hoping I'd never have to make this choice, to become responsible for choosing the kind of baby I'd get, the kind of baby we'd accept. But everyone, my doctors, my parents, my friends, everyone urged me to . . . have amniocentesis. Now, I guess I'm having a modern baby. And they all told me I'd feel more in control. But I guess I feel less in control. It's still my baby, but only if it's good enough to be our baby, if you see what I mean. (Nancy Smithers, white lawyer, 36)

To tell you the truth, I had a sonogram with my first one at eight weeks, and it changed my ideas about abortion. We all say it isn't a human being, but that's no longer true. This pregnancy, I waited for the sonogram till the amnio. At sixteen weeks, when you see it, everything is there. The heart is beating, the fingers are separating, the spinal chord is closed. It's your decision, it's your body, and you must do whatever is right because you must raise whoever you have. But it's a human being. You can't have this test without thinking about it like this. (Amana Owasu, Nigerian-born hospital attendant, 35)

The anxiety described by pregnant women and their supporters as they view sonograms invokes dread because it confronts the issue of choice. A diagnosed fetus is potentially an aborted fetus. And the fear of taking the responsibility for ending a pregnancy that one has desired is substantial. Prenatal diagnosis forces a confrontation between attitudes toward abortion in general and the concrete fears, fantasies, and phobias any woman and her supporters may hold about particular disabilities, or disability in general. Notions of normalcy, damage, disappointment, guilt, or grief may thus be invoked by the possibility that the particular fetus under surveillance will fail a eugenic test. Accepting or refusing a particular pregnancy in which a disability has been diagnosed requires an active consideration of whether there are limits to voluntary parenthood, and if so, what those limits might be. Notions of maternal responsibility—for causing the disability, for accepting or rejecting the fetus—are sharply etched into the uses of this technology.

Late abortion is thus the hidden or overt interlocutor of all amniocentesis and ultrasound stories. Surely this notion of abortion choice brings a thoroughly contemporary and American idea of individuality and its rights into intersection with older, quite Protestant and philosophical discourses of responsibility and free will. A modern technology turns every user into a moral philosopher as she engages her fears and fantasies about mothering a fetus with a disability. While this disabled fetal imaginary is properly the focus of another essay, its presence haunts prenatal diagnostic technologies. This dilemma thus interpolates women as gatekeepers of the limits of "the private" in the midst of the continuing US abortion controversy.

But women do not come to the anxiety of this "individual" choice unassisted; there is a representational politics which plays itself out on maternal-fetal terrain. Sonograms are constructed on a ground shot through with power differences in both personal and institutional terms (Duden 1991, 1993; Hartouni 1991, 1992; Pechesky 1987).

> As a counselor, I consider it my job to accompany my patients to every-thing. The sonographer here at City is a right-to-lifer; when he knows someone has a positive diagnosis, that they are going to abort, he hands them a photo of the fetus. Imagine, being forced to take it, to take that picture, when you know the pregnancy is Down's, you know you're going to abort! (Felicia Arcana, South American–born genetic coun-selor, 50)

Our political memory should include attempts by antiabortion representatives to place both a medicalized discourse on the sanctity of the fetus, and fetal imagery, into the *Congressional Record* during the early–Reagan-era hearings on a Human Life Amendment. "The decade [of the 1980s] began, need we remind ourselves, in a flurry of antiabortion, antigay, anti-ERA, profamily, prolife, pro-American rhetoric" (Hartouni 1991:33). Fetuses visualized as persons loom large in the contested politics of feminism and antifeminism that have become a central aspect of our social landscape (Cartwright 1992; Hartouni 1991, 1992).

As I have hinted throughout this essay, interventions into pregnancy have a public, visual, and political history. Fetal images currently are imbricated into an ongoing political struggle over abortion rights. They are thus located and interpreted not only within biomedicine and among women and their supporters, but on our public air waves, in the halls of Congress, in Presidential primaries, and in the marketplace as well. The anxiety about which I have been writing is multilayered. It feeds upon older, more existential fears engendered in the liminality of pregnancy and bears the discursive weight of prior generations of commentary and blame. Now it is given a technological assist by representations of the fetus that are available ever earlier in pregnancy and contribute to powerful, emerging notions of personhood and separation of fetuses from the women who carry them.

While this anxiety is experienced as deeply personal and embodied, its new salience and medicalization are in large measure *socially* constructed through the politics of technological intervention. If fetal images are produced in real time, inside of women's bodies, they are also and simultaneously remediated visually, disembodied, free floating, over the politicized and commodified fault lines of our culture. A visual object at once personal, domestic, and intimate, the real-time fetus also indexes women's obligations, responsibilities, and choices in a world where highly stratified access to health care, child care, and the resources on which stable parenting depend makes the role of "mother" a contested terrain. What are feminists and their allies to make of a technology that tunes in both the visual pleasures and the anxieties of our individual "choices," disembedding them from the social matrix within which they will be actualized? We may be better able to answer this question once we subject the sonogram to multiple interpretations by the diverse women now exposed to it. Surely it is they who can best articulate the aspirations and fears that a new, routinizing reproductive technology calls forth as it intersects old and familiar scripts of inequality.

Notes

The research on which this essay builds was conducted in New York City over the course of the decade 1984–1993. At various times my research has been funded by the National Science Foundation, the National Endowment for the Humanities, the Rockefeller Foundation's "Changing Gender Roles" program, the Institute for Advanced Study, the Spencer Foundation, and a sabbatical leave from the New School for Social Research. I thank them all for their support and absolve them from any responsibility for what I have made of it. Above all, I thank the hundreds of pregnant women and mothers of young children, and their supporters, as well as the medical service providers, who believed in the importance of this work and participated in it. All names have been changed for reasons of confidentiality. My colleagues at the School of American Research seminar greatly enriched my understandings of the literature on social studies of science, especially medical imaging. Their questions and suggestions inspired important revisions of my argument. Paul Rabinow's comments when I presented this paper at the University of California–Berkeley in 1995 inspired a political updating. And Faye Ginsburg applied her usual, quite extraordinary editorial energy to the penultimate draft of this paper; I am grateful, as ever, for her intellectual and personal commitment to the development of my work.

1.   The number of ultrasound examinations performed on pregnant women in the US is not reportable or monitored. While hospital records reveal the number ordered in their facilities, ultrasound equipment has rapidly diffused into the offices of private obstetricians and group practices, where it is in frequent use. There are thus neither national nor local data to support exact figures on the number of pregnant women and their fetuses who undergo ultrasound examinations. One recent collaborative study suggests 50 percent as a national baseline figure, but there are strong indications that the rates in urban areas are significantly higher (LeFevre et al. 1993).

2.   Janelle Taylor (Department of Anthropology, University of Chicago) is currently writing a dissertation on the cultural construction of sonograms. The role of paternal "bonding" and the creation of ties of kinship through imaging are among the many questions she is investigating.

3.   A range of Jews interviewed for the larger study from which this chapter is drawn expressed an unambivalent relationship to technology in general, in marked contrast to those drawn from other religious groups. From the most orthodox to the most secular, Jews seemed comfortable with the idea that God provides the technology, just as he provides the problem; both must be interpreted as aspects of His will. For further elaboration, see Rapp 1998.

AIDS, KNOWLEDGE, AND DISCRIMINATION IN THE INNER
CITY / An Anthropological Analysis of the Experiences of Injection
Drug Users / **Emily Martin, Laury Oaks, Karen-Sue Taussig,
and Ariane van der Straten**

I N WHAT WAYS ARE THE EXPERIENCES of being at high risk for HIV infection,
being infected with HIV, or being sick because of HIV understood and given
meaning by injection drug users in the inner city?[1] From 1989 to 1991
we conducted an ethnographic study in Baltimore designed to illuminate
people's cultural concepts of the body, health, and illness. Using a variety of
methods (participant-observation in community groups, volunteering in clinics, open-ended interviewing), we explored whether people were interested in
the "immune system" and, if so, what they might take the "immune system" to
mean.[2] The study was designed to include people representing a wide variety of
socioeconomic positions and ethnic identities. In the third year of the project we
were able to include a group we had had virtually no access to before: former or
current injectable drug users (IDUs) in inner-city neighborhoods.

The Research and Writing

Through the ALIVE (AIDS Link to Intravenous Experiences) study, we carried out
over forty extended interviews, using as a guide the same topics for discussion
we had already used in approximately 200 interviews in various urban clinics
and neighborhoods.[3] The ALIVE study is a longitudinal study of the natural history of HIV-1 (human immunodeficiency virus type 1) infection among IDUs. A
cohort of former or current IDUs was enrolled in 1988–89 and has been followed
through semiannual visits at a freestanding clinic.[4] Of the 2,921 individuals
recruited, 24 percent were HIV-seropositive at enrollment. The annual rate of
seroconversion in the study population is approximately 4 percent. Injectable
drug users were recruited by word of mouth from community agencies and
housing projects.

The individuals who make up the ALIVE study population are primarily African American men drawn from the surrounding neighborhoods, and therefore the sample of people we interviewed reflects this ethnic and gender composition. Once we had gained permission from the ALIVE researchers, project director, and board of directors, our study became a way station on the round of research sites that each study participant passed on his or her way through the clinic. In one office, he (or she) would give a sample of urine or blood, in return for a $10 fee; in our office he would, if he chose to, give a tape-recorded interview in return for a $10 fee.[5] Written permission to publish quotes was obtained from everyone who agreed to be interviewed.

This positioning of our project placed us squarely within the camp of the scientists studying the health of poor people in Baltimore's inner city. Our window into the lives of inner-city residents was built differently from that of the public health experts, but it was no less narrow. For reasons of confidentiality that ALIVE participants were guaranteed, we made no attempt to follow them into their home territory to observe how AIDS/HIV issues functioned in their daily lives. Also for reasons of confidentiality, during the interviews we did not ask questions about anything not directly related to health, including occupation and family. These limitations on our research closed off many doors we would have liked to walk through. But another door our position opened provided some unexpected insights. The scientists themselves confounded many stereotypes about the clinical detachment and overly rational blinders that non-scientists often suppose confine scientific medical research.

The ALIVE study was conceived with the help of street-corner conversations between its co-PI (project investigator), David Vlahof, and men like those who eventually participated. Vlahof's question to them was, How could a study of HIV be designed that would make long-term participation worth their while? The answer, put simply, was to provide long-term, all-round care for all their health needs, whether related to HIV or not. This insight, obtained from ethnographic encounters carried out by a scientist across a social and cultural divide, informed the way ALIVE works and has probably been largely responsible for the extremely low dropout rate of its participants. Many of our interviewees mentioned how important their access to general health care had been to them. Mostly they obtained this through the various departments of the Johns Hopkins Hospital, facilitated by ALIVE staff.

The headquarters for our larger project was the main campus of Johns Hopkins University, surrounded by the affluent neighborhoods of north Baltimore. Traveling to the headquarters of the ALIVE study on the edge of Hopkins' medical campus meant moving to a different world. The main campus is composed of stately, red-brick, Jeffersonian-style buildings in a groomed, parklike setting with trees, flowers, and grass. The ALIVE clinic, just twenty minutes away, faces an inner-city landscape. On the edge of the tightly clustered buildings of the medical school, it looks out over poor neighborhoods, housing projects, and the turrets of the city jail.

Coming from north Baltimore, one is struck by the presence of uniformed police and security guards on the streets around the clinic. Their presence increases periodically, usually after news of violent crimes in the area. During the period of our fieldwork, security was increased following the kidnapping of a physician and the sexual assault of a female medical student. A security guard sits inside the clinic doors and the staff regularly worry about safety concerns as they walk to and from the clinic.

The writing of this paper as well as the research itself were organized by both collaboration and division of labor. The forty individual interviews were conducted entirely by Karen-Sue Taussig, a Ph.D. student in the department of anthropology at Johns Hopkins, and Ariane van der Straten, a Ph.D. in molecular biology, who at the time was planning to begin graduate studies in public health. This work, by turns tiring, frightening, exhilarating, and fascinating, was transformed by the labor of Jackie Nguyen (the secretarial assistant for the project) into typed transcripts which all of us read and studied. Laury Oaks, also a graduate student in the department of anthropology at Johns Hopkins, carried out participant observation in HIV/AIDS related contexts elsewhere in the city, helping us understand the ALIVE study more broadly. Maintenance of our administrative connections to the Johns Hopkins Medical Institutions, to the director of the project, and to the co-PI was primarily in the hands of Emily Martin, by means of her faculty appointment at the Homewood campus of the university.

All of us who were to be authors wanted the writing of the paper to be collaborative in inception and execution. To ensure that everyone would be first writer on part of the manuscript, we conceived the framework of the paper in group discussions, divided up the writing into sections, and assigned some sections to each person. The resulting manuscript was then discussed and edited, edited and discussed by each of us singly and in groups. Our goal was to facilitate intellectual exchange among scholars at different points in their careers.

Discrimination

Discrimination and marginalization are integral to the experience of participants in the ALIVE study. They attributed the low priority accorded HIV/AIDS in the United States, as many in the gay community do, to their marginalized social status. Many informants speculated that HIV was developed by the United States government, and that no cure is available because the government wants to "kill off certain groups" of people. Some felt that "the government wants to cut down on the population" (Mary Wilson). Seemingly sharing this view, Joe Simpson said, "There's probably a cure. It's just the system, waitin' for more people to die." Earl Johnston expressed little doubt about the possibility of a genocidal purpose of AIDS. He referred to the Tuskegee experiment that the United States Public Health Service conducted for forty years on the effects of untreated syphilis on African Americans in Alabama (Jones 1981):

> I would hate to believe that it was a manmade disease put out here to eliminate certain people in our society, but I wouldn't put it past 'em. Simply because a long time ago they had—what do you call that—syphilis—and they actually gave syphilis to these guys to see how they would act, you know.[6]

Carina Fisher called HIV/AIDS "germ warfare," explaining, "I think the population was over-booming. And the heads of the states got together—I do believe that . . . we were infected by our own government." Gladys Jones expressed her feeling that there is no cure for AIDS "'cause a lot of black people die from it . . . that's what I think." [7]

The participants in the study acknowledged that race and class mark their marginalization and discrimination in American society. Some spoke eloquently about their own attitudes toward racism. Sam Kantor, who is in his thirties, said,

> I try to live and let live and . . . these people walkin' around the street prejudiced, I mean hell, people are people. This thing with white/black, I don't wanna hear that. People are people . . . Everybody should start realizin' that. The world would be a lot better place to live.

It makes sense that Sam Kantor would see discrimination as "a bunch of crap" because, as many informants told us, color and class are irrelevant when it comes to HIV and AIDS. Even if people discriminate, HIV does not. AIDS is perceived as something anybody can get:

> Anybody can catch it . . . the virus does not discriminate. But sometime, you know, it's like people . . . think that money can buy everything, [but] once people get the virus they'll find out, you know, your money don't mean nothin' . . . A lot of people just say, "Well, you shouldn't have been doin' this thing," . . . [but] it doesn't make a difference—we're human, we make mistakes. (Joe Simpson)

The theme that HIV does not discriminate is extended to people with power and public figures: "You can get it, a doctor can get it, anybody can get it. Even the President can get it" (Carina Fisher). For William Porter,

> Ain't no set timing or nothin' . . . A lotta people . . . figure it would never . . . happen to Magic Johnson! . . . A lot of important people . . . caught AIDS . . . It don't make much difference how much star you are, how much money you have, how poor you is . . . You got it, you got it . . . Magic Johnson is just an ordinary person. He a human being, he bleeds the same way everybody else bleeds . . . He's a human being just like everybody else.

We asked participants specifically about their reactions to the news of Magic Johnson's HIV infection because his serostatus was made public the week before we began interviews at the ALIVE clinic. Many participants used Magic Johnson as an example to illustrate their view that anybody is susceptible to the disease. The fact that anybody can get HIV/AIDS translates into the belief that those who have the virus should not be marginalized.

Although the gay community, other "high-risk" populations, and medical and civil rights organizations abhor the idea of mandatory testing for HIV because of the possibility of discrimination against those who test positive, most of our informants at the clinic viewed mandatory testing as something beneficial. "I think it's a good idea," said Frank Stevens, when asked about mandatory HIV testing for health care providers and patients. Jimmy Currant replied,

"I think it's good. I think it's a good thing to do." William Porter supports the idea of testing health care providers and patients as "a beautiful program" because, as he explained, "You don't know who's got what, see, so it's best for everybody to at least have that test done." Sam Kantor responded, "I agree with it totally . . . I mean why put yourself in a position where you might be infecting other people—innocent people . . . I think it should be no problem . . . I feel as though if you don't want to take a test then you have somethin' to hide."

Many activist and legal advocacy groups perceive mandatory testing for HIV as inappropriate state intervention into private life. The participants of the study frequently experience state intervention in the form of police presence in the inner city, which serves as a constant reminder of their marginalized status. One informant told us that he became part of the study population while he was in prison. He was tested for HIV during his processing in the criminal justice system. Because this information was on his criminal record, he became involved in AIDS research studies in Baltimore. Another informant told us that he thinks mandatory HIV testing for health care workers and patients is a good idea, adding that he thought law enforcement officials should be tested as well. He spoke about violent arrests in which blood could pass between police officers and people being arrested. Mandatory testing for HIV, then, can be seen as a social leveler, a process that emphasizes the message that AIDS does not discriminate.

Many of the people we spoke with said that the desire to know their sero-status brought them to the clinic for the first time.[8] A number of participants also told us that the ten dollars they were paid in exchange for their blood prompted their participation in the study, suggesting the dire financial circumstances of a majority of the people we interviewed. Ongoing participation in the research at the clinic, however, became related to feelings about acceptance and about both scientific and self-knowledge. Patrick Williams said,

> Well, I've [been] coming I guess for over two years and first [it] was for the ten dollars, and that was a way of getting high . . . then I fell in love with the people here and their interests and how they made you feel as a person, how they don't mind touching you, talking to you. I look forward to coming in and seeing them now because they have a very warm, positive smile. They offer any kind of help possible and not only offer. They give it to you . . . I like the clinic.

Leroy West said, "When I first came . . . they just amazed [me]. They test you for the AIDS virus and they paid me . . . ten dollars to come to the clinic . . . I can't understand . . . being paid to tell you about your health." Carina Fisher said that she came to the clinic because she was "tired of wondering . . . tired of guessin' whether or not I was positive or negative . . . Curiosity." Donald Jackson described his participation at the clinic as a learning experience. He said "I talked to the lady [social worker] pretty much the way I talk to you. I learned. It was an education for myself also."

This valuing of knowledge makes the participants' giving of themselves for long-term clinical study a gift toward the development of scientific knowledge and to the society that will benefit from the creation of increased knowledge about HIV/AIDS. Furthermore, the worth that participants attached to scientific knowledge suggests they have a certain faith in science and medicine.

Faith in Physicians and Scientists and Its Limits

Study participants' faith in physicians and scientists seemed to be built on their belief that the depth of knowledge these experts possess gives them legitimate power and authority. The imbalance of power and knowledge clinic patients experienced between themselves and health care providers often led them to feel powerless. Many participants resisted answering questions about their own perception of disease processes. Others imagined the immune system as an extension of the modern medical system: "I guess the body has its own little doctor in there" (Horace Miller); "I'd say it [the immune system] is like a twenty-four-hour hospital. Just helps you going" (James Sargent).

Faith in the power of doctors was extended to a general trust in the range of modern medicines available to cure diseases. Homer Stone vividly described the medical era in which we are living:

> My grandmother, right, she had a lot of home remedies . . . the cod liver/castor oil . . . The real nasty medicine, right? You know, if you got sick you didn't even want to go home, but you know it was effective. Now, we are in the age of the pill, somethin' serious. Back then you had aspirin, but it wasn't a lot of pills and things like that.

Or, as Joe Simpson put it, "I mean, some of the same [home] remedies work, but I just go in to the doctor, and the technology has like a pill for this, a pill for that." Treatments can also be perceived as tricks pulled out of the doctor/magician's hat. Erwin Coulter described the workings of a vaccine as "magical," and although he said he did not understand how a shot would protect him from a disease, he trusted his doctor's word on the matter.

Though the participants clearly experienced the doctor-patient relationship as one of unequal power, they also revealed that there are ways to resist the control that doctors wield. Resistance was expressed in different ways. A unanimous belief in the importance of self-help and maintaining a positive attitude illustrates their attempt to regain control over their bodies and lives: "You know, God's gonna help you, and the doctor's gonna be there to help you, but you got to help yourself" (Earl Johnston). Keeping a positive attitude is empowering. Words such as "strength," "energy," "winning," "self-confidence," and "self-esteem" were used by the participants to describe what a positive attitude means to them. For Luther Scofield, even being HIV-positive could be turned into a positive experience:[9]

> And it doesn't do any good to have a negative attitude . . . I think that helps tear down the immune system, too . . . You know, [AIDS is] a negative thing all the way. But then again, some negative things bring out positive things.

Asserting control over his life and having a sense of his personal responsibility and dignity have led Joe Simpson to openly express resistance to his doctors' recommendations:

I ought to keep a positive frame of mind . . . When I first found out, and I was seeing a lot of people that was takin' AZT, and I see what it did to them. I doubt if most of them people died from the virus, or complications of the virus. I think they died from complications from the drugs that they were given. And I have been totally against it . . . I'll be honest, I don't participate in a lot of studies that want to give out medication. I'm not a guinea pig . . . I'll participate in studies where I have to give blood and answer questions . . . I have refused to be in studies where they want to inject me with this and inject me with that.

Increased knowledge and awareness about AIDS brings participants a greater sense of control over their own lives. This knowledge is augmented by understandings of the immune system, which counter fear and confusion about AIDS. Jim Black discussed his thoughts about the importance of comprehending processes surrounding HIV/AIDS:

I believe some people don't even want to know about it [HIV/AIDS] . . . I think they are stupid. But right now I feel more comfortable about it because, y'know, I'm more aware of how to prevent getting AIDS . . . I think about my immune system now more than ever. Since I started this program. I mean really more. I'm more aware or I'm more cautious or I'm more . . . I just, you know, I take care of myself better now then I used to.

For Joe Simpson, becoming seropositive has brought about a drastic change in his life in terms of knowledge and attitudes:

Before I found out I was positive, I had no idea what an immune system was. It was of no big concern . . . it didn't do nothing, it was just like a toe or a foot or an arm . . . and now it's like, okay, this is my life force. You know, this is my body, and everything else works from my immune system and my brain . . . and it's like . . . now I'm lookin' at the body as one big mechanism that all works together. If one thing breaks down, then something else breaks down . . . My outlook on things is a lot different. It's like I understand that I can't live in this world alone, so if I expect somebody to help me in my time . . . I have to do the same.

Finding a Cure

Many of the participants in the ALIVE study believe that a cure for AIDS will eventually be found and place a great deal of hope in scientific research and modern medicine. Derek Hedley says, "I choose to believe" that a cure will be developed. Patrick Williams, who is seropositive, believes that "with the research goin' on today there's a possibility of a cure. That's what my hope is built on." Mary Wilson also expresses a faith in modern medicine: "It might take fifty more years, but it'll be found. With all the great technology we have today they'll find something."

For others, though, optimism is tempered by doubts about the speed at which a cure will be found:

I believe that eventually they find a cure for every . . . any kind of disease if they study long enough and somebody smart enough comes along. I don't know if it's just around the corner though. 'Cause look how long they've been trying to find a cure for cancer. (Earl Johnston)

Although clinic participants acknowledged doctors' ability to control other diseases, many informants recognized the powerlessness of the medical system when faced with AIDS:

Y'know, but once you got it, you got it, you know. I mean you got it. There ain't no ifs around about it, it's not like you get it and like cancer where you, they might cut a limb off, right, you know, of that part of your body and . . . they caught it before it spreads, right? With AIDS, you got it, it don't spread. You can't stop it, it's there. (William Porter)

Comparing syphilis and AIDS, Porter goes on to describe AIDS as unusual because it is "something now they can't get rid of."

Several people expressed convictions about the limitations of science and research, suggesting there will not be a quick solution for AIDS. Robert Gregory believes that "if we do [find a cure for AIDS], it'll be years from now . . . the same way it took years to find out about AIDS, it's gonna take years to find a cure." Many informants, like Homer Stone, are caught between their faith in technology and the apparent failure of science to find a cure: "I'm very pessimistic . . . But, you know, like I think that you know, I hope that scientists are incredible. They can send a man to the moon, right? I hope they can find a cure for AIDS."

Sam Kantor is even more concerned, believing that "eventually [a cure will be found]. The way it is going now, I don't know. I would like to think so, yeah. I would like to think so." He also expressed a general concern of many clinic participants regarding financial issues—how money is spent by the government, the need of money for research, and the cost involved in finding a cure:

I think if you gonna put money on here for like the stadium, hell with the stadium, you put that money to research.[10] I mean that's just my opinion. The hell with this stadium. Why not put that money into this research and try to cure this thing before . . . They gotta find out some kind of cure to kill this thing. I mean if you wanna spend all this money on all this other crap like stadiums and plazas. They cuttin' everything. Take some of that money that ya'll jivin' the government and all this, take some of that money and put the thing in use . . . If you gonna sit there and spend a whole lotta money, it's a good cause for it right there . . . for the dead.

As much as some clients believe that there is a genuine desire among scientists to find a cure, their perception is that the government's attitude toward the epidemic involves greed or even genocidal racism. Some participants believe an existing cure is being held back, while others suggest that finding a cure is simply not a priority for the government. Frank Stevens said, "They probably got a whole bunch of, like, cures, or so-called cures. The government—I guess they can't get no money from it—I guess they ain't going to put it on the market." Similarly, Ed Vanross said, "The Food and Drug Administration, I mean realistically . . . you know, the bottom line is money. You know, if people can't get their money's worth out of something, then they don't want to be part of it."

Although a few informants believed that a cure will only come from God, most felt sure that science and technology would eventually produce a cure: "I think man searches for something deeper, something above it, for a miracle—a miracle drug" (Patrick Williams). Describing the form that the cure will take, words like "pill," "injections," "shots," "inoculation," or "vaccine" were often used. For Ed Vanross, the cure for AIDS would be "possibly an inoculation . . . as with most of they [sic] 'cure-all' drugs, you catch something and they inject you with another needle, and it cures it." There were also those who believe in more science-fiction, high-technology approaches: "It might be a form where as though you have to walk through this machine with a laser might be hittin' your body. Somethin' like that. You know. Or it might be a shock treatment" (Robert Gregory).

Origin of HIV/AIDS

Whereas the possibility of finding a medical cure someday seemed to be a hope most sustained, on the subject of the origin of HIV/AIDS, most ALIVE participants did not refer to medical knowledge. When we asked what the origin of the disease might have been, participants mainly focused on the government, environmental pollution, chemical warfare, and religion. Jim Black, like many participants, felt that AIDS has been around for quite a while but was not identified or recognized as such:

> I heard it started in Haiti, I heard it started in Africa, I heard the CIA invented it—I heard the KGB invented it . . . I heard somebody got it from a monkey. Homosexuals came over from America to Africa, and they had sexual relations, and then the homosexuals came back to the United States and spread it to the other homosexuals . . . I really don't know what the hell to believe but I will say this: I do believe that the virus is out a long time ago, but nobody just had, didn't have a name for it.

The AIDS pandemic was often linked to "bad changes" in the environment, in society, or in people themselves. These included physical changes such as pollution, changes in aspects of the cities, or changes in individual bodies. Jimmy Currant said, "The government made it [AIDS] probably as far as the waste materials and the toxic waste and all that." Earl Johnston also cited environmental causes: "It's probably through chemicals and stuff like they say . . . 'cause I can remember my grandmother sayin' when they used natural fertilizers, they didn't have these things. When they started these chemical fertilizers, then they started things." Others, like William Porter, viewed AIDS itself as a poison, "just like somethin' done leaked out in our body, right, that wasn't supposed to." Joe Simpson mentioned his grandfather's theory that the virus was brought back from the moon. Thus, for many, the HIV virus was depicted as an intruder, an alien, coming from an outside source.

The HIV virus was often believed to be man-made. As mentioned above, some perceive it as having been intentionally developed, an attempt on the part

57

of government to eliminate "undesirables" such as African Americans, homosexuals, and IDUs. Others felt it was a mistake, an accident that "leaked out" from a scientific or military experiment.

The origin of HIV was also frequently related to war. "My theory," said Erwin Coulter, "is it came from Vietnam, they were spraying a little bit orange, and it ain't worth it. Orange—AIDS—Agent Orange, the chemical." Some thought of HIV as a by-product of chemical or germicidal warfare or the atomic bomb. Luther Scofield said, "It's gas release from germicidal warfare in some of those storage containers."

Still others saw a religious aspect to the origins of HIV/AIDS: "Sometimes I think it's God's punishment for the wrongness that we do" (Patrick Williams). For Homer Stone,

> Even though they said it was man-made, I think everything arise from God and I think that it's a reason for the disease to be here . . . I believe that the disease was an enhancer to enlighten the human race because of the way things were going. I think that God had a lot to do with it.

The Meaning of AIDS

A number of clinic participants discussed the significance of the AIDS pandemic in terms of holy scripture. They conveyed a sense of apocalypse, of the end of an age:

> The world's outta control. And we got—well, things are happenin' . . . Some prophecies all comin' to pass and they said in Revelation like, "There will be plagues in the last season," and I think this might be a plague. This might be one of the plagues. (Homer Stone)

For Derek Hedley, AIDS was a way for society to purify itself: "Mankind will probably find a cure for it. But for right now, things are like filtering, filtering out the undesirables. The people that's doing that's not what's right."

In spite of the religious and moral dimensions ascribed to AIDS by some participants, for most, AIDS was related to violence and death. "It's a deadly disease that I know I'll die from. It's scary. It's scary and when I hear AIDS I hear death, and that's what I relate it to" (Derek Hedley). "For me . . . it's a very strong disease that can kill you . . . It's a known-in-fact-killer" (William Porter). Luther Scofield expressed his perception of AIDS as a dangerous outsider from whom one needed protection, saying, "I guess society on a whole is used to being able to combat intruders whether it be human or disease-wise."

Participants described both devastating and beneficial changes in their lives. AIDS can strain families and cause dramatic deterioration in physical appearance:

> I think it's just eating them up, just deteriorating their body. Like my friend, I've seen the way he lost weight. I guess it just eats up the immune, you know, your system, your immune system, whatever you're supposed to have, it eats you [alive]. (Sam Kantor)

> HIV is cancer to me. It's eatin' away your body. (Steve Jones)

It makes you aware that life is really short and that you have to live each day to the fullest now . . . I mean, I think what it do, it like scares you into reality . . . what it means, it means stayin' in reality and living your extent to the fullest. (Joe Simpson)

Body Images

The most common image participants used to explain how a person stays healthy was a strong immune system, and this system was almost always described in explicitly militaristic terms. For some the battle was offensive, and the individual saw himself or herself as identified with the offensive army. As Homer Stone said, "Yeah, it's like a[n] army inside of us are [in] a constant battle to defend our body from all invaders. The combat that our body has . . . Yeah, combat, just instead of like defending. I've always been a hero-man myself."

For others, the battle went on inside the body, separate from the person, and the body relied on the person's life activities for help. Ray Green said, "Your body don't protect itself. It's you that protect your body, you know. Your body don't do nothin' unless you do it yourself. It's your mind." When asked what he does, he answered "See, what I do, I eat very cautious, more careful about a lot of things. That's one thing."

When asked to imagine what happens inside her body when she starts to catch a cold, Carina Fisher answered, "Well, I can imagine they're fighting in there. The reds and the whites and all. Decidin' who's gonna take over. You know. Then I come along and give 'em a little assistance to help alleviate the problem, and to give the winning team some extra help." When asked what kind of help she referred to, she replied, "All types of medicines. Some cough medicines. I might increase and take an extra vitamin pill. And drink a lot of juices."

For some participants the action that matters takes place at the edges of the body's territory, as if once the walls of the body are breached, the consequences are entirely uncertain and not likely to come out well:

Okay, it's like our defenses. It's like we got our own army. Yeah, in our body, right? And automatically when something—aliens!—comes into it, it doesn't belong there, our army attacks because it has no business being there—unless the alien is too strong, we cannot. (Homer Stone)

He sees the immune system as

a[n] entry control . . . I imagine it like radar . . . As soon as the alien hit, we have an automatic radar system that sets off our soldiers or white blood cells that automatically light on instinct . . . It's like it's hard for a disease to just creep up in the body . . . It's like you got your own defense mechanism like an alarm system that goes off.

For James Sargent, the immune system is "more or less like our police force. Protect, you know, the people. You know, hey. 'No, you cannot come in or we do what we can to fight you off!' "

For others, there are interior responses the body can make even if its outer defenses are breached. When asked what happens inside his body when he is exposed to some contagious illness and does not catch it, Donald Jackson answered:

> If most of the germ-fighting cells are finally gone, then the body has a way of putting up another defense against these—it may not be as strong as the cells that's necessary to fight them, but it has enough to ward them off. I don't think it will be able to withstand several attacks, but I think one or two minor attacks, yes . . . The body, I mean, I don't know what it is, but I think the body is equipped with a second line of defense to ward off, just for a while though because eventually it's going to be worn down . . . Whatever cells are strongest, once they run out, once they're gone, then you have a second line of defense, you know what I mean? Somewhat for a pretty small time and then after that eventually it would be overrun.

HIV infection is seen as damaging the immune system irreparably and these men and women developed a variety of images to express this. In articulating the bodily processes they associate with HIV/AIDS, participants' descriptions included ideas about deterioration, viral takeover, military abandonment, exhaustion, and morality.

Deterioration

When she described what is happening in someone who has AIDS, Sandra Lyon said, "It's deterioratin' . . . To me the body's just rotten, you know. It's just everything inside is just, just, somethin' is eatin' at it. Stuff like that." For Mary Wilson, "Oh, it's constantly eating away at them. They just deteriorating inside. I don't know whether AIDS hurts, you know, physically, but looking at them and the pain in they face, it looks like it hurts real physical."

Viral Takeover

Well, the immune system is your defense mechanism . . . It's once, you know, once the defenses are down, you're subject to any blows that might come your way, you know. And without medical assistance your days would be far fewer. [The immune system is] made of, like I said, white cells which combat foreign, foreign intruders into your, you know, your normal system. And they are like a . . . mini-computer. You know, they have memory banks. If they have been invaded by a certain thing prior, then they know exactly what to do. You know, they call the troops out and they get busy and eliminate this. But I was reading somewhere in *Reader's Digest* the AIDS thing, you know, they take on the same shape. They get up on the cells, and mimic, you know . . . And they alter all your cells . . . That's their main objective. They just take over your whole body . . . you're helpless. And your body is confused because, you know, there's a . . . [guy over] here's supposed to be helpin'. And he's floored, right, so. The main, you know, function of the T cells is, is to retain . . . and store . . . memories of different illnesses . . . like childhood illness, measles, mumps, [influenza]. If you come in touch with it later on, your body has an immune system because it has developed as a defense to attack. But if something new comes, it'll still combat it and develop a defense for that for future reference. But with AIDS rendering it and destroying the immune system . . . it'll activate, but it just doesn't have the strength to ward it off, you know. And as it gets progressively worse, you know, there's no resistance at all. And that's the cause of death. Diseases just happen to luck up and join the crowd. (Luther Scofield)

Military Abandonment
What do you imagine is happening inside the body of somebody who has AIDS?

Well, most of the germ-fighting cells are being eaten up by the disease . . . They just not there anymore, whereas they would have forces one time, they won't be and they gradually leaving now but you know, on a scale one to ten they be gone completely and something like a common cold be able to just wipe the body out. It's like a army deserting . . . going over the hill of life. (Donald Jackson)

Exhaustion
You see because AIDS is a disease that all your antibiotics are being used to attack and defend against that disease so if you get enough of it, you have no defense or anything else. That's why most people that have AIDS you don't hear them really just dying of the disease, they die of pneumonia and you know, other illnesses because the defenses are so totally exhausted and broken down that they have no defense system to really combat the other diseases that's coming into the body. (Homer Stone)

Lived Experiences

Talking with participants in the ALIVE study not only gave us insights into their body images but also vividly conveyed to us the nature of their daily lived experiences. Questions about how much they think about HIV/AIDS and what HIV means in their lives were often answered with detailed descriptions or theories. HIV/AIDS is a reality that most must confront personally on a daily basis. Each person's reactions to HIV/AIDS was different, but almost everyone shared a concern for the present and future effects of HIV/AIDS on self, others, and society in general.

The seriousness with which most participants treat the disease is clear; HIV/AIDS is described as "deadly," "a known-in-fact-killer," and "slow death." The vast majority demonstrate an understanding of which activities can lead to HIV infection. The knowledge that some activities increase the likelihood of infection resulted in clear-cut divisions between things one should and should not do. Statements about how the virus is transmitted often referred to "good," "bad," or "dirty" activities. Sam Kantor described dirty needles, sex, and "messin' with dirty people" as likely ways for a person to become HIV-infected.

William Porter explained that HIV/AIDS comes "from a person that have bad blood . . . And it comes through dealin' with drugs . . . y'know, know bad news by not taking care of your needles . . . an' it come from bad sex. Not usin' condoms right."

Through gaining medical knowledge or, as Jim Black put it, the "medical things . . . that let me know what to do, what not to do," ALIVE participants can assign "good" and "bad" to specific actions. "Bad" actions, such as sharing used needles and having sex without condoms, are "dirty" and can lead to HIV infection. "Good" actions, then, are supposedly "safe." William Porter emphasized that, despite guidelines, many do not follow safe practices: "And for society . . . they need to know about it [AIDS] because . . . it's to me, it's the number one killer. Because it's so many young people out there, right? And not only young people but you got a lot a' old people doin' drugs too, right, you got a lot a' old people and young people practicing unsafe sex."

A knowledge of what is "good" or "safe" and what is "bad," "dirty," or "unsafe" can encourage people to focus on activities that do not transmit HIV. William Porter advocated "getting people involved [informed]" and telling people, "If y'on drugs, clean the needle. And if you havin' unsafe sex, start havin' safe sex." With concentration on the role of individuals' activities, decreasing the spread of HIV/AIDS lies in changing behaviors from "bad" to "good." At a broader level, however, it is obvious that such changes are not merely individual. Relationships with others—whether friends who share needles, sexual partners, or family members—become involved. Many have had to confront challenges to acting in a "safe" manner as individuals and in the context of relationships with others. Many ALIVE participants say that being HIV-positive or being fearful of becoming infected causes strains and changes in intimate relationships in particular.

For example, having safer sex often involves the question of using condoms with a partner. While AIDS researchers have paid increased attention to women's negotiating powers, little has been said about the complexities of men's active roles in demanding safer sex practice outside of gay communities (Kane 1990). Sam Kantor's comments indicate that male insistence on using condoms is also cause for stress on a relationship:

> I've been more prone likely to use condoms better than I've been. That's kind of got her [his girlfriend] a little shaky right now. She wonders now why I'm using condoms now but I tell her I'm trying to have no more babies. I'm trying to look out for her benefits as well as mine. I don't, as a matter of fact, I know I don't have the virus or whatever, but who knows, you know, anything could happen. So in that aspect I'm trying to protect her as well as myself.

Those who have tested HIV-positive tell of changes in their relationships also. Efforts to be sure that one does not place others at risk of infection often result in avoiding speaking about one's serostatus, avoiding sexual intimacy, and avoiding "casual" physical contact.

Regarding his bouts with AIDS-related illnesses, Leon Walker said, "But nobody know what I went through but me and a few people. A lotta people don't know, I just . . . you know, I really haven't never opened up and told anybody."

Luther Scofield pointed out that he only told one other person, someone he had "shared [needles] with." Patrick Williams and Leroy West say they have only told their families about their HIV-positive status. Patrick Williams explained that "most of the time I'd be hiding and being ashamed of sharing it [his sero-status]."

Changes in relationships were expressed by some participants as one-sided, permanent decisions. Negotiating with others about changes such as always having safer sex is not apparent in most descriptions. Homer Stone, who told us he is seronegative, said, "I have actually told my girlfriend that if I had it, I wouldn't have nothin' to do with her, you know . . . Condoms are not 100 percent. I wouldn't do that." Patrick Williams stated plainly, "I refuse to have a girlfriend," and Ed Vanross admitted, "I've become celibate." These men and others mentioned the broader consequences of living with HIV and fears of transmitting it. Even casual physical contact is avoided by Patrick Williams, who said, "I don't kiss my niece," and Leroy West explained, "I told my sister, . . . 'Well I don't wanna hold your baby.' And she was shocked."

Virtually all of the people we spoke with are well informed about current public health knowledge concerning how HIV is transmitted and which activities are best avoided. Some even take further precautions. Abstaining from all sexual contact and avoiding "casual" physical touch are examples of attempts to be absolutely safe. A grounding in scientific evidence and knowledge does not necessarily provide much confidence when one is faced with actions that do not seem totally safe in the face of a fatal disease.

Other people also expressed disbelief that there are a limited number of ways HIV is transmitted. Patrick Williams stated,

> I think there are more ways that people can make contact with it . . . there's no safe way. Well like with HIV for instance I had a cut on my hand and you know passing money, coins, that is constantly change into other people's hands, their handshakes. The person with HIV and they have a cut and you have a cut or open wound, and just the contact. I think, you know, it's being passed a lot of different ways.

Ed Vanross noted, "Statistically it's said that they [HIV/AIDS] can't be transmuted any other way. I'm really not too sure of that, I mean, if you kiss your husband and he's . . . you know, the saliva, it enters into your blood stream."

Homer Stone stated that intercourse and IV drug use are the most likely ways to be exposed to HIV, adding that transmission through kissing, due to saliva, may also be a possibility. Chris Stevens too mentioned the possibility that saliva transmits HIV, and went on to say, "I don' know about all this 'toilet seat,' I don't know about all that, you might can, you might cannot [be infected with HIV this way]." Joe Simpson thinks that all the ways HIV is transmitted are not yet discovered: "The virus finds ways that I guess we don't think about. To transmit itself to other people. And then sometimes, it's not successful . . ."

The reluctance of so many participants to believe in and act solely on scientific or medical knowledge can be seen as connected to the pervasiveness of HIV/AIDS and the threat of infection in their lives. The virus and AIDS-related illnesses confront these individuals at the most personal levels; this is true for those who are HIV-positive as well as those who are not. Leon Walker, who is

HIV-positive, said that he thinks about AIDS everyday. Similarly, Jim Black, who is seronegative, commented, "I don't think there's a day that's passed that I haven't thought about AIDS."

HIV/AIDS was described by various participants as "very, very contagious" (Jimmy Currant), a "very contagious killer" (Michael Langford), and "contagious to the point where anybody can get it" (Homer Stone). Still, "safe" behavior is recognized as effective by some. Frank Stevens noted, "You can avoid it. Practice things that you know could prevent it. You know, it's as contagious as you want it to be." Luther Scofield echoed this: "It [AIDS] is no more contagious than the individual makes it . . . It's very low on contagion if, if, if, the guidelines are followed, you know."

To these people, HIV/AIDS was not present merely in distant or abstract ways. They might themselves be infected, have friends or family who are HIV-positive or who died due to AIDS-related complications, or they might be afraid of becoming infected with HIV. Still, not all descriptions of what life is like for someone who is HIV-positive are constituted in negative terms. Those who are immersed in a situation where HIV/AIDS is "everywhere" may have particularly clear perspectives on what it is to be a person living with HIV/AIDS.

Being HIV-positive does not mean inactivity and constant hopelessness. Rather it involves increased attention to aspects of life not privileged before, such as health. Care for one's physical state, bodily "maintenance," and a focus on positive emotional attitudes about life are all a part of changes described as associated with being HIV-positive. Michael Langford noted, "I know a couple of guys who are HIV-positive, you know, when they found out they had it, you know, they didn't give up; they got their bodies in shape and eatin' healthy, you know. Keep active, that helps." Ed Vanross, who is seropositive and has "lost several friends to this disease," said that taking care of one's body and being strong mentally are important. Joe Simpson, who is seropositive and says he does volunteer work with people whose "bodies are deteriorating," admitted, "I sometimes wonder, 'Am I gonna be like that?' And then I have to come down to reality and be grateful that I'm not payin' yet. And that I'm still healthy and life is goin' fairly well."

Conclusions

Our conversations with ALIVE study participants reveal that their understanding of HIV/AIDS is congruent with their personal experience of extreme discrimination. Many of the participants live in a world where HIV/AIDS is ever present, yet the disease (or threat of the disease) is but one of many uncertainties they face daily. Economic and social concerns converge under the shadow of HIV. The possibility of sporadic or continuous unemployment and the exorbitantly high costs of medical care (both financially and in terms of time spent dealing with bureaucratic systems such as Medicaid) are compounded by the social stigma attached to HIV/AIDS.

Among the participants of the study there is a tension between faith in science and medicine (a belief in the possibility of a cure developed through scientific and medical research) and the suspicion that a cure already exists and is being concealed. The possibility that a cure is being withheld fits participants' beliefs that the virus originated in a government or scientific project gone awry and is reinforced by knowledge of at least one documented case in which an existing cure for syphilis was withheld from a part of the African American population (Jones 1981).

The idea that the virus might have come from the government and that an existing cure is being held back in order to "kill off certain groups" may serve to explain another example of the violence that appears so present in the lives of the study participants. The visibility of police in the area around the clinic suggests the proximity of the state in the daily lives of the local people, while the suggestion that police officers be tested for HIV demonstrates that state violence is an integral part of the experience of clinic participants. The military and battle metaphors for disease frequently used by ALIVE participants occur very widely in the US, as our and others' research has shown (see, for example, Haraway 1991a; Martin 1994; Treichler 1987; and Waldby 1996). But ALIVE participants' strikingly consistent use of these images, to the exclusion of any others, to describe biological and bodily processes hints that for them violence is understood as an immediate reality that has—apart from any particular illness—already been incorporated inside the body.

One important role ethnographers can try to play in any society is to increase the visibility of those granted little legitimacy by the society at large and to increase the audibility of their concerns. Listening to ALIVE participants talking about their daily lives may help the scientific community to gain insight into the distinctiveness of their situation, as complex human beings emerge from behind the statistics. Listening to them talk cogently about why they believe they experience systematic discrimination at the level of their bodies and their communities may help overcome common middle-class stereotypes about the lesser mental acuity of poor people. Listening to them describe the positive actions they take on behalf of their own health might help overcome common stereotypes about the passivity of the poor. In these cases, ethnography gives a conversational voice to people who have been silenced in discussions about health and the body.

Another role ethnography can play is to broaden lines of communication between different groups. Our research unexpectedly made a small contribution to that effort when we presented some of our findings to the clinic staff—the nurses, social workers, and office managers who keep the study organized, the participants in the right place at the right time, and the office functioning smoothly. Although these employees see ALIVE participants every day and have obviously warm and friendly relationships with many of them, they were not accustomed to hearing the outpouring of deeply felt appreciation that was often expressed in our interviews. The sentiments behind Patrick Williams's remarks were repeated many times: "I fell in love with the people here and their interests and how they made you feel as a person, how they don't mind touching you, talking to you. I look forward to coming in and seeing them now because they have a very warm, positive smile. They offer any kind of help possible and not only offer. They give it to you . . . I like the clinic."

When we read this and other selections from our interviews at a clinic meeting, several staff members had to brush away tears from their eyes. The conduit of information between study participants and staff was opened a little wider, allowing the intensely positive feelings on both sides to flow more freely. Whether mitigating the distance associated with inequality or simply opening ongoing conversations to new topics, our research makes it plain that ethnography can function as an intervention in complex social processes involving knowledge and power.

Notes

The research for and writing of this paper were collaborative. Authors' names are listed in alphabetical order.

1.   See Feldman and Biernacki (1988), Grund, Kaplan, and Adriaana (1991), Kane (1991), Kane and Mason (1992), and Sibthorpe (1992) for other ethnographic studies of IDUs.

2.   For other publications from this research, see Claeson et al. (forthcoming) and Martin (1994). The research was funded by the Spencer Foundation. In addition to the authors of this article, Bjorn Claeson, Monica Schoch-Spana, and Wendy Richardson helped with research for the longer study. Jackie Nguyen carried out the essential work of transcribing the interviews used for this paper with accuracy and care.

3.   We owe this opportunity to the gracious and generous help of David Vlahof, co-investigator of the ALIVE study, and Liza Solomon, project director and co-investigator.

4.   For detailed information about the ALIVE study, see Vlahof et al. (1991) and Vlahof et al. (1990).

5.   We would like to acknowledge with thanks the help of the ALIVE staff, who arranged the interviews and handled the paperwork associated with our visits.

6.   All names attributed to participants are pseudonyms. Remarks in italics were made by the interviewer; ellipses indicate brief omitted material. Words in brackets within a quote were added for clarification.

7.   Four of our interviewees were women. By the happenstance of who came to the clinic on the days we were there, and who agreed to the interview, women are more highly represented in our interviews than in the ALIVE study itself.

8.   For reasons of confidentiality, the clinic did not reveal anyone's sero-status to us. We asked in the course of the interview, and everyone told us.

9.   One organization, Street Voice, has taken a particularly active role in prompting further action on the ways HIV affects life in inner-city Baltimore. Each week this nonprofit group distributes a two-page, brightly colored newsletter to over 7,000 people "face-to-face on the streets of Baltimore." The practical information it contains is meant to be used as a "tool for education and survival" and covers a wide array of topics such as finding short-term housing and low- or no-cost medical programs, advice from doctors on afflictions like herpes and syphilis, and resources for locating employment, job training, and literacy programs. Positive messages from former IDUs are regularly featured, along with quotes of motivation and inspiration.

10.   The City of Baltimore erected a brand-new, state-of-the-art baseball stadium for the Orioles, which opened in the spring of 1992.

BODIES, ANTIBODIES, AND MODEST INTERVENTIONS /
Deborah Heath

SINCE 1992 I HAVE BEEN DOING FIELDWORK on the cultural practices surrounding contemporary genetics, spending extended periods in two US research laboratories and following the networks that link laboratory life to wider worlds. My interest is both in comparing particular variants of local, embodied technoscientific practice and in what emerges from translocal encounters. One of the labs is in a molecular biotechnology department, where I found my participant niche working as a DNA sequencing technician. Preparing DNA samples for automated computerized sequencing, I worked with an array of tools ranging from bacteria and enzymes to robotic workstations—a cyborgian network of organisms and machines.

The second lab, the focus of much of this paper, is part of a research unit located in a children's orthopedic hospital. Working as an apprentice cell culture technician, my tasks included learning to collaborate with mice, lymphocytes, and tumor cells and to harvest and purify the monoclonal antibodies that they/we produced. The unit's overall focus is research on connective tissue. The principal focus of my lab group has been on the characterization of a connective-tissue protein called fibrillin, discovered by Dr. Lynn Y. Sakai, the head of the lab (Sakai, Keene, and Engvall 1986). Mutations in the fibrillin gene have been shown to be the key factor in a heritable condition known as Marfan syndrome.

In November 1992 I followed members of the lab to the Second International Symposium on the Marfan Syndrome, a gathering held primarily for scientists and clinicians. In August 1993 the National Marfan Foundation, the US lay organization for affected individuals and their advocates, held its ninth annual meeting in the city where the lab is located, in conjunction with a scientific workshop on the Marfan syndrome. The scientific meeting was chaired by Lynn Sakai and organized with support from members of her lab, including its resident ethnographer. Like the marketplace, the fair, the pilgrimage site, or the E-mail conference, meetings such as these are terrains where boundaries of identity and difference are mapped and contested, stretching the limits of local cultural practices.

This essay joins a conversation weaving together the interpretive and critical threads of cultural studies, feminist theory, and social studies of technoscientific knowledge and power (cf. Haraway 1994, 1997; Rouse 1993, 1996a; Traweek 1993). The paper explores the heterogeneous networks of association (*pace* Latour 1987) that infuse everyday laboratory practice and link it to other cultural-material domains, including the annual conventions that bring together those affected by Marfan syndrome with researchers, clinicians, their patrons, and others, including filmmakers, public relations officers, and ethnographers. Tracing the traffic in and out of the lab underscores the heterogeneity of technoscience and the permeability of its borders. At the same time, it draws attention to asymmetries, differences, and contestation among producers and consumers of technoscientific knowledge.[1]

The main characters of this essay comprise a spliced community, a multiplex alignment of human and nonhuman players linked through medical, molecular, and personal embodiments of Marfan syndrome. The stories presented here are neither the master narratives of disembodied subjects nor transparently descriptive anecdotes. Instead, these tales are devices in what I'm calling *modest interventions*—translocal engagements that reveal, perturb, and perhaps transform the constructed boundaries between local, situated knowledges.

In their account of Robert Boyle and the emergent culture of modern science in seventeenth-century England, Shapin and Schaffer (1985) describe the figure of the modest witness, the self-invisible gentleman-scientist who aimed to mirror nature while revealing not a trace of his own history.[2] The originary practices of the Early Modern modest witness, as Elizabeth Potter (n.d.) and Donna Haraway (1997) have compellingly shown, undergird canonical gendered notions of objectivity. The accounts in this essay speak to an alternative mode of witnessing, based on modest interventions and achieved not through holding objects at a distance but through partial connections and intermittent engagements among different constituencies. Unlike the "view from nowhere" (Bordo 1990; Nagel 1986) that has been the legacy of the modest witness, modest interventions recognize—and make use of—both the local, contingent character of scientific practice and the traffic that connects different locales.

Appropriating Latourian actor-networks (cf. Latour 1987; Callon 1986) for use in the poststructuralist toolkit, my approach is to chart the networks of association that link laboratory practices to other domains, while at the same time attending to the engagements, disjunctures, and constructed boundaries that disrupt what might otherwise appear to be a seamless web of linkages. Taking ethnographic liberties, some of the interventions that I will describe are my own; others are made by those with prior claim to native status in these technoscientific milieux. In both cases the effect of these border incidents is to make connections visible, at best transforming contradiction into a resource, a field of possibility.

The Mindful Body at the Bench

Lynn Y. Sakai is a protein biochemist, well regarded in her field. She is sansei, a third-generation Japanese American. Before beginning her scientific career she did graduate work in political philosophy. I had worked briefly in the lab where she is PI (principal investigator) before beginning my research in the DNA sequencing group. She had invited me back in part, she said, so that I would have a broader understanding of contemporary biology. "It's more than just what the DNA gene jockeys do," she said. "You should learn something about what DNA expresses: the proteins." Sakai and I also share an interest in critical theory, though apparently from opposite sides of a modernist divide: she regards me, usually with a kind of gracious scientific curiosity, as a "nonlinear thinker." Since beginning our work together we have maintained a running dialogue about power and technoscience. In the following exchange I asked her to explain why she had left philosophy for science.

> *LYS*: It was because I came to think that philosophy, theory, had no place in the modern world. It used to be that philosophy was related to political activism, to what went on in the world. These days, my old mentor used to say, theory has gone mad. There's no unified theory to account for the complexities of the modern world. In science you work with your hands; this activity is what Marx said makes us uniquely human. You have a direct impact on things in the world in science, with less chance of being alienated from your work.
> *DH*: Unless you're a technician.
> *LYS*: [laughs] Yes, that's right. Most of the time technicians don't get credit for the work that they do. I think that's wrong. Often what happens is that the post-doc is handed a project that a technician had started. The post-doc just puts the icing on the cake and then gets credit for the work, usually as first author on an article. Of course, the PI is last author; they still own the work.
> *DH*: So really, the scientific labor process works much the way the industrial labor process does.
> *LYS*: Yes, that's why scientists cling to the distinction between mind and hands; it has to do with how credit for work is allocated. The PI is the mind; the technicians are the hands.
> *DH*: So what would make the hierarchical order of things in science change?
> *LYS*: I don't know. I try to do things differently in my own lab. My technicians get credit for the work that they do; they're listed as authors on my papers, and I have them give presentations. I can run my lab the way I choose to. [She laughs.] Of course, I'm still the last author. Labs are like independent fiefdoms; they're really premodern.

On one level, the bench laboratory is the territorial domain of a particular PI. Still, no lab is wholly independent or self-contained. Its autonomy is mitigated by dependence on patrons for funding, space, and equipment, as well as by interdependence with collaborators and reliance on an infrastructure of technical, administrative, and maintenance personnel shared with other laboratories.

Lynn Sakai's laboratory is located in the research unit that occupies the fifth floor of a children's orthopedic hospital in Portland, Oregon. Both hospital and research unit are supported by the Ancient Arabic Order Nobles of the Mystic Shrine, the Masonic fraternal organization known as the Shriners that was founded in the US in the 1870s. On my way to the lab I pass a display case in the first-floor lobby of the hospital that houses a collection of red fezzes with black tassels, part of the orientalist ceremonial garb of the organization. Patient care is free at all Shriners Hospitals for Crippled Children, and the funding for researchers is comparable to that available through US government sources. On clinic day the lobby is filled with children, many of them in wheelchairs or moving down the hallways using walkers or crutches. Ignoring the elevator, I pass a room marked "Prosthetics and Orthotics" before I enter the stairway to the fifth floor.

The perceived boundary between the world of the research unit and the activities of the rest of the hospital is monitored and reinforced by the unit's spatial segregation. As I enter from the stairway, I am met by large signs on the exterior doors that read "Warning: No Unauthorized Personnel." The floor of the research unit is divided into individual laboratories, each one allocated to a particular principal investigator and his or her staff of technicians, graduate students, and post-docs. In the hallways adjoining the labs hang a series of framed photographs, enlargements of pictures captured by an electron microscope. Several are images of the protein called fibrillin.

It is one of my first days in the lab where I will be working as a cell culture technician. I have successfully passed the initial induction requirements, drug-screening and TB tests, and have been given a photo ID card that identifies me as a "research tech." I am now in the cell culture room in Lab Three with the head of the lab, who is showing me how to make the medium that is used to feed cells. I am wondering if the skills I acquired working as a DNA sequencing technician in another lab will help me out. I am concentrating, holding an electric pipetter fitted with a long disposable plastic pipette tube, trying carefully to measure one of the ingredients for the medium. It feels awkward in comparison to the smaller plastic-tipped pipetter that I had grown accustomed to using for DNA work. "No, no, not like that." The head of the lab shakes her head, laughing. "You don't have to be that careful. It's not like molecular biology!"

This encounter, among many others, taught me something about how technoscience is grounded in everyday practice and how specific, often mundane, practices help to distinguish one field of endeavor from another. It reveals both the local, embodied materiality of technoscientific knowledge and its translocal heterogeneity. The body-knowledge that I had brought with me from my other field site served in this new setting as a cultural boundary marker, revealing my time spent with a different disciplinary clan.

Recent studies in the sociology of science have pointed to the importance of local or tacit knowledge in technoscientific practice and knowledge production (cf. Collins 1987; Knorr-Cetina 1981, 1992; Lynch 1985). As Cambrosio and Keating (1988:249) put it, "Much of what is important to the understanding of an experimental protocol is not contained in the instructions but is incorporated in the various visual and corporal movements that make up the actual practice." Yet this work has often erred, in my opinion, in describing tacit, experiential, or nonverbal knowledge as inarticulable or unconscious. For example, Knorr-Cetina (1992:121) insightfully portrays the local, holistic approach that benchworkers in a molecular genetics lab use to optimize laboratory procedures, drawing on knowledge that is "implicit, embodied, and encapsulated within the person." However, her discussion takes for granted the mind/body dichotomy that grounds Western notions of objectivity, as well as the cultural-ideological distinctions between technology and science and between technicians and other laboratory workers:

> It is a knowledge which draws upon scientists' bodies rather than their minds. Consciousness and even intentionality are left out of the picture. And there is no native theory as to what this body without mind is doing, or should be doing, when it develops sense. (Knorr-Cetina 1992:121)

As I understand her argument, Knorr-Cetina (1992:119) relates what she sees as the unconscious aspect of embodied knowledge to her claim (which I find otherwise persuasive) that scientists and technicians are "methods," that they are part of a field's apparatus of knowledge production.

As in Knorr-Cetina's account, discussions with my laboratory colleagues, along with my own hands-on experiences in the lab, reveal the persistent division between mind and body in technoscientific practice. Yet the same interlocutors also present critiques of the dominant paradigm. These counterdiscourses (might we call them a "native theory"?) accord significance to an intuitive, corporeal knowledge that, while imbedded in practice, is nonetheless conscious and socially transmissible. Terms like "body-knowledge," "art," "magic," and "good hands" are frequently used to describe this alternative way of knowing.[3]

In the course of one of our conversations I asked Lynn Sakai to comment on the assertion that embodied knowledge is unconscious. She said,

> Boy, is *that* a Cartesian argument! . . . It [the work you do at the bench] is about body-knowledge, not cerebral knowledge. But, no, it's not unconscious. It's like having good hands. There are scientists who have cerebral knowledge without the body-knowledge, and they're no good. Those who have good hands know it, the way a gardener knows he has a green thumb.

Sally Hacker's (1989) term "techno-eroticism" aptly describes the deeply pleasurable sensation of being in sync with certain technological extensions of our mental-physical selves. As my proficiency as a cell culture technician increased, I came to find a kind of kinetic pleasure in the steady cadence of carrying out a repeated task, handling the once-foreign accoutrements of the laboratory with growing dexterity. The comments of native members of the lab confirmed my perceptions about knowledge that is embodied in material practice, not held at a distance by a disembodied mind.

One of the researchers was showing me how to do a procedure for the first time. Doing what ethnographers are inclined to do, I interjected questions at each step in the procedure. She seemed to grow increasingly impatient with my queries and finally said, "Just watch. You don't have to understand, because there's a lot that you don't understand. You have to be mindless hands before you can be mind and hands." The ethnographer snorted skeptically. "You do," the researcher insisted. "You have to be hands first."

I initially read the exhortation to be "mindless hands" as a move to reinforce my low status as a neophyte technician, and this may have been partly true. However, subsequent conversations and my own experiences at the bench led me to see that my guide, in urging me to stop asking questions and learn by doing, was also attempting to initiate me into the body-knowledge of the craft of cell culture. I began to have a sense that being "mindless hands" was, on a phenomenological level, about being a "mindful body," entering into the flow of a series of interconnected activities.[4]

Contradictory ideologies—both of which are present in the cultural practices of contemporary biology—inform these contrasting readings of my colleague's words. The first describes the dichotomous cultural-material world of the modest witness, with a line clearly separating mind from body and mental (scientific) from manual (technical) labor. The division of technoscientific labor is characterized by an apprenticeship system in which individuals are expected to work their way up from the "manual" labor of benchwork through graduate training and post-doctoral fellowships, eventually attaining the credentials necessary to do the "mental" labor of the PI (and to be the "mind" that controls the "hands" of others). This privileges the role of rationality, while claiming to limit its distribution. It also relegates the career technician permanently to the status of nonmind.

Although the legacy of the modest witness—including the hierarchies that it supports—predominates, it coexists with an alternative epistemology that recognizes the corporeality of technoscientific knowledge and the ways that the mindful body engages the world. The latter perspective supports the possibility of modest interventions like those of Lynn Sakai, whose understanding of the importance of "body-knowledge" in the technoscientific enterprise is linked to her critique of the alienation of technicians' labor and shapes her laboratory practices.

Monoclonal Antibody Technology: Works of Art in the Age of Cyborgian Reproduction

> Hybridomas are permanent cell lines with the potential for unlimited proliferative capacity . . . The hybridoma technique makes it possible to obtain virtually unlimited quantities of homogeneous antibodies with specificity for any desired antigen. (Hood et al. 1984:20)

At the heart of the circulation of materials and information both within and between labs is an experimental technology that "treats natural objects as processing materials, as transitory object states . . . decomposable entities from which effects can be extracted" (Knorr-Cetina 1992:126). Organisms—singly and in combination with one another—and the products or reagents derived from them become part of the experimental apparatus of the lab. This includes the human benchworkers, with the day-to-day work of the lab resulting in an ongoing reconfiguration of the network of associations that we might call cyborg (Haraway 1991a), actant (Latour 1987), or "self-others-things" (Knorr-Cetina 1992; Merleau-Ponty 1945).

In the language of immunologists, an antigen is identified as "self" and its antibody is known as "other." Monoclonal antibody technology is a collaborative self-other recognition system joining the capacities of mice, tumor cells, antibodies, and the benchworkers who set the process in motion.[5] Immunized mice produce antibodies to a chosen antigen. The mice are "sacrificed" and then "immortalized" as the lymphocytes from their spleens are "fused" with myeloma tumor cells. The result, a chimeric organism called a hybridoma, can produce countless copies of the same antibody indefinitely. The organisms become machines—or supernatural entities.

Developed in 1975 by Georges Köhler and Cesar Milstein, who later won the Nobel prize for their discovery, the technology's discovery is regarded as a watershed event in the new era of genetic engineering and biotechnology. In 1977, working in an immunology lab in her first job as a technician, Lynn Sakai taught herself the technique, using the original article by Köhler and Milstein (1975). At the time no one else in her lab knew how to do the procedure. She says the deep satisfaction of figuring out how to successfully execute this new technology helped give her a sense that science was a creative, empowering endeavor.

My interlocutors' accounts of the work involved in monoclonal antibody technology convey the sense of what might be seen as *shared embodiment*, a co-performance that involves having "a feeling for the organism" (Keller 1983). As one of the researchers in the lab put it, describing the process of caring for hybridoma cells,

> It's about rhythm. You have to be *in sync with your cells*; you have to be able to feel the flow of the experiment. It's not just a matter of mechanically feeding your cells every so many days. You have to really look at them, and have a feeling for when they need to be fed, or they'll poop out on you.

Sakai describes monoclonal antibody technology in animated tones as "the industrial revolution come to biology," with hybridoma cells harnessed to create "factories" for the continuous production of a particular antibody. She laughs and says, "It really is cyborgian, isn't it?" She contrasts the technology with the polyclonal antibody technique that preceded it, in which a particular rabbit produced antiserum: "This was like preindustrial craft. The antiserum was the distinctive creation of that rabbit; when it died, there was no other source." As the process has become routinized, Sakai says, the situation of the typical technician has changed. There are now graduate programs that train students specifically in monoclonal antibody technology; the skill is now acquired as received knowledge to a much greater extent than before. Still, Sakai says, a technician gets the thrill of discovery when she sees a newly produced antibody for the first time, "something that no one else has seen before in the history of the world." She continues:

> It's wrong for that sort of creative work to be alienated from those who produce it, which is what typically happens for technicians when they are denied credit for what they do. When I said that monoclonal anti-bodies were like the industrial revolution in biology, the workers I was referring to were the hybridomas. Now what's happening, with large-scale automated operations coming to biology, is that technicians are being turned into industrial workers.

Sakai says she doesn't mind harnessing cells in order to effect mechanical reproduction, but she doesn't want people to be treated the same way. "But I'm a Buddhist," she says. "I still think we should live in harmony with Nature, at the same time that we harness it. [She laughs.] I feel attached to my hybrido-mas; I created them, they work for me. They're kind of like my pets."

The benchwork laboratory of contemporary biology is, as Karin Knorr-Cetina (1992:129) says, "a *link between internal and external environments, a border* in a *wider* traffic of objects and observations" (original emphasis).[6] Following the initial discovery of fibrillin, Sakai's lab began a collaboration with two groups of medical researchers who had been conducting research on Marfan patients, collecting blood and skin samples and the kinship charts that geneti-cists call pedigrees. As a result of these collaborations, mutations in the fibrillin gene have been identified as the cause of Marfan syndrome. The association be-tween fibrillin and Marfan syndrome was established through, and has contin-ued to expand, an international network of collaborations and exchanges.

Among the most highly valued trade goods in this circuit are Ab 201 and Ab 69, two of the antibodies that Sakai originally used to identify fibrillin. The electronic and postal conduits between her lab and the worlds beyond it bring in a steady stream of requests for the antibodies, as well as DNA probes and clones, from clinicians, graduate students, and other basic researchers. When she travels to professional meetings, Sakai will often carry centrifuge tubes in her pocket containing allotments of her reagents, the fibrillin antibodies, to par-cel out to selected colleagues.

The term *reagent* is generally applied to the materials used in experiments. A reagent is not, however, simply an element that occurs naturally. It is some-thing that has been produced, purified; it is the product of someone's labor. Reading the agency back into our understanding of reagents raises questions of ownership and of control over circuits of exchange. The traffic in Lynn Sakai's antibodies reflects her ownership of them; she controls the networks of relations that her exchanges engender. A technician's labor, along with the surplus value that it produces, may be one of the "decomposable entities" from which such ownership claims are extracted. The fruits of intellectual-manual labor in the laboratory are also subject to claims from patrons. For instance, monoclonal antibodies have been determined to be patentable, which means potentially profitable (cf. MacKenzie, Keating, and Cambrosio 1990). The terms of most sci-entific funding specify the funding agency's right to profits from any patentable discoveries. Monoclonals are among the myriad commercial reagents available, sold as proprietary ingredients. The fibrillin antibodies are not patented, although Sakai says, "I'd be a fool not to patent any future discoveries." She is clear, though, about their value in creating and expanding alliances.

> LYS: Science is moved along by the individual trades that go on. When a reagent is first developed, it's passed around. If it yields good results, it leads to collaborations.
> DH: So it's about networks of reciprocity.
> LYS: That's *exactly* what it is. It's about meaningful exchanges. Buying a commercial reagent isn't meaningful. It's just a purchase.

As anthropologists working in many other settings have observed, trade goods such as the fibrillin antibodies lose social value when they enter the cash nexus. With commercialization, the scientist whose lab has produced a particular reagent is no longer able to use the reagent directly to initiate or sustain trading relations that extend her networks of association.

The professional meetings discussed below are one of the arenas where trade occurs, not just between individual scientists but also across disciplinary and occupational divides and between scientific researchers and nonscientists. The goods exchanged are symbolic as well as material and, as with most trading relationships, some exchanges are asymmetrical.

Marfan Embodiments

> Feminist embodiment . . . is not about fixed location in a reified body, female or otherwise, but about nodes in fields, inflections in orientations, and responsibility for difference in material-semiotic fields of meaning. Embodiment is significant prosthesis; objectivity cannot be about fixed vision when what counts as an object is precisely what world history turns out to be about. (Haraway 1991b:195)

Three women, one in a white lab coat, stand in the hallway outside Lab Three looking at the electron micrographs of fibrillin that are hanging on the wall. One of the two women who have come to tour the lab is over six feet tall. She is a bus driver and founder of the local support group for people with Marfan syndrome. Her height is one of the signs that she is affected. The two visitors are on the planning committee for the annual convention of the National Marfan Foundation (NMF), which will begin later in the week. Lynn Sakai, the woman in the lab coat (a uniform generally reserved for encounters such as this with outsiders), points to the images of the protein that she discovered. One of the visitors asks if they can see mutations in the images of the fibrillin molecule. Sakai explains that the resolution isn't great enough to see the genetic components where mutations occur.

It is these mutations in the gene that codes for fibrillin that cause Marfan syndrome, which affects the connective tissue in different parts of the body, characteristically the eyes, the bones and ligaments, and the heart and blood vessels. The results include acute nearsightedness, lens dislocation, and

75

scoliosis, as well as above-average height. The most life-threatening manifestations are cardiovascular, in particular the stretching or dilation of the wall around the valve of the aorta, which can result in unexpected rupture and death. In the past this condition was considered untreatable, but developments in pharmaceutical treatment from the 1970s onward, and in open-heart surgery since the 1980s, have substantially improved the prognosis of affected individuals. Most notable is the creation of the composite aortic valve-graft, a prosthetic heart valve sutured onto one end of a composite graft, a woven cloth tube (usually made of Dacron®) that replaces a section of the aorta.[7] Playing a supporting role are the imaging technologies echocardiography and MRI (magnetic resonance imaging), which permit monitoring of the size and function of the aorta both before and after surgery.

If advances in cardiovascular interventions have the highest profile in ongoing treatment of those affected with Marfan, hopes for the future are focused at the body's molecular level, on the fibrillin gene and the as yet unrealized possibilities for gene therapy. These two therapeutic approaches to the Marfan body correspond roughly to two professional communities, clinicians and scientists. There is a perceived division between the two domains that is borne out in practice, despite the traffic between them. The articulation of this boundary in terms of a tension between "applied" and "basic" research can be traced historically to shifts in the post–World War II political economy of research funding, with attendant effects on the relative autonomy of academic science vis-à-vis the biomedical establishment (cf. Wright 1994).[8]

The Second International Symposium on the Marfan Syndrome was held in San Francisco in 1992. Intended primarily for scientists and health care professionals, the conference was divided into moieties, with sessions of interest to clinicians and basic researchers held on separate days. Like the oral presentations, the posters that visually summarize late-breaking research results were also segregated by being displayed in two rooms, one for each group. A full day was devoted to cardiovascular concerns. A second day and a half emphasized research on fibrillin, billed as the "Marfan gene." Although some of the key figures in the research community—physicians actively pursuing research programs—span the divide, clinicians not involved in research seemed reticent to participate fully in the scientific portions of the symposium. For instance, during the question period following one of the sessions, the chair asked, "Are there any orthopedists left?" A tentative hand went up at the back of the room, and a physician said, "I don't really have anything scientific to contribute."

The main sessions were bracketed by an opening speech and a final panel discussion that addressed links between biomedical research and the wider concerns of affected people. The theme of the keynote address, entitled "The Joining Circles," was the integration of "the four frontiers" of Marfan syndrome: research, clinical medicine, social support, and life experience (Gasner 1993). It was delivered by Cheryll Gasner, a nurse-practitioner in her mid-30s who heads a university clinic for Marfan patients. One of the founding members of the NMF, Gasner has Marfan syndrome herself and has had five open-heart surgeries. In addition to her clinical work, she participates in the laboratory research program of a medical geneticist who specializes in Marfan syndrome. She embodies the connections she described in her talk and is a committed advocate for strengthening them in ways that make a difference for those who are affected, as she made evident in a subsequent interview:

I have to keep clear which hat I'm wearing. There's my position as a nurse-practitioner, my work with the Northern California chapter of the National Marfan Foundation, my position as a patient. It's hard sometimes . . . I try hard to work with the physicians; it's my job. But I also know that they're human, that they're fallible. Many patients think that they're infallible. I try to pass on the sense that patients need to develop self-sufficiency. I teach people from Day One to take personal responsibility, to press for what they want done. I work to get people empowered. We've got a saying, "Either change your doctor, or change doctors."

Although those affected with Marfan were not the principal participants in the conference, members of the NMF attended an evening reception along with representatives of an organization called Tall Clubs International (TCI), a federation of social clubs for "unusually tall" people (defined in their bylaws as a minimum of five feet, ten inches for women and six feet, two inches for men). TCI has selected the NMF as its chosen charity; during the reception one of its members, a woman who had been elected Miss Tall San Francisco, presented a check to the head of the NMF. TCI's literature reveals that 60 percent of its membership is female, which may speak to the stigmatization of tall women in US culture. Negative stereotypes about women of above-average height appear to carry over into the medical treatment of Marfan patients, with some clinicians recommending that girls in particular be given hormones to speed the onset of puberty and thus limit adult height.

At the opening session of the Ninth Annual National NMF Conference one of the first speakers was a woman on the organizing committee, the six-foot, two-inch bus driver who had visited the lab earlier in the week. Her remarks, which were about using humor as a defensive strategy, drew attention to gender biases about height. The focus was a mean-spirited co-worker who had started a rumor that the speaker had had a sex change; why else would she be so tall? Her response was to join with a male colleague in starting a counter-rumor. Their story was that the two of them had previously been married, and had separated when they both decided to have sex change operations. They had since reunited, the story continued, and he was now carrying their child. The audience, composed mostly of people with Marfan, met her account with hearty laughter. When she had finished her narrative, she joked about having to lower the microphone for the next speaker, Lynn Sakai, who is more than a foot shorter than she.

Sakai spoke both as the organizer of the scientific meeting running concurrently with the NMF conference and as a representative of the Shriners Hospitals, one of the sponsors of both meetings. The centerpiece of her brief presentation was a slide that showed an electron micrograph of the fibrillin molecule, much like the one hanging on the wall outside her lab. Addressing the question that one of her visitors had asked earlier in the week, she noted that mutations were not visible. When the slide first appeared, someone in the audience hissed at the villain molecule responsible for Marfan syndrome. Sakai told me later that even though she was sure it had been done in jest, she still felt somewhat hurt that the protein that she had discovered would elicit such a response.

Other opportunities for contact with NMF members gave scientists and clinicians at the conference a human dimension to their understanding of genetic variability. One scientist said that, although she has collaborated on all but one of the articles describing different Marfan mutations, it was only when she visited the clinic on the day before the NMF meeting that she understood what diversity in the expression of the syndrome actually meant. She said that one of the attending physicians at the clinic, a medical geneticist who has seen individual Marfan patients for years, reported having a similar reaction. As she put it, "Seeing a whole collection of people with Marfan syndrome in the same place, and seeing how they all look really different, gave a new meaning to the diversity of phenotype. It was striking relating the mutations to these whole people."

For those affected with Marfan syndrome, the physical and cultural signs of the condition and the medical crises and interventions that they endure are part of the shared life experiences that foster a sense of identity at gatherings such as local meetings and the annual national NMF conferences. In a playful performance in the closing session of the NMF meeting, a group of several women calling themselves "The Melodic Marfettes" sang their own rendition of Woody Guthrie's song "So Long, It's Been Good to Know You." It captured some of the feeling of solidarity that I had heard expressed in many other ways throughout the meeting, both during workshops and in the hallways between sessions, much of which focused on markers of difference in the world dominated by those who are unaffected. Here are two of the verses, which contain references to both the significance of storytelling and the shared bodily experiences of people with Marfan syndrome:

> The day I arrived I felt lonely and shy.
> I said to myself, "Let's give it a try."
> I heard people's stories and people heard mine
> About shoe size and lenses, aorta and spine.
> Chorus: Singing so long, it's been good to know you, etc.

> I came here this week with a lot on my mind,
> I came here not knowing what I might find.
> I looked at these Marfans and what did I see?
> A whole brand-new family that looked just like me.

Hillary Rose (1983, 1994) has written that a feminist epistemology for the natural sciences would resolve the mind/body dichotomy by insisting that heart be linked to head and hands. This is a lesson that the Marfan activists who inhabit the borderlands of technoscience already know. Given the consequences of untreated cardiovascular problems and the extent to which medical intervention has extended the lives of many people with Marfan, it is not surprising that the heart has become a focal point for the efforts of both lay advocacy groups and clinicians. The National Marfan Foundation uses the heart as a symbol in its fundraising campaigns and has designated February, when the annual campaign takes place, as Have-a-Heart Month because of Valentine's Day and the birthday of Abraham Lincoln, who is thought to have had Marfan syndrome. Items distributed by the NMF include heart-shaped Post-it notes and T-shirts that read, "The Progress Is Heartening." At the Ninth Annual National NMF Conference, one of the fundraising events was the raffling of a quilt covered with hearts. Beneath the trappings of public relations schemes and consumer capitalism, the heart is an icon for a politics of truth and caring grounded in a kinship of affliction and a sense of shared embodiment.

Partial Connections and Modest Interventions

> The knowing self is partial in all its guises, never finished, whole, simply there and original; it is always constructed and stitched together imperfectly, and therefore able to join with another, to see together without claiming to be another. Here is the promise of objectivity: a scientific knower seeks the subject position not of identity, but of objectivity; that is, partial connection. (Haraway 1991b:193)

Operating in an experimentalist mode, I organized two roundtable discussions during the NMF meeting, inviting researchers, clinicians, and advocates to engage one another in open-ended discussion. I was curious to see how representatives of these different constituencies—whose domains are interdependent yet largely distinct from one another—would interact, and how both commonalities and discontinuities might become apparent.

In the course of one of these conversations, a patient-advocate pointedly conveyed the frustration that many people with Marfan feel about the pace of research results and an apparent lack of focus on concerns of immediate importance to those who are affected. She capped her remarks to the clinicians and researchers at the table by saying, "It's been two years [since the partial sequence of fibrillin was published]. What the patients want to know is: 'Where's the beef?'"

After the discussions I asked Lynn Sakai, who had attended both sessions, for her reactions. She said that she appreciated such contacts with patient-advocates for giving a human dimension to her research. At the same time, she said, some of the advocates' comments, such as their focus on therapeutic concerns, had been disturbing. "I'm a basic researcher," she said, "not a clinician." The tension she felt between the power of the patients' perspective and the high value she places on her professional autonomy seems indicative of the contradictory connections and divisions that describe the networks linking this laboratory researcher to wider worlds.

In the 1992 annual report of the Shriners Hospitals medical research programs, Lynn Sakai extols the virtues of "pure" research, expressing her conviction that it provides the firmest foundation for clinically significant discoveries:

> I believe that the story of fibrillin and the Marfan syndrome is instructive. In the Portland Unit, scientists were performing research on the connective tissue, without any prior idea that their work would lead them to a specific result. In Baltimore, clinicians were actively studying the Marfan syndrome and collecting patient samples. The cause of the

malady was unknown for almost one hundred years. In 1991, the time was right; the groups in Portland and Baltimore got together to collaborate, and the cause of the Marfan syndrome was discovered. Research is like that: it is difficult to predict the outcome of research, which needs only a free and open, and well-funded environment, but our belief is that, since there is so much about what makes our bodies work that is unknown, clinical progress can only be made through basic research. (Shriners Hospitals for Crippled Children 1992)

Working within a hospital system in which the research units are generally headed by MDs, Sakai is well aware that the notion of "pure science" is an ideal type—though a compelling one—and that scientific knowledge production is dependent on shifting power relations within wider networks of patrons and allies. "Twenty years ago," she says, "scientists were seen as the handmaidens of the MDs. That's begun to change." I ask with a smile, "Is that because now you're the handmaidens of industry?" She laughs and nods. "Yes, the rise of the biotech industry has had something to do with it. Scientists are much more powerful now. But we still have to compete with MDs for funding." Although Sakai's words in the Shriners annual report can be read as the rhetorical appeal of a research scientist addressing her patrons, they also convey her deeply held beliefs about the importance of scientific autonomy to the successful pursuit of new knowledge. This is linked to her concern about the targeting of federal funding for the life sciences toward particular clinical problems, constraining the resources available for basic research.

A week or so after the NMF Conference, Sakai and I watched a videotape of local television coverage of the event, which the head of public relations for the Shriners Hospital had put together. The clips, from two local stations, each contained a short interview with Sakai concerning the scientific aspects of the disease. There were also interviews with people affected with Marfan syndrome, including the head of the local chapter. Other shots included the clinic day, workshops, a speaker at the scientific sessions, panelists at the medical presentations, and stock footage of one of the technicians in Sakai's lab "doing science," that is, pipetting while wearing a white lab coat. The latter footage was taken in 1991, when Sakai received national coverage following the publication of articles in *Nature* (Dietz et al. 1991; Maslen et al. 1991) about the fibrillin sequence and the characterization of mutations in the fibrillin gene in Marfan patients.

I complimented her on how well she had done in both interviews, kidding her a little at the same time. I told her that she always did a good job communicating with lay people when she actually did it, that it was only before the fact that she groused about having to do it. "In practice, you're a populist; it's only in principle that you're an elitist," I joked. "I'm always an elitist," Sakai snarled, and then laughed. Then she said, "The patients really don't know very much about what the research is really all about." "Well," I countered, "are there many opportunities for patients and researchers to come into contact with one another?" Sakai conceded that there weren't, and then praised Priscilla Ciccariello, head of the NMF, for pushing researchers to make a commitment to Marfan syndrome and those affected with it. As she had said during her introductory remarks at the NMF meeting, Ciccariello's efforts had given her own research "a human face." At the same time, some of the contact Sakai had had with members of the affected community had been "unsettling." "The patients really think that I'm responsible to them," she said.

She stopped, thought for a moment, and then said that with a just a little reorientation, a little tinkering here and there, she could push parts of her own research agenda in directions that could provide diagnostic or therapeutic insights. We discussed what some of those possibilities might be. Soon thereafter, however, she said, "But science is supposed to be pure. The data are supposed to follow their own course." "But science is a human product," I replied. "The data don't just invent themselves." "No, of course not," she retorted, "but it's not good when research is dictated by these applied concerns; it's misguided." At this point she seemed quite irritated, and said, "I don't want to talk about this anymore." We stood in silence for a moment, and then I said, with a tentative smile, "You're mad at the Marfan patients, aren't you?" She paused, then laughed and nodded. "You're right; I am. They've made a difference in how I think about my work."

Throughout this conversation it struck me that Sakai alternately advanced and retreated across the boundary between an insular science and its larger context. This is precisely the site of her practice; it is both circumscribed by its local boundaries and pulled to reach beyond them. Having left philosophy because it divorced mind from action, in search of a world of activity where head and hands are joined, she still lives with contradictions, as, of course, we all do. Some she engages directly, doing her best, for instance, to mitigate the effects of laboratory hierarchies that minimize contributions of technical labor. Others are more problematic; the privilege of pursuing "pure" science is closely linked to the relative structural autonomy that permits her both to run her lab largely as she chooses, and to pursue, and sometimes achieve, profoundly satisfying mental-corporeal pleasures.

If cultural studies of science are "politically and epistemically engaged" (Rouse 1993:20) in ways that implicate its practitioners in the practice of technoscience, the boundaries of anthropology are no less permeable. As an ethnographer of technoscience, I have found my own interpretive and epistemic practices shaped by the encounters with my interlocutors "in the field" as we participate in, observe, and critique one another's practices. Lynn Sakai's work and my own are both anchored in the privileged pursuit of curiosity. My curiosity about the local knowledges of laboratory practice has taught me both about the embodied pleasures of participant performance and about body-knowledge as a locus of critical discourse on the nature of technoscientific knowledge. My encounters with Marfan advocates gave me a different sense of shared embodiment at the intersection between engagement and the kinship of affliction.

Like my interlocutors, I am committed to an "itinerant territoriality" (Deleuze and Guattari 1987) that traverses and perhaps destabilizes the institutionalized boundaries between Science and Not Science, aiming to make the partial connections between them matter more. We share the interstitial spaces of what Donna Haraway (1997) calls the *mutated* modest witness, where received boundaries between the knower and the known are critically contested. The modest interventions that bring together scientists and clinicians with border denizens like Marfan activists and ethnographers combine local knowledges in order to build a differently situated—but never disembodied—translocal knowledge and practice.

Notes

1. Within the laboratory setting I want to highlight the viewpoint of the technicians and of certain nonhuman benchworkers. Beyond the lab I want to draw attention to the experiences of people affected by Marfan syndrome. I am also concerned with conveying a sense of alternative hegemonies as well as counterhegemonies, a sense of the heterogeneity within the "view from above," with rivalries and interdependencies marking distinctive yet mutually constitutive cultural domains. The recent history of research on the Marfan syndrome, for example, has been contingent on developments in biotechnology, influenced by the lobbying efforts of lay advocates, dependent on the support of public- and private-sector patrons, and carried out by both basic researchers and clinicians from a range of disciplines and subfields, each with its own constituencies and characteristic practices.

2. The masculinist asceticism of the modest witness can be seen as originating earlier, as David Noble (1992) argues, in the emergence of an ascetic culture among Christian clerics in the late Middle Ages, which excluded women from science and institutions of higher learning.

3. See Heath (1992) for a discussion of the notion of good hands in a DNA sequencing lab.

4. See Scheper-Hughes and Lock (1987) for an insightful account of the notion of the mindful body.

5. For a social history of the art and science of hybridoma technology, see Cambrosio and Keating (1992).

6. While Knorr-Cetina limits this wider landscape to connections among other laboratories, I want to extend the terrain of the present discussion to include the traffic that links the lab to (among others) the worlds of clinicians, organ donors, people with Marfan syndrome and their advocates, and the patron institutions that fund biomedical and basic research.

7. Depending on where or how severely the aorta is weakened or torn, larger sections may be replaced. As a pamphlet on the Marfan syndrome published by the NMF puts it, "[I]ndeed, a few people with the Marfan syndrome have had their entire aorta converted to Dacron®!" (Pyeritz and Conant 1989:15).

8. Susan Wright's (1994) study of genetic engineering policy provides a detailed and revealing comparative account of the political economy of research funding in both the US and the UK since World War II. She examines the consequences of the shift from the postwar boom in relatively unrestricted science funding toward increasing demands for "targeted" research and "accountability" among researchers to produce results with demonstrable applications. This pressure to articulate basic research agendas in terms of biomedical concerns originates in dependence for funding from both public and private sources. At the same time, the direct involvement of some basic researchers in, for example, biotech firms has provided them with a new measure of partial autonomy from public and academic biomedical institutions.

12345678910 11

A DIGITAL IMAGE OF THE CATEGORY OF THE PERSON /

PET Scanning and Objective Self-Fashioning / Joseph Dumit

> Probably one of the most important initiatives we have ever undertaken [at the National Institute for Neurological Communicative Diseases and Stroke] is our support for positron emission tomography (PET), an intriguing new research technique . . . With PET we will be able to examine what happens functionally, in the living human brain, when a person speaks, hears, sees, thinks. The potential payoffs from this technique are enormous.
> —Donald B. Tower

> We must learn to distinguish it [the body in which I live and experience, just as I live and experience it] from the objective body as set forth by physiology. This is not the body which is capable of being inhabited by a consciousness.
> —Maurice Merleau-Ponty

MARCEL MAUSS AND OTHERS FOLLOWING HIM argued that the basic human unit, "the person," is a cultural category with different attributes—rationality, agency, participation, gender divisions, and so forth—for different cultures in different times and places (Mauss 1985; see also Carrithers, Collins, and Lukes 1985; Geertz 1973; Strathern 1988, 1992). For Mauss and his successors, "the person" is a category stuffed into a physical body but independent of the body's physicality. They argue as if each culture or historical period has its own category of the person. Other anthropologists have been more troubled by the findings of medicine and neuroscience. Victor Turner, for instance, once expressed great difficulty in keeping up with the latest findings:

> This is because I am having to submit to question some of the axioms anthropologists . . . were taught to hallow. These axioms express the belief that all human behavior is the result of social conditioning. Clearly a very great deal of it is, but gradually it has been borne home to me that there are inherent resistances to conditioning. (Turner 1983:221)

Turner is describing how new facts from medicine and neuroscience disturb his notion of personhood and personal behavior. What kinds of biological limitations are built into our brains—limitations on, for example, personality, sexuality, violence, mental illness—that might resist being changed by society?

Facing these facts requires reimagining what kinds of persons humans are. How do we as anthropologists and other scholars understand our bodies?[1] How do we put together the facts of science and medicine that we read in the *New York Times* and receive from our doctors with the role of culture in our constitution? In anthropological terms, I am interested in how facts come to play a role in our everyday category of the person.

Medical anthropologists have long faced the relation between what Merleau-Ponty called our objective body and our lived body, or our person, with a variety of more subtle analyses. For clinical medical anthropology, oriented around the question of efficacy, the lived body (cultural) and the objective body (physiological) have initially different causes but mutually influence each other throughout development. For example, physiological diseases are often inseparable from cultural variables like political violence, discrimination, housing conditions, poverty, and diet (Desjarlais 1995; Farmer 1992; Kleinman 1986; Kleinman and Good 1985; Rhodes 1991; Romanucci-Ross, Moerman, and Tancredi 1991). In spite of this flexibility, each culture is ultimately assigned its "body" that is lived and explained in relation to an objective body, which provides the touchstone of cross-cultural comparison and criticism. *Change* in the category of the person is not well attended to. Instead, categories are often explained as a reflection of changes in other spheres of society: economics, politics, colonization, and religion (see Carrithers, Collins, and Lukes 1985).

Other medical anthropologies, some sociologies of medicine, and the history of science and medicine take a different approach. Instead of viewing the *experience* of health and illness as variable, the "objective body" is understood to be culturally and historically contingent, the object of a scientific and medical gaze that changes with the times and according to discipline, site, culture, and circumstance (Farquhar 1992; Manning and Fabrega 1973; Saunders 1989; Taussig 1992, 1993). These approaches regard the objective body as varying with the development (positive or negative) of technoscientific culture and examine how the historical-cultural category of the person (via politics, economics, etc.) influences the evaluation of the objective body (Canguilhem 1989; Foucault et al. 1988; Gilman 1988; Terry 1989). The objective body and the experienced body remain side by side, both variable, but analytically separate.[2]

Focusing on brain theories and brain images, this paper begins to explore the way that facts and categories of persons are produced, proved, contested, and lived—in other words, how they are at stake in social interactions. It treats the emergence and maintenance of categories of persons as a dialectical process involving expert-researchers, mediators (such as science writers, anthropologists, popular psychologists, and mass mediators), and laypersons.[3] I am interested in the question, How do mutations in our categories of the person happen?

Given the unevenness of scientific knowledge and our dependence on its authority for self-knowledge, is it possible that local mutations in categories of persons take place daily, that they are contested within American cultures because they are lived and not just known? Building up a dynamic notion of the category of the person—objective self-fashioning—I first examine a best-selling account by a psychiatrist of how his patients learned to reconfigure their notions of core personalities through attending to how the drug Prozac altered their physiology. Next I turn to a popular movie in which a brain-imaging technology, PET scanning, is used to decide whether or not a murderer is insane. Finally, I look to the operation of a small PET center where images of mental illness as located in brains, and therefore biological, are sought out by sufferers and their families. In each case I am concerned with understanding how the circulation of evidence—first-hand experiences, reports, newspaper articles, movies, interviews—helps to form and reform human possibilities and probabilities. Each of these examples proposes a biomedical answer to uncertainty and anxiety over human nature by evoking the citadel of scientific certainty. One question I would like to answer, but cannot, is what would constitute final certainty with regard to human nature.

Living with the Facts: Prozac

In his recent nonfiction bestseller, *Listening to Prozac*, psychiatrist Peter Kramer begins with the following story. Kramer is visited by a patient, Sam, who suffers from a brooding depression following the death of his parents. Kramer first prescribes an antidepressant that does not seem to have an effect. He then proposes Prozac, which Sam agrees to try.

> The change, when it came, was remarkable: Sam not only recovered from his depression, he declared himself "better than well." He felt unencumbered, more vitally alive, less pessimistic. Now he could complete projects in one draft, whereas before he had sketched and sketched again. His memory was more reliable, his concentration keener. Every aspect of his work went more smoothly. He appeared more poised, more thoughtful, less distracted . . . Though he enjoyed sex as much as ever, he no longer had any interest in pornography. He experienced this change as a loss. The style he had nurtured and defended for years now seemed not a part of him but an illness. What he had touted as independence of spirit was a biological tic. In particular, Sam was convinced that his interest in pornography had been mere physiological obsessionality . . . This one aspect of his recovery was disconcerting, because the medication redefined what was essential and what was contingent about his own personality—and the drug agreed with his wife when she was being critical. Sam was under the influence of medication in more ways than one: he had allowed Prozac not only to cure the episode of depression but also to tell him how he was constituted . . . Though I had never taken psychotherapeutic medication, I, too, seemed to be under its influence. (Kramer 1993:xi)

Sam becomes more alert, attentive, happy, adjusted, and "successful" than ever before in his life. Observing this, Kramer realizes that both he and his patient now understand the "real Sam" to be the one that Prozac revealed, and the "former Sam" to be a biological sickness. Sam and Kramer have "listened to Prozac" rather than to Sam's previous three decades of life. Because Prozac is a biologic drug, Sam must in some sense have been cured by it, freed at last from his strange psychophysiological disease and able to be his true self—and his true self becomes something that is perhaps revealed only with Prozac.[4] I want to note the account of Sam taking Prozac and then behaving differently (and better) as a "fact-in-the-world," a reminder that facts don't just pop into our consciousness. Facts have to find us; we have to hear of them or read them, and we have to incorporate them as facts.

Sam's story is not just an anecdote but an apparently objective account made as part of a psychiatric case history. We only know this "fact" about Sam through the story told by Peter Kramer, MD. I almost want to name this "fact" a "factoid" to call attention to the specific ways that we *learn* the fact, that we attend to all of the cultural aspects of our learning: the objective voice, the authorship of a psychiatrist, doctor, scientist, book-writer, the way in which Kramer's discussion of his own disconcertedness and surprise allows us to share these feelings as part of the novelty of this fact.[5] This story is a challenge for us. We deny it or fit it into our categories of persons. These are ways in which the story makes sense to us, seems possible, even as it shifts our notions of what is possible. We are the sorts of people who take facts seriously. But *how* do we take them so? How do we incorporate them into ourselves, especially ones that shape who we are but that we ourselves are not equipped to properly verify?

Facts typically imply relationships between things that are not bound to time and space and culture; they simply are. But facts are not untethered. They are facts-in-the-world. One task is to understand how the meanings of facts change—how we are never simply handed facts but are continually faced with facts-in-the-world and continually judge their status and relative worth for ourselves. Facts are bits of mastery in expert culture. Expert culture is about being extremely knowledgeable about a very few things. We each know very little about most things, and in their entirety the facts are beyond reach. The very category of the person, it seems, has become parceled out among expert discourses. All facts contain, imply, or exclude categories of persons. Calling the case of Sam a fact-in-the-world is an attempt to mnemonically maintain the perspective that a particular category of the person is at stake in the "fact," and that this fact has traveled.

We must ask ourselves, however, why this Prozac story can be so compelling, and why we might consider it authoritative. One objection to Kramer's description might be that Sam *experienced* a new self, and it was so compelling that he simply adopted it as his true self. Kramer and Sam's friends followed suit because they too experienced a different Sam. But this does not account for my feelings and others' upon hearing about Sam. In discussing this case I have been struck by a double response. On the one hand there is a desire to have it not be true, to deny the fact of the transformation and assert a less mutable category of the person. On the other hand is a desire to know more about the story, to begin to play with the fact of Prozac changing personalities and call into question one's own category of the person. My sense is that the fact exploits the incompleteness of our categories of persons, that there is much that is either unaccounted for or contradictorily accounted for in our categories, and that each fact provides material "good to think with," in Lévi-Strauss's (1963:89) memorable coinage.[6]

What makes *Listening to Prozac* fascinating reading is that Kramer is well aware of the middle-class American cultural boundedness of his understanding of Sam's self, and he is both frightened and eager to work with it. He goes on to consider more borderline cases, for instance, a woman who has been "spacey and flaky" all her life. When on Prozac, she becomes a faster and more articulate speaker. A businessperson on Prozac becomes less sensitive to the possible problems in proposals and therefore more risk-taking and successful. These examples raise the dilemma of what Kramer calls "cosmetic psychopharmacology," people who are taking Prozac in order to become better than their "normal" selves (1983:244–49). At stake in these stories are categories of persons: flakiness and eloquence, risk-taking ability and self-deprecation, as neurochemical on/off switches. These in turn alter how we feel about the drugs qua controlled substances: "Once these medicines have colored our view of how the self is constituted, our understanding of related ethical issues inevitably will be affected" (249).

Kramer's work illustrates how, at least in the US, expert scientific and medical facts play a key role in how we experience our selves, our bodies, and others. In other words, there appear to be many objective bodies that we inhabit consciously, in part through adjusting our categories of persons to account for compelling facts. Of course this is not a one-way imposition of science upon laypersons. Scientific facts affect us, but we are not, as Roger Cooter (1984) has pointed out, passive laypersons. We participate in the instantiation and legitimation of facts. In the next sections I will consider our role as social scientists in the business of producing and maintaining facts.

Mediating Facts

> The years passed. I continued to treat ritual essentially as a cultural system. Meanwhile exciting new findings were coming from genetics, ethology, and neurology, particularly the neurobiology of the brain. I found myself asking a stream of questions more or less along the following lines. Can we enlarge our understanding of the ritual process by relating it to some of these findings?
> —Victor Turner

We, as scholars and laypersons, are involved in the midst of science and in the midst of facts; our persons are built into them. Every fact involves asserting a particular view of human nature. Scholars have studied the use of facts in abortion debates (Hartouni 1991; Pechesky 1987; Strathern 1992) and in research on purported biologic differences between homosexual and heterosexual individuals (Bayer 1981; LeVay 1993; Terry 1989) and between races and sexes (Fausto-Sterling 1985; Gould 1981; Lawrence 1982; Stepan and Gilman 1993). In Stanley Fish's terms, "disagreements are not settled by the facts, but are the means by which the facts are settled" (1980:338).[7]

Who takes facts up? Who does not? How are they produced and distributed? These are critical anthropological questions in another sense as well. New facts-in-the-world, for instance, literally make Turner reconsider his notions of personhood. Turner goes on to examine, find fault with, and then propose his own theories of how brain topographies might make sense of cultural rituals. Like Kramer and Sam, Turner listens to the facts propounded by neuroscience and physiology and wonders how to refigure what he knows so as to make sense with them. His response, quoted above, is instructive. Accepting the importance of these facts about the brain, he discovered that they could not account for the fundamentally important concepts of religion and play. Rather than using this insight to discount the value of these studies built on flawed and deficient theories of human nature, he instead reworked them into his own theories of the brain that could account for religion and play. I suggest that these neuroscientific facts compel such reworking because they provide authoritative starting points along with combinatory possibilities. Like Lévi-Strauss's totem animals and Turkle's computers, they are good and solid and fun to think with, lively facts with provocative connotations.[8]

Turner's implications regarding neuroscience are twofold. First, he makes it clear that merely to ignore contemporary neuroscience is to risk building an outdated (wrong) neuroscience into our categories of the person.[9] Second, he makes it clear that anthropology has a lot to contribute to neuroscience, especially with regard to human specificity (how we are different from animals) and human differences (cultural differences among humans). Of course, anthropology already does this. Konner's *Tangled Wing* (1983), for example, one of the more forceful and eloquent defenses of sociobiology, depends upon cultural anthropological facts to substantiate the relative determinism of human nature by biology.

Other studies have concerned themselves with popular categories of persons that are taken up into scientific theories as they are developed. Evelyn Fox Keller's studies of gendered and capitalist subjects built into biology, and Sahlins' examination of sociobiology's roots in possessive individualism, are excellent examples (Keller 1985, 1992; Sahlins 1976a). Feminist studies of science have concentrated on gender bias in scientific practice as well as patriarchal presuppositions in good science (Bleier 1986; Haraway 1989, 1991a; Harding 1987, 1991; Longino 1990; Martin 1987). Cultural studies of science and technology have been especially active in tracing the profound role the media have played in shaping the development of scientific facts (Hartouni 1991; Martin 1987; Pechesky 1987; Treichler 1991). And historians and sociological studies of science have traced the political orientations of scientific and technological research (Shapin 1979b; Shapin and Schaffer 1985).[10] If categories of persons are built into facts, and facts are mutually borrowed between disciplines such as anthropology and neuroscience, human nature at least at this level is quite dynamic and dialogic.

Objective Self-Fashioning

> Given the explosive rate at which the fields of molecular genetics and neurobiology are expanding, it is inevitable that the perception of our own nature, in the field of sex as in all attributes of our physical and mental lives, will be increasingly dominated by concepts derived from the biological sciences.
> —S. A. LeVay

Within this broad sketch of three symbiotic actors—experts, laypersons, and mediators—each drawing upon and reconfiguring the presuppositions of the others, I am going to concentrate my attention on the aspect I call objective self-fashioning.[11] The objective self is an active category of the person that is developed through references to expert knowledge and invoked through facts. The objective self is also an embodied theory of human nature, both scientific and popular. Objective self-fashioning calls attention to the equivocal site of this production of new objective knowledge of the self. From one perspective, science produces facts that define who our selves are objectively, which we then accept. From another perspective, our selves are fashioned by us out of the facts available to us through the media, and these categories of persons are in turn the cultural basis from which new theories of human nature are constructed.

Kramer provides an excellent illustration of this by relating how both he and his patients incorporate the fact that Prozac makes some people "better." Out of this fact Sam fashions a new objectively true self and a new history (of a self that was defective until Prozac), while Kramer goes on to experiment with Prozac and draw upon other human and animal facts to propose a new set of theories of human nature, packaged for a popular audience and read by psychiatrists and other neuroscientists.[12]

Objective self-fashioning is thus an acknowledgment of local mutations in categories of persons highlighting the active and continual process of self-definition and self-participation in that process. Objective self-fashioning is how we take facts about ourselves—about our bodies, minds, capacities, traits, states, limitations, propensities, etc.—that we have read, heard, or otherwise encountered in the world, and *incorporate* them into our lives.[13] As anthropologists and other scholars we are, like Turner, most often in the mediator role, casting theories of objective selves out of our own categories of the person.

These cases point to two interrelated meanings of objective self-fashioning: (1) How we come to understand ourselves as subject to the scientific, medical, and technical discourses of objectivity, and (2) How these discourses choose "us" as their object of study. The difference between the two meanings is a matter of point of view. On the one hand these cases point to the ways in which we fashion our selves—person, body, brain, and mind—out of ready-made objective types, and therefore subject ourselves to the disciplines of science and technology, expertise and machines. This kind of self encounters objectivity in the form of resistance; who we are is a product of discourse networks and technologies over which we have little control (Kittler 1985). On the other hand the practices of science, technology, and medicine fashion selves as objective facts through scientific experimentation, subject selection, and medical taxonomic exercises. This latter case emphasizes social and disciplinary production of selves, while the former emphasizes cultural presuppositions built into concepts and practices.

Attending to the categories of the person built into facts and attending to facts-in-the-world as facts enables us to see more clearly how medical and scientific claims, along with our own, are as much about dividing persons as they are about describing them. Here, along with Emily Martin, I believe we should also "acknowledge the varieties of ways in which experience resists science and medicine" (Martin 1987). More specifically, the question of objective self-fashioning raises the issue of creativity with regard to facts. Rayna Rapp, for instance, has followed the different ways in which people incorporate the possibilities and results of amniocentesis into their lives—for one mother, the fact of a genetic defect means a decision to abort, while for another it means preparing to take proper care of a challenging baby (Rapp 1990, 1993). Martin and Rapp are both calling for a reader-response analysis of our relation to science, medicine, and other facts of life.[14]

PET Scanning in Courts and at the Movies

> With PET imaging, we can begin to explore the degree to which biological and social factors affect brain chemistry. Perhaps one day we will speak of an individual's brain chemotype as well as his or her genotype and phenotype.
> —Henry N. Wagner, Jr., and Linda E. Ketchum

I am concerned with objective self-fashioning as a result of my work in the field of positron emission tomography (PET) scanning, a brain-imaging technique that promises to provide images of the living brain in action as it thinks, worries, adds, gets sad, and goes mad. I have been examining what might be called, following Stone (1992), the "virtual community" of PET scans—the heterogeneous community of people who interact with these scans and each other. In addition to fieldwork among those who work with the injection and imaging of radiopharmaceuticals, I have interviewed graduate students, imaging technologists, and PET researchers. I have also followed how PET scans have appeared on TV, in newspapers, and in Hollywood movies. In particular, my attention has been drawn to the use of PET scans as authoritative facts in claims about how the world and people objectively are—that is, what attributes and properties our objective bodies have, and what this means for the rest of our persons: our lived bodies, subjective souls, and/or our selves (see, for example, Begley 1991).

The PET scan, produced in university research laboratories, is one of the iconic centerpieces of the 1990s' "Decade of the Brain."[15] This recent technology produces images of living brain and body functions through the use of radioactive tracers.[16] Unlike CT (computed tomography) and MR (magnetic resonance), which provide images of the tissue and structure of the brain, PET produces high-resolution, functional images of blood flow and glucose consumption within the brain. Producing PET images is an extremely capital– and expert–labor–intensive process. PET requires an infrastructure—including interdisciplinary personnel, a cyclotron, a nuclear chemistry lab, high-speed computers, and a scanner—that costs upwards of six million dollars to install and one to two million dollars a year to operate. Government fears of high-cost medicine have contributed to the fact that PET did not enter regular clinical medicine as CT and MRI did, but has remained an experimental science into the 1990s (Dumit 1995).

After an experiment is designed and representative subjects selected, a small cyclotron is used to produce radioactive isotopes.[17] These isotopes are short-lived, with half-lives that range from two minutes to two hours. They are immediately "tagged" or attached onto other chemicals to form radio-labeled substances, or radiopharmaceuticals. Flourine-18, for instance, can be tagged onto glucose, and Oxygen-15 can be tagged onto water. The radiopharmaceuticals thus formed either mimic or are analogs of substances regularly circulating through the brain.

The next step is to set up the experiment, inject the human subject with the radiopharmaceutical, and place him or her in the scanner. While the subject carries out some task (such as looking at words) or attempts to maintain some state (such as rest or anxiety), his or her brain is assumed to be using energy differentially in those regions involved in that activity or state. Scans can be taken quickly for a "picture" of blood flow during a thirty-second period, or they are taken after forty minutes for a "picture" of the glucose utilization up to the scan.

The scanner depends on the physics and biophysics of positron emission. As the radiopharmaceutical decays in the brain, it emits positrons that travel a short distance, run into an electron, and burst into two photons or gamma rays that fly off at almost 180 degrees. The scanner consists of a ring of detectors connected to a computing system that reacts when two detectors are hit by gamma rays at about the same time. The computer then assumes that there was a positron along the line between the two detectors.

After collecting hundreds of thousands of data points, the computer attempts mathematically to reconstruct the approximate spatial density of the radiopharmaceutical, a process involving many assumptions about brain biochemistry and metabolism. The result is a simultaneously simple (in the sense of transparent) and complex image of a human brain at work. In addition to appearing in popular magazines and newspapers, these images are increasingly being used in court cases to argue for incompetency and insanity as well as neurotoxic damage and head trauma (see, for example, Stipp 1992).

To begin considering the role these images can play in our own lives, I would like to present an example of PET as depicted in a popular film about schizophrenia, violence, and insanity. The following is my own transcription of part of the final four minutes of a 1989 movie, *Rampage*, directed by William Friedkin. I believe it represents the first use of PET in a Hollywood movie. At this point in the script, Charles Reese has committed six grisly murders and is about to be found guilty of them by a jury.

> *In a courtroom.*
> *Defense attorney*: [*whispering to his client, Charles Reese*] We still have a shot to save your life. We can still show the jury that you weren't responsible.
>
> *Cut to the Judge's chamber.*
> *Defense attorney*: Your honor, I'm going to request that a PET scan be performed as part of a defense to show the jury that he is mentally ill, during the penalty phase.
> *Prosecutor*: A PET scan purports to show only a patient's brain chemistry at a certain moment of time. In this case it is after the crime is committed.
> *Judge*: A PET scan is a form of medical imaging which is used in the diagnosis of epilepsy, some Alzheimer's, as well as mental deficiency. Depriving Mr. Reese of putting this in front of the jury . . .
> *Prosecutor*: [*interrupting*] It's only another gadget to hide Mr. Reese's responsibility.

Judge: [*pausing, contemplating*] Well, we're going to err on the side of caution. I'm going to order the test. We'll let the jury evaluate it. Nobody knows what it will show.

Cut to medical laboratory. Charles Reese is put in the PET scanner. A computer-generated rotating skull is shown, peeled back to reveal a rotating brain in red, then green.
Two scans come up side by side. One is labeled "Normal Control," the other "Reese, John." The scans look significantly different.
Medical Doctor: [*pointing to Reese's scan*] These are abnormal patterns, without a doubt.
Defense Attorney: What does that tell you?
Medical Doctor: Well, this yellow-green area here is consistent with schizophrenia. What you are seeing is a computer-enhanced image of the chemistry of the brain. And what it shows is a picture of madness.

Cut to the courtroom again.
Jury Foreman: Your honor, based on the new scientific evidence, We, the jury, find that the defendant should go to a state mental hospital.
At the end of the movie, text: "Charles Reese has served four years in a state mental facility. He has had one hearing to determine his eligibility for release. His next hearing is in four months."

In the microcosm of this movie, a convicted brutal murderer is not put into prison but is treated as a mentally diseased subject who may be released in the near future. The sole element presented to account for the jury's decision is a PET scan.[18] The words of the doctor—"This is his brain . . . These areas are definitely abnormal . . . consistent with schizophrenia . . . a picture of madness"—concatenate a history of struggle and controversy within the medical and legal communities regarding a host of relationships: PET scan to brain, brain to schizophrenia, schizophrenia to insanity. In the movie, the PET scan stands as *the fact*, the linchpin referent, which holds the chain of connections together, convincing a jury that an abnormal brain scan is an abnormal brain is an abnormal person who does not bear responsibility for murder.

Not one of these connections, however, is settled in the scientific and medical community, in the legal community, or in my own mind.[19] Medical anthropologist Horacio Fabrega discusses the reluctance of Anglo-American society to accept a theory of illness-caused deviance. He suggests that this is primarily due to a need to have the will be socially or rationally motivated: "In essence, mental illness as a defense of homicide requires a suspension of our attribution of personhood if the latter is equated with willful symbolic behavior" (Fabrega 1989:592). Although I think that this argument makes sense in general when comparing societies, I am interested in the ways in which the attributes of personhood in the US are continually contested using batteries of facts. *Rampage* is an intervention into the facts of PET and the facts of life, presenting as it does a definition of PET, a set of presumptions about imaging and mental illness, and a possible scenario of PET's use in a court. Watching the movie, one confronts these facts of PET and is drawn into the virtual community of its images.[20]

What is the status of these "facts" proclaimed via Hollywood? Are they true? These questions trip me up as I watch a world of biotechnopower where technology judges who is responsible/sane/rational and who is not. This is a "view of the world that might be different from my current one" (Martin 1985:195). Like Emily Martin, I often find myself stumbling "over accepting [these] scientific medical statements as truth" (1985:10). But Hollywood reframes the question of truth, calling for an examination of the ways in which new facts, worlds, and persons are produced, distributed, and incorporated. For example, *Rampage* mediates between experts who presumably provided the details of PET, brains, and schizophrenia, and us lay viewers.

Though some might want to claim that there is a set of accepted medical truths, the purpose of this paper is to work with a notion of uneven flows of knowledge and contradictory versions of acceptability and legitimacy. We don't know how much we don't know about medical truths. Hollywood movies, along with best-selling novels written by physicians and our own doctors' advice, help to shape our notions of "accepted medical knowledge" and thus help shape our categories of the person. As part of my ethnography I follow this shaping process, examining how facts travel in the world, but also how they never travel alone. Instead they are packaged in the form of stories, explanations, and experiences, as authorized or unauthorized accounts, and they necessarily include definitions of human nature. Faced with novel facts, we may indeed stumble over accepting them.

When I have shown the movie clip from *Rampage* and pictures of PET scans during talks, some people with social constructionist tendencies and some with strong feelings about the social or psychodynamic nature of schizophrenia have been upset over the biosocial totalitarian implications of this apparently seamless presentation of clear difference between "them" and "us." I want first to note that despite constant work on PET and schizophrenia over the last twenty years, there is still much disagreement over whether PET is ready yet for clinical work with mental illness. In addition, over 90 percent of the PET community furiously opposes the use of PET for the insanity defense (Mayberg 1992; Rojas-Burke 1993). In spite of this unreliability for regular clinical work, in some places PET has nevertheless been heavily supported, including financially, by mental illness activists, that is, organized families of people with mental illness. Here another set of contests emerges. Should researchers look for biological correlates of schizophrenia, and how should such correlates be interpreted? What do the facts mean? Surprisingly, the meaning of these facts does not emerge solely from the research community; the whole virtual community must be examined.

In order to examine this story I have to back up forty years to the beginning of the "biological revolution" in psychiatry. During the 1950s and through the early 1960s, new pharmacological agents—drugs such as thorazine (chlorpromazine), lithium, and valium, which helped reduce symptoms in mental patients—were discovered and allowed many patients to live at home for the first time (e.g., Andreassen 1984, 1989). In the 1960s and 1970s, however, mental illness treatment critics organized to reform institutionalization practices. These critics created an uneasy alliance with psychotherapeutic psychiatrists who were invested in talking cures, and together they campaigned heavily for the notion that schizophrenia and the affective disorders were psychogenic. These "antipsychiatrists" argued that mental illness was socially constructed and therefore in need of social cures, not drugs (Laing 1967; Szasz 1970).[21] Their argument drew in part on the fact that there were no known biological mechanisms for mental illness. Perhaps, the antipsychiatric camp argued, drugs only affected the symptoms, not the cause.[22]

In the late 1970s and 1980s the increasing availability of new diagnostic techniques such as computed tomography (CT) scanning and PET scanning changed this perspective. These techniques offered different ways of examining *living* brains (Pardes and Pincus 1985). The medical imaging advantage was measured in two ways. First, it allowed correlation between brains and diagnosis among living humans, thus permitting anew the equation of "brain = illness." Second, medical imaging promised to provide early warnings of the onset of mental illness, one of the largest problems in its treatment and prevention.

PET in particular was hailed as significant because it promised to provide functional images of the brain in action. Early on, it was realized that many head injuries, strokes, and epilepsies leave the structure of the brain relatively unchanged but show up with different degrees of clarity on PET scans. In biological psychiatry, such proof of pathology was talked about as a Holy Grail. One biological psychiatrist, for instance, began a review of PET with the statement, "In the 1970s, the antipsychiatry movement almost had us (Szasz 1970), but now we have proof" (Kuhar 1989). For this subdiscipline, eager to demonstrate the physiology of mental illness, images of brain differences between mentally ill patients and non–mentally ill controls were facts that implied that a full biological explanation of mental disease was only a matter of time.[23] This technique thus functioned as a promise that mental illness was not "in the head" but in the brain.

Patients, Victims: On Seeing Oneself in a Brain Mirror

> I find a tremendous interest in PET scanning everywhere I go. I do a lot of public speaking, and I find that people are very interested in this. And they are always appreciative of the first ten minutes where I go through how positrons decay into gamma rays and the coincidence detections. They follow this, they understand it, they have a concept of how they whole thing works, and they are terribly fascinated with the whole idea. People are tremendously interested in the brain. You know, almost everybody thinks they are going to get Alzheimer's disease. If for no other reason, they want to know what is going on.
> —Richard Haier

To illustrate the ongoing negotiation of personhood and illness and call attention to the wider virtual community of objective self-fashioning around PET, I turn now to one site of my fieldwork, the Brain Imaging Center at the University of California at Irvine (UCI). This center was unlike most PET centers in two important respects. First, it was located in a psychiatry department, not in a chemistry, nuclear medicine, or radiology department. Second, for a PET center, it was extremely underfunded. Other major PET centers have received either Department of Energy or National Institutes of Health program grants to support the multimillion-dollar costs of laboratories in nuclear medicine or radiology. UCI's program was started in a psychiatry department and purchased its scanner and then its cyclotron with bank loans. Monthly payments were dependent upon an external fee schedule that dampened free operation. In the words of one researcher,

> *YY*: We were sort of a shoestring operation. I think we were sort of an upstart in some sense, because other places that have PET centers are much better endowed than we were. We were sort of the scrappy, come-from-behind, shoestring budget kind of guys. And we did things on a budget that is probably one-tenth of the budget that Hopkins or UCLA has for their PET centers. They are very well endowed and they support their PET centers in a maximum way. I think that we have a much more sort of guerrilla-type operation. We are unconventional in that we did so many things on our own, but I think we were fairly productive. We've done a fair amount of work even though we are on a shoestring budget, relative to a lot of other facilities.

This PET center operated from such a precarious financial position that its researchers spent much time doing local community outreach. They found a ready alliance with the mental illness community in Orange County, especially with families who had schizophrenia among their children. As Haier details below, the psychodynamic approach, while supporting the social nature of schizophrenia, often localized this causation into the family and specifically in the mother.

> *RH*: The Lockhardts contributed $250,000 to help pay for our scanner . . . By that time, the scanner had arrived and we were making pictures, they had schizophrenia in their family, and they were very interested in it. And they knew our emphasis was going to be on schizophrenia. We always approach it that in the long run, the main help will come through research. Probably not for people who currently have it, but because there is a genetic component, there are still the grandchildren to worry about. And families find this compelling. Remember, even in the late '80s, the public was just coming out of the idea of the schizogenic mother, that schizophrenia was somehow induced because the mother was doing something wrong. Virtually every set of parents that we talk to now, when schizophrenics are now in their twenties and their thirties, almost every parent has had the experience of going to a psychologist early on and getting the idea that somehow they were at fault. So it is all in their memory. And the idea that it is biological has caught on real fast over the last five or eight years. Family groups have organized around this to support biological research, and imaging is obviously at the heart of that. So it is kind of a natural sequence of events.

Supporting PET research became a means for these families to empower their participation within science, stay informed, and come to understand their role as accountable to, but not responsible for, the fact of familial schizophrenia. Along with the National Alliance for the Mentally Ill (NAMI), these families advocated a biological redefinition of mental illness and actively helped to produce facts about the nature of personhood and mental illness (OTA 1992). Objective self-fashioning is here a strategy without which such research might not get done.

Within the daily practice of clinical psychiatry, these brain-imaging techniques have also helped sufferers deal with the fact of mental illness symptoms. The following excerpt is from an interview with Dr. Joseph Wu, a psychiatrist at UCI.

JD: Do you show the patients their PET scans?
JW: Oh yes. We try and show them the PET scans, and then some of these patients will refer them out to people. I have a part-time private practice with some of them, and they may like to continue with me.
JD: Does it help them overcome part of the stigma of mental illness?
JW: I think so. I think that definitely. One of the intrinsic messages is that the depression isn't something to be ashamed of; it is an illness which needs to be understood. And it is not something that is their fault.

I think that there is a destigmatization that occurs with the biological emphasis. It is a fine line, because there are some arenas of personal responsibility that people can and should assume for their feelings. But I think it is a very narrow and tricky balance. It is important not to think that it is all biology; that can lead to a certain eschewing of what is appropriate for one's own role in understanding one's emotions. On the other hand, I think that people can go overboard, and say, "Gee, I'm entirely at fault for how I feel." [It is important] to try and understand one's role in helping to monitor one's emotions without being unnecessarily harshly judgmental of oneself.

The reconfiguration of mental illness as biological through the use of PET scans becomes part of a personal reconfiguration of one's own category of person. A strict division between the biological self and the personal self is not at issue here. Rather, the relations between the two selves are redistributed so that, although the patient must continue to experience the illness and live with it, she or he no longer has to identify with it. The diseased brain, in this case, becomes a part of a biological body that is experienced phenomenologically but is not the bearer of personhood. Rather, the patient who looks at his or her PET brain scan is an innocent sufferer rationally seeking help.[24]

Other researchers who have also shown patients their scans have agreed that, especially in cases of neurological and mental diseases, which are often accompanied with self-disgust or a sense of failure, both the scan and the process help legitimate the problem. They make it something that can at least be explored.[25] These patients (and their families) want schizophrenia and depression to be medicalized, to have a single cause or explanation, even if there is no solution or cure for them.

Anthropologists of medicine have long explored this kind of effect as a crucial aspect of every health care system. Jean Jackson discusses the failure of culture to come to grips with chronic pain (Jackson 1994; see also Good et al. 1992).

The tension Jackson describes involves mental versus physical pain. Chronic pain sufferers seek out, even hope for, positive test results, even cancer, because then there would be something to point to and work on to solve the problem. Regarding depression, Dr. Wu concurred with this interpretation when I asked him about the history of psychiatry.

> JD: Dr. Wu, Nancy Andreasen has written about the biological revolution in psychiatry.[26] You were in medical school during this time. Did you also get the other side of psychiatry?
>
> JW: Oh, very much so. I would say that most of the psychiatrists in this department are still analytically, dynamically focused. I would say that biologically oriented psychiatrists still make up a minority of the faculty. Maybe 30 to 40 percent, as opposed to the psychodynamically oriented people [who] are 50 to 60 percent.
>
> JD: Do both of these sides come into play in your work?
>
> JW: Somewhat. For me, when I do a study of depression, there is a part of me, a whole human dimension, that really tugs at my heart. Part of me feels moved by the pain of the patient that we work with. I am also moved by the courage and the willingness that many of these people have to participate in this study, even with the depth of their emotional pain and anguish. I think we try to offer to them the gratification that comes with knowing that they are contributing to the fund of knowledge that will eventually help to, we hope, eliminate depression or mitigate it. And that is something that many of these people find appealing, because there may be some greater purpose to their suffering. It is a way of reconnecting in some sense with the broader community. It is a way of making a personal meaning out of the emotional pain that they suffer from. For me, I see the whole biological aspect as not being contradictory or mutually exclusive from the psychodynamic aspect. I really see it as complementary and synergistic with the dynamic aspect. There are some people that see it as either/or. I see it more as a both/and type of proposition.

PET research into mental illness has thus become an area of study worthy of community support and patient contribution. The both/and approach to psychiatry, popularized by writers like Peter Kramer (1993), involves realizing that the brain can be altered by the social environment *and* by genetic development and drugs. The kindling theory, for instance, suggests that repeated abuse during childhood can build up depressed reactions until the depression is neurologically self-sustaining (Post and Ballenger 1984, cited in Kramer 1993:110–18, 334). The brain becomes "rewired" as if the person had been born that way. In the same vein, both psychodynamic talk therapy and psychopharmaceutical drug treatment can change brain chemistry and rewire the brain toward freedom from depression. Note that the brain remains the bearer of mental illness, but has now become an intersection for social and biological influences.

Dr. Wu's "both/and" approach to psychodynamic and biological explanations of mental illness arises, I suspect, from taking patients' perspectives into his account.[27] Patients are able to participate in social and medical reform by participating in research that might produce facts implying a category of person who suffers from a physiological rather than a psychological disturbance.

If we see that responsibility and causations are part of our categories of persons, this example demonstrates the flexibility and contestibility of these categories. Patients and activists are actively getting together to support and promote research on the shared biological nature of mental illness because of their desire to see the results and their hope for cures. Paul Rabinow has called this grouping on the basis of biological commonality "biosociality" (Rabinow 1992). A key point to remember here is that the facts of biology around which these groups are organizing are not necessarily fully decided within the scientific community. Yet they provide the means for social action, justifications for support of certain kinds of research, and arguments for a biological understanding of mental illness. The facts enable the groups to further promote a category of the objective person that does not, in their view, prejudge them and condemn them to blame and guilt. This involves understanding the many very different ways facts (science, technology, nature) and experience (subjectivity, personality, culture) are constantly shaping and tripping over each other. These people are working creatively to refigure responsibility for mental illness, in this case to biology, in an attempt to gain control over this part of their world.

The challenge here isn't just to the social construction of mental illness. This is not a simple story of the gradual emergence of the right view of depression, schizophrenia, and PET scanning. Biological psychiatry, for instance, often leads to deinstitutionalization, which burdens lower-income communities more than upper-income ones. But this story is not one of victims and blame. By tracing facts-in-the-world throughout the virtual community of PET images, I hope that responsibility for these situations might be multiplied—that accountability might adhere to experts, mediators, and laypersons alike for their participation in objective self-fashioning.

Cyborg: Machines My Eyes and Ears

> My present research with PET scanning is concerned with investigating chemical reactions constantly taking place inside the human brain, and how these reactions affect how we think, feel and act . . . how they affect whether we are afraid, violent or destructive . . . Perhaps we will be able to learn enough about the brain chemistry of fear, violence and destructiveness to save ourselves from the problems of interpersonal violence and war.
> —Henry N. Wagner

When PET researcher Henry Wagner (1986:253) says that "in PET, we now have a new set of eyes that permits us to examine the chemistry of the human mind," he is pointing to a particular kind of humanoid: a cyborg whose experience of vision includes the physiology of the brain as witnessed through PET scanning. Kramer and Sam listen to Prozac to hear Sam's true self speak. Some of us may shudder at the alienation implied in selves mediated by radiotracers, new pharmaceuticals, and multimillion-dollar bioscience. Others may breathe a sigh of relief at not being blamed for personally constructing schizophrenic children, at finally being respected as having wonderful children who happen to have a visible and therefore real brain dysfunction. Still others may wonder when and how they will be classed as normal or abnormal, or if the binary categorization will finally prevail.

In conclusion, I have tried to point out some of the ways in which contemporary biomedical and scientific practices are culturally situated. These practices are participating in ongoing negotiations not just of specific brain-behavior-mind links but also of the nature of human nature and the significance of human differences. I have tried to show both the complexity of the process of producing contemporary neuroscientific facts and images as well as the numerous ways in which practical considerations often build in assumptions about human nature with undesirable and socially unequal consequences. My purpose is not to point a finger at any particular sets of people or techniques. I think it is necessary to recognize the social and cognitive benefits of these practices for many, many people. Rather, I am seeking to find a language to talk about multiple accountabilities between the diverse communities engaged with PET.

The challenges of how to understand the continuing and increasing presence of biotechnopower require close attention not only to the multiple uses and arenas of facts-in-the-world but also to their deployment within discourses of objectivity and to the ways that they have built-in, presupposed notions of human nature. The point is that science and medicine turn out to be our business on a daily basis. We are involved in them, they involve us, and they draw upon the ways in which we configure the person. My hunch is that this process will reveal much about the multiple circuits of theory transfer from laypersons to experts and back again to laypersons via all kinds of mediators—movies, magazines, personal physicians, and anthropologists. These circuits of fact distribution and presupposition are worth understanding if we want to play a critical role in our own understanding of our selves.

Notes

1. I'd like to acknowledge the comments from the SAR Advanced Seminar on Cyborg Anthropology, and especially the critiques of Wendy Belcher, Bruce Grant, John Hartigan, David Hess, Lorraine Kenney, Kim Fortun, Anjie Rosga, Gail Sansbury, and Sylvia Sensiper.

2. "The body is not the object of study but the subject of culture" (Csordas 1990). Culture entails science for us; therefore our bodies have to take science into account.

3. The etymology of "layperson" extends beyond its recent application to a nonscientist to any form of nonexpert. Its original use for noncleric hints at the search for truth, certainty, and redemption through expertise (Barzun 1964). In interdisciplinary neuroscience, for example, each researcher is usually expert in only part of an experiment and an informed layperson with regard to the rest. Complete interpretation of the results are therefore continually deferred to "other experts." See Ademuwagun (1979) for anthropological discussions of lay knowledge.

4. There are other perspectives on Prozac as well (Breggin 1994; Elfenbein 1995; Norden 1995; Wurtzel 1994).

5. I use "factoid" in the CNN sense of an apparently nonscientific, even pseudoscientific claim. In a *New York Times Magazine* article, William Safire (1993) describes

many current definitions of "factoid." He eventually declares that though he prefers another definition, CNN's popular media power and its constant use of the term means that CNN's notion of it will in the end prevail.

6. "When one says in connection with totemism that certain animal species are chosen not because they are 'good to eat' but because they are 'good to think', one is no doubt disclosing an important truth. But it must not lead one to neglect the questions that then follow: why are some species 'better to think' than others; why is one pair of oppositions chosen over all the other possible pairs offered by nature; who thinks these pairs, when and how?" (Castoriadis 1984:19). See also Turkle (1984).

7. Fish, cited in Rabinow (1986:255). Science and technology studies researchers have a tradition of examining the contested production and stabilization of facts-in-the-world. See the work of Harry Collins (1985) on replication as well as David Hess's article in this book.

8. See also Fischer (1990), Harrington (1992), and Star (1989, 1992) for analyses of the "brain" as a particularly inspiring object to think with.

9. Other disciplines also make use of this conceit. De Lauretis (1984), for example, uses the then contemporary neuroscience of vision to point out flaws in previous theories of the specular gaze in the cinema. This argument actually takes up where previous neuroscientific studies of vision left off regarding the persistence of vision phenomenon (de Lauretis and Heath 1980).

10. See also Fujimura (1988), Gerson and Star (1986), and Griesemer and Wimsatt (1989) for careful sociological attention to the framing of scientific questions within a conceptual environment.

11. With an acknowledgment to Stephen Greenblatt (1980).

12. My attention was first drawn to Kramer's book when a psychiatrist I was interviewing referred to it as something he had spent a lot of time thinking about.

13. On incorporations see Casper (1993) and Farquhar (1992) and also the magnum collection *Incorporations* (Crary and Kwinter 1992) and the much more modest *Cartographies* (Diprose and Ferrell 1991). The work of Douglas (1966), Douglas and Wildavsky (1982), and Ong (1987) has built the groundwork for this kind of approach.

14. Particular inspirations for me are Cheng (1993), Long (1985), and Radway (1984), who each examine the ways in which different groups of readers incorporate texts (science fiction, reading group selections, romance novels) into their lives creatively and powerfully.

15. Ordered by Congress and signed by President George Bush in 1989. See *Decade of the Brain: Answers through Scientific Research* (National Advisory Neurological and Communicative Disorders and Stroke Council, 1989). The full declaration and a summary of previous and promised projects including a national database of brain research can be found in *Mapping the Brain and Its Functions* (Pechura 1991), sponsored by the Institute of Medicine. The cover of the book features four PET scans with labels reading, "Seeing Words," "Hearing Words," "Reading Words," "Generating Words."

16. This paper emphasizes PET as a brain research technique and not as a clinical or whole-body diagnostic technique. For many researchers and clinicians, the latter concept of PET is the only meaningful one. PET is used to image the spread of cancer in the body, to localize focal epilepsy for surgery, to characterize heart conditions, and for many other uses, both in the brain and in almost all other organs.

17. The following description is provided as a guide to the process of PET scanning. A more developed and explanatory description is provided in Dumit (1995).

18. It should be noted that *Rampage* (Wood 1985), the book upon which the movie is based, does not mention PET scanning at all. It also has no pictures.

19. There is quite a lot of research on mental illness and violence. The legal category in the US is "dangerousness" (a danger to self or others can be grounds for in-

voluntary commitment). Mestrovic and Cook (1986) and Monahan and Shah (1989) provide histories and evaluations of the dangerousness standard. Many researchers have argued that mental illness is actually *not* predictive of violence (Monahan 1988; Mulvey and Lidz 1984; Pollack 1990).

20.   Movies and other media are often the direct route for new sciences like PET to enter the courtroom. Consider for instance the case of Barry Wayne McNamara, who killed his parents, his sister, and his niece in 1985. When this case was brought to trial McNamara's attorney, Santa Barbara deputy public defender Michael McGrath, thought his client was not sane and sought proof beyond psychiatrists diagnosing schizophrenia: "We know there was a trend in the law which is hostile to psychiatrists in the courtrooms. . . . what we needed was objective evidence." McGrath learned of PET scanning through a PBS television series, "The Brain," and contacted Monte Buchsbaum at UC–Irvine's Brain Imaging Center. Buchsbaum reports that McNamara received a life sentence and not the death penalty "partly, perhaps, because of the ameliorating circumstances of a brain which was not entirely normal" (Black 1989).

21.   There is ample documentation concerning abuse of drugs and electroshock therapy in many state mental institutions during this period. See also Lovell and Scheper-Hughes (1986) for an overview of this history of deinstitutionalization in the US and in Italy.

22.   In the anthropology of medicine, much work has attended to the cultural construction of illness, especially mental illness. Kleinman's notion of Explanatory Models (EMs), for example, has been useful in helping to understand the many ways in which illness can be defined, explained causally, treated clinically, and respected socially. Further work has extended this to look at the cultural construction of the body in relation to EMs (Scheper-Hughes and Lock 1993). Gaines (1985) begins a serious critique of prevailing approaches to relativity by showing that the concept of the person underlies and crosscuts the very boundaries of our thoughts on the world:

> Kleinman et al. (1978) and Kleinman (1980b) have found that patients have cognitive models of their illness episodes; they refer to these models as 'Explanatory Models' (EMs). They recognize that not only do patients have EMs, but so do healers. What I will suggest here is that EMs are in part reflections of larger cultural conceptions; in particular conceptions of the person. I suggest that the key conception of person *organizes* cultural knowledge which gives rise to the EM of patient and healer. That is, a non-medically focused notion, that of person, lies behind and organizes patients' and healers' thinking about sickness episodes. Put another way, we may say that a cultural or folk theory underlies and gives shape to cultural knowledge and direction to cultural thinking about sickness. (Gaines 1985:230–31)

23.   See, for instance, Frackowiak (1986:33): "Much optimism has been generated in those who feel that the only reason organic bases for the various psychiatric syndromes have not been elucidated has been the lack of a suitable investigative tool . . . It is probable that PET is the investigative technique of choice for research of such hypotheses in man."

24.   The psychiatrist in this instance is using the PET scan to "therapeutically emplot" the patient as well. That is, by helping the patient to see mental illness as a physiological phenomenon, physiological intervention is also facilitated. See Good et al. (1994) and Mattingly (1994).

25.   One researcher noted during our conversation that he should list this feeling of validation as a confounder in future PET studies of depressed patients.

26.   "These advances [in PET and SPECT] have dramatically changed the training of psychiatrists, as well as their clinical practice and their research. They now see neuroscience as their primary basic science, modification of brain chemistry and metabolism as one of their primary modes of treatment, and the brain as the organ that they are treating. (Good psychiatrists also recognize that they are treating people, and that sensitive counseling and psychotherapy are also a fundamental part of their specialty)" (Andreasen 1992:842). See also Andreasen (1989).

27.   There has also been much controlled research on the usefulness of combining psychodynamic treatment (psychotherapies) with drugs (pharmacotherapies) (DiMascio et al. 1979; Elkin et al. 1985; Pardes and Pincus 1985).

ICONIC DEVICES / Toward an Ethnography of Physics Images /

Sharon Traweek

PARTICLE PHYSICS HAS BEEN HEAVILY FUNDED by all the richest countries of the world since World War II. Throughout most of the twentieth century almost all scientists in the United States, Japan, and several European countries considered particle physics the most fundamental science because it identified the basic elements and processes of the phenomenal world, elements and processes that were the building blocks of all the more complex aspects of our universe. During the past twenty years the sciences of complexity, artificial intelligence, and artificial life and the mapping of the human genome emerged to challenge the foundational status of physics. Today most science students focus on biology and computer science, whereas in the 1960s half of all undergraduate science students in the United States studied physics.

Currently there are fewer than ten thousand people worldwide who are active in particle physics. So why is it that since the mid-1980s books by and about these people (Dyson 1988; Feynman 1985; Hawking 1988) have appeared on the world's bestseller lists for months at a time? In the popular cultures of wealthy countries, physicists have become our guides to the galaxy, the purveyors of doom, our priests of time and space, and the saviors of civilization. A big job; a heroic job. Like all scientists they also are ridiculed as socially inept, ugly automata and called "nerds" or, somewhat more politely, "brains" (implying that their bodies are insignificant and useless). The physicists are themselves, of course, a part of the cultures that make those popular images, which in turn have helped to fashion the physicists. In this essay I will be considering the physicists' favorite images of physics and of themselves. My interest is in what their pictures of Albert Einstein and their legends of Richard Feynman might have to do with physics charts, tables, timelines, and so on, and in how the physicists' world of images shape their strategies for getting governments to support them.

Asking these questions is part of my ongoing study of the production, reproduction, consumption, and revisioning of knowledge among scientists and in popular culture. I am studying an international community's strategic practices (visual, narrative, mathematical, mechanistic, financial, computational, institutional, pedagogic, governmental) for doing physics and for fashioning themselves. I am studying how all their strategies shape and are shaped by each other. I am studying colonialist practices, engendered practices, and generational practices. I am studying unstable ecologies that are simultaneously personal, local, regional, national, transnational, and global. I am asking such questions as, Do physicists have their own kind of common sense? Are aesthetic pleasures and thinking connected in physics? Is physics profitable? Do machines crafted by physicists make science? Are simulations desirable? Is physics popular culture? What kind of subjectivity makes the best physicists now and fifty years ago? Is physics political? Traditionally it has not been acceptable for scholars to think and write about all this at once, but I have been querying just such a messy world in a half-dozen recent articles (Traweek 1996a, 1996b, 1996c, 1995a, 1995b, in press).

Concerned with the performance and enactment of cultures, for many decades anthropologists have been studying the ways visual, aural, oral, and written media are composed, consumed, circulated, and revised through cultures. For the last decade, at least, many anthropologists, including those of us who study the social production and allocation of knowledges, have had a special interest in the discursive, rhetorical, genre, and narrative strategies being deployed in these media. Today we are especially alert to how differences and knowledges are constituted together in multiple arenas. We are also focused on diurnal, mundane practices, habits, and common sense, even when we study the most privileged forms of knowledge a society chooses to support. Increasingly, we study how knowledges circulate through and between cultures, how those knowledges are adopted, revised, or repudiated as they travel, and who gets to go along for the ride. We are interested in how all this is changing in the midst of the massive shifts in global political economies over the past fifty years.

Charting High Energy Physics

Among all the global traffic in images circulating incessantly within the high energy physics community, one very noticeable set of static visual images is frequently seen on the walls of physics laboratories, classrooms, and their adjacent corridors. The same images are found around the world, suggesting that they have particular significance in the high energy physics community. That consistency is remarkable to me, given these physicists' acute interest in stylistic differences in experiment design, detector design, data analysis, and troubleshooting, and their transmission of those stylistic differences from one generation to the next in certain research groups and schools.

These pervasive images include photographs of important physicists, locally used research equipment, and particle interactions; charts of particle/field relations and laboratory timelines of the universe and discoveries in physics about the universe, as well as the construction of research laboratories; timetables illustrating laboratory schedules of operation; graphic displays of data, usually on Cartesian coordinates, but sometimes designed as bar graphs; diagrams of particle actions and research equipment; lists of particle data and research equipment; cartoons, usually drawn by physicists and engineers at the labs; and posters, usually announcing future conferences.

I am especially interested in the forms employed in these images and in their meanings for the viewing physicists. With rare exception the images hung on laboratory walls are rendered in black and white on heavyweight, glossy paper. In some cases text is printed on the reverse side. Almost none of the images (with the exception of cartoons) are original, and many are finely (and expensively) printed. The designers, illustrators, photographers, and artists are almost never identified; often the copyright marks note publishers and dates of publication. I have asked many, many physicists if they know who made these images. All were startled by the question and had no idea of the answer; nonetheless, everyone could describe the images from memory.

Little Boxes

The charts, tables, timelines, and graphs commonly found in physics labs are very simple. Almost all of them consist of sets of rectangular boxes segmented into smaller rectangular boxes and arrayed in columns. Each box is labeled, sometimes with letters (Roman or Greek, upper- or lowercase), sometimes with numbers (now and then with the symbols for positive or negative, often with superscript, occasionally with the symbols for "less than" or "greater than," at times with the symbol for multiplication). One very well known chart also has sets of circles segmented into smaller circles, with arrows pointing from one set to another. In this case color is used to link separate columns and rows of boxes and different clusters of circles. But the use of black and white is the most salient feature of these images, with significant areas remaining blank, usually white, and in one case black.

That description of the shared elements in these images underscores the overwhelming impression they present of order: symmetry, balance, simplicity, and clarity. The images remind me of mandalas, those designs of the universe used as aids to meditation in some South Asian religious practices. Like the mandalas, though far less visually complex, the charts, tables, and graphs used in high energy physics are devices that serve the initiated as codes or pathways for understanding the relationship of everything to everything. These charts mime the world of high energy physics. A lengthy education in high energy physics enables the adept to follow the iconography of these images much better than the novice. I want to emphasize what most physicists reported to be the raison d'être of these stylized images: like Mendeleyev's periodic table of the elements, a chart should enable adepts to predict new elements of the system by noting the characteristics on the coordinates pointing to any blank boxes on the chart.

Particle physicists believe the world is composed of a limited number of elements with a limited number of characteristics. Sets of elements are related to each other, and the elements can interact in a limited number of ways. Thus, a chart that depicts the "standard model of fundamental particles and interactions" may be read as a kinship chart that maps the endogamous and exogamous relations of the world's basic elements.

A History of Words: Charts, Graphs, and Classifications

Many terms may be used to describe the graphic images employed and displayed by particle physicists: charts, tables, maps, timelines, graphs, displays. A glance at their meanings indicates a system of social values, an ordered world which these devices visually represent. My dictionary (Morris 1979), for example, defines "chart" as "an outline map on which special information . . . can be plotted" and "table" as "an orderly written, typed, or printed display of data, especially a rectangular array exhibiting one or more characteristics of designated entities or categories" or "an abbreviated list, as of contents; a synopsis." In the plural, "tables" means "a system of laws or decrees" or "a code." A "display" can be "printed matter that is set off prominently." An "array" is "an orderly arrangement, especially of troops," and "a rectangular arrangement of quantities in rows and columns, as in a matrix." "Chart" is from Latin *charta*, for papyrus leaf or paper, and "table" is from Latin *tabula*, for board or list; "display" is from the Indo-European root *plek*, meaning to plait and to fold, to weave and entwine. "Array" is from the root *reidh*, meaning to ride, to provide, and to prepare (Morris 1979:73, 227, 380, 1307, 1535).

The same dictionary defines "classification" in biology as "the systematic grouping of organisms into categories based on shared characteristics or traits," and as "taxonomy." Taxonomy is "the science, laws, or principles of classification"; the suffix *-taxis* means order or arrangement. "Classification" is from the Indo-European root *kel*, meaning to shout or to call out; taxonomy is from the root *tag*, meaning to set in order, to arrange (Morris 1979:248, 1319, 1545).

A graph is defined as "a drawing that exhibits a relationship, often functional, between two sets of numbers as a set of points having coordinates determined by the relationship," and as "any pictorial device, as a pie chart or a bar graph, used to display numerical relationships." As a suffix, *-graph* "indicates an apparatus that writes or records," as in telegraph, or "something drawn or written," as in monograph; one can personify the suffix by adding an "er," as in *-grapher*: "one who employs a specific means to write, draw, or record," such as an ethnographer. These "graphic" words are all drawn from the Indo-European root *gerebh*, meaning "to scratch, cut, carve, draw, and write," and indicating a piece of writing, a picture, and a written letter (Morris 1979:574, 1516–17).

These clusters of definitions point us to the notions of displays, rectangular arrays, and coordinates; we are led to lists, outlines, and synopses; we focus on orderly, functional relationships and categories; we notice techniques, devices, and apparatus; we are given laws and decrees; we are in the realm of writing and making pictures, plotting, and drawing; we began by scratching on leaves and cutting on boards. We also have voices and we act, making patterns and order, weaving and folding, preparing and providing. The etymology of all these nested words of Indo-European origin has given us a social world and a world of practice; it has given us a world of law and order, a world of differences that make a difference.

I am familiar with only one non–Indo-European language: Japanese. The Japanese words for chart, diagram, graph, and map all share one root, the ideogram pronounced "zu" or "to." One of my Japanese dictionaries says the ideogram is a simplification of one that originally included the ideograms for both fence and village; it was once used to mean a map and later came to mean "to draw or to project on paper," as well as a plan, a sketch, and "a matter written on paper" (Osaka University of Foreign Studies 1969:180–81; see also O'Neill 1973:65, 75). The same root is used in the word for library. The word for chart, *zuhyo*, includes another ideogram, pronounced "ara" and "omote" as well as "hyo." Derived from two earlier ideograms for "fur" and "clothes," it once meant clothing but came to mean surface and outside, as well as "to show, reveal, and express." Words using this ideogram include facial expression, superficial, table, list, the cover of a book, representation, and to make public (Osaka University of Foreign Studies 1969:182–83). From this set of ideograms we get an ordered social world (villages with fences and social expression), a world of practice (drawing and projecting onto paper), and a tactile world (clothing). We also get something new: the notion of surface and its associated meanings.

A Cultural History: Charting European Knowledge

My search for the history of charts and graphs in the intellectual history of Europe and Japan has just begun. Certainly the practice of charting knowledge was widespread in Europe's medieval period and in the Renaissance. Furthermore, the practice of representing knowledge visually was well established in the same periods; an entire sector of the history of European art is devoted to the decoding of that iconography. (Its best-known practitioner, Erwin Panofsky, was the father of two sons who studied physics; one became internationally known in high energy physics.)

We also know that knowledge was displayed iconographically on many surfaces other than paper. For example, the architecture, statuary, and windows of a cathedral were meant to represent visually both theology and the events depicted in the Christian Bible. Stage direction in Shakespeare's time oriented actors toward certain architectural features so that they could use them as mnemonic devices for their speeches (Yates 1974).

More recently, anthropologists of cognition have claimed that the visual array of knowledge into atemporal, discrete classifications is characteristic of what Claude Lévi-Strauss (1966) has called the "primitive mind" (non-European thinking). Michel Foucault has argued that the construction of "grids of knowledge" is especially characteristic of what he calls the "classical episteme," that is, the seventeenth and eighteenth centuries—the period we associate with the emergence of experimental science. Speaking of natural history, Foucault (1970:131) describes such grids of knowledge in the following terms:

> The documents of this new history are not other words, texts or records,
> but unencumbered spaces in which things are juxtaposed: herbariums,

collections, gardens; the locus of this history is a non-temporal rec-
tangle in which, stripped of all commentary, of all enveloping language, creatures present themselves one beside another, their surfaces
visible, grouped according to their common features, and thus already
virtually analyzed, and bearers of nothing but their own individual
names. . . . What had changed was the space in which it was possible
to see them and from which it was possible to describe them. . . . The
natural history room and the garden, as created in the Classical period,
replace the circular procession of the "show" with the arrangement
of things in a "table." What came surreptitiously into being between
the age of the theatre and that of the catalogue was not the desire for
knowledge, but a new of way of connecting things both to the eye and
to discourse.

Foucault goes on to declare that such grids of knowledge confine knowledge
of their objects of inquiry to just four variables: "the form of the elements, the
quantity of those elements, the manner in which they are distributed in space
in relation to each other, and the relative magnitude of each element" (Foucault
1970:134).

These grids of knowledge are also significant for what they lack. As Foucault
(1970:133) noted, they exclude all knowledge derived from taste, smell, and
touch, and they privilege the visual; nonetheless, the visual is restricted to lines,
surfaces, and forms, usually in black and white. The epistemological exclusions
are more subtle. Change over time is erased. The naming and tabulation of visually accessible characteristics is the defining method of the grid system of
knowledge. The grid form forces researchers to accept that "the system of variables has been defined at the outset [and so] it is no longer possible to modify
it"; they also must assume that all the entities in the same space on the grid
"have the same relation of resemblance in all and each of their parts" (Foucault
1970:143, 140). Perhaps the cheerful acceptance of this last constraint accounts
for the fascination I have found among physicists for visual representations of
fractals, along with their deep aesthetic appreciation of the resemblances they
perceive to be permeating "powers of ten."

I am not claiming that particle physicists are either "primitive" or lost in the
archives of the seventeenth and eighteenth centuries. I think that Lévi-Strauss
fetishized Europe as well as the distinction between ways of thinking into two
geographically situated types; I also think that Foucault neglected to consider
the survival of certain modes of thought in some communities long after they
had been dropped in the groups that had developed them in the first place. I am
proposing that we ask why knowledge in physics came to be and is still arrayed
in grids borrowed from natural history (botany in particular), why such grids
are so unproblematic to physicists, and why these grids of controlled difference
are so intellectually and aesthetically satisfying to physicists.

Laboratories as Charts

Elsewhere I have argued that the spatial array of scientific laboratories can be
read iconographically (Traweek 1988:chap. 1). That is, the labs themselves are

massive mnemonic devices for determining who people are and what they should be doing. To control space in a laboratory shows the other researchers that one also controls the scarcest commodity of all: access to operating time on the most prized equipment.

The built environments at two North American laboratories I have studied bring the Capitoline Plaza and Lenôtre's gardens to mind. The architecture of both labs reinforces and constructs privileged positions, places where one properly stands to survey one's domain, which recedes to the horizon like perspectival lines in paintings of the European Renaissance. In Japan the lab environment is arrayed on an entirely different grid: If a Japanese box had been gift-wrapped and then unwrapped, the diagonal, off-centered lines left on the paper would correspond to a map of the lab. The lines also resemble the paths of a classical Japanese stroll garden of the sixteenth- and seventeenth-century intelligentsia, which focused the eye on what was near and unexpected, always avoiding grand views. The North American laboratories' array cultivates a dominating gaze, the Japanese lab attends to the glance. Both situate the practitioners in a spatial array of controlled differences where surfaces are all.

Living on Timelines

Grids are static; they are not images of development, of phenomena changing through time. I have argued that the regulation of time permeates laboratories (Traweek 1988). There are six kinds of time every experimental high energy physicist constantly calibrates: whether or not the accelerator is running, how much beamtime—the number of pulses of the accelerator beam—is allotted to which research group, the expected lifetimes of the group's detector and the lab's accelerator, and the lifetimes of ideas and careers. All these are ephemeral, all are slipping away. Only beamtime can be replenished and accumulated. I claimed that what can be replicated is desirable for physicists and what is ephemeral, slipping away, is a constant source of anxiety. It is at the laboratory, and especially at the detector, that time is most dramatically annihilated and resurrected. Now I would say that there are other devices in the laboratory for annihilating and resurrecting time. First are the computers, with their capacity for the endless repetition of simulations of data. More pervasive are the charts and timelines that also annihilate and resurrect time.

My dictionary defines *timetable* as "a schedule listing the times at which certain events, such as arrivals and departures at a transportation station, are expected to take place" (Morris 1979:1346). Accordingly, physics timelines are usually plotted by powers of ten, starting with the origin of the universe in the putative big bang, the formation of our galaxy much later, going on to the recent appearance of our solar system, and ending very soon thereafter in the present. Charted along with time we find energy and mass, lest we forget that they are interchangeable, and along the bottom, like the mythical figures in the margins of old maps, the "constituents" are listed: leptons and quarks, gauge bosons, and photons. The only mundane bit of our universe plotted on one chart is the "rest energy of flea."

My current favorite timeline—although it is now outdated, its provocative challenge having not been met—is titled "Scientific Knowledge: 500 BC to

109

AD 2000" and subtitled "The Road to Waxahachie." Referring to the timeline as a road is certainly consistent with the notion that these charts and timelines, as objects of meditation, are meant to lead the viewer somewhere, to understanding and to action. The narrative road begins on the left side of the page:

> The road to Waxahachie began 2400 years ago in the town of Abdera, Greece, when a philosopher named Democritus stated: "Nothing exists except atoms and space; everything else is opinion."

This timeline is in color and large, about 24 inches high by 48 inches wide. The top three-quarters of the page is blue, navy in the center and a lighter blue at the sides. From the left to the center there are yellow stars; to the right there are mostly clear skies, but with some light and dark clouds. The bottom quarter of the page is green; where the green meets the blue there are hills and a few buildings. At the left stands an acropolis, in the center little houses, and to the right a set of high-rise structures, one of them quite phallic. Running across the width of the image is a bright ribbon of red; on the left it runs through the blue, in the center it is lower, in the green, and on the right the red ribbon reaches abruptly to the top right corner. In that corner there are dashed black lines intersecting the wide stroke of red.

At the bottom of the page is a timeline from 500 BC to AD 2000, with columns of scientific, technical, and medical discoveries listed above the appropriate dates. In the period labeled "Dark Ages," the church bell, bed, crossbow, linen and wool manufacture, and plowshare are mentioned. The columns are taller during the "Greek Empire" and "Roman Empire" eras and ascend again during the "Scientific Revival & Renaissance," "Industrial Revolution," and "Space Age." The red ribbon undulates across the page at the height of the columns of discoveries. Just beneath the dashed line across the ribbon the text reads, in part:

> The quest for knowledge has brought us, at last, to Waxahachie, where scientists and engineers are building the next great instrument of discovery—the SSC [the superconducting supercollider accelerator]. They toil in the midst of controversy. Strident naysayers demand to know *now* what benefits the SSC will bring. They want to ignore the pattern: *curiosity leads to discovery; discovery leads to invention; invention leads to benefit.* (Emphasis in the original)

Beneath the dashed line that bisects the red ribbon of soaring knowledge is written, "Cut here if the SSC is defeated." There is a great deal of text on this timeline, mostly in white print; the challenge about the funding for Waxahachie is in black ink. (Altogether there are thirteen paragraphs of text and twenty-four columns of discoveries and inventions on this timeline.) In the middle of the page, over the "Dark Ages" on the timeline, this passage appears in white letters against the navy blue nighttime sky:

> A warning: *knowledge is perishable* [emphasis in the original]. From AD 530—when the emperor Justinian closed the Academy and Lyceum in Athens—until AD 1453—when Greek-speaking scholars fled west following the fall of Constantinople—scientific activity in Europe virtually

ceased. According to Hellemans and Bunch in *The Timetables of Science*, "large cities disappeared, roads and aqueducts disintegrated, and trade became more limited. The general infrastructure in which learning and science could develop had ceased to exist."

Altogether this timeline presents quite an ominous message. When asked by a reporter for the *New York Times* why he thought members of Congress might not support the SSC, the then-director of the lab replied that it would be "the revenge of the 'C' students." The barbarian hordes went to college!

It is not difficult to make light of this doomsday iconography. Nonetheless, almost every North American high energy physicist I know endorses this representation of the history and future of knowledge (although most of the European and Asian physicists I have asked find the chart both silly and illustrative of the tactics certain North American physicists used to get funding for the SSC). This timeline is no benevolent mandala, leading the adepts to the right path for discovery; it is a narrative of the history of a fragile but beautiful and beneficent science, always vulnerable to ignorance. Science "nerds" are always vulnerable to the revenge of the "C" students; the "background noise" has suddenly became uncontrollable again.

Einstein's Hair, Feynman's Drums, and Repeating Voices

Almost all autobiographies of scientists follow the same genre conventions. As a child the protoscientist realizes that he is different from the other children. (With notable exceptions, women are almost never the subject of autobiographies and biographies of science.) That difference translates eventually into a social isolation that is both painful and a form of pleasure as the young scientist-in-the-making learns to enjoy the advantages of privacy, using the time to explore the world and ideas or to dismantle and rebuild one artifact after another. Upon entering college he begins to find others like himself and suddenly feels normal in the newfound world of science. Meanwhile these young men are sorting through engineering and the sciences and locating their true intellectual calling, usually in physics. Gradually a few begin to realize that they remain different from even their peers. Within a few years they have the ideas that enable others to understand how exceptional, how original they are; finally, they are invited to live and work at the most special places among their true peers.

In their most rudimentary telling, our traditional notions about science in popular culture are usually recited in the cultural forms developed by the medieval European Catholic church: a list of saints (geniuses), their miracles (discoveries), and holy sites (laboratories). These reverential stories can be found easily on television, especially on The Discovery Channel and public broadcasting stations, invoked in documentaries on science and technology, complete with an authoritative male voice-over offering instructive and amusing examples of Derrida's notion of "absent presences" (Derrida 1976). What is intriguing in this context is how closely exceptional scientists use the same genres and narrative strategies as the medieval European Catholic Church when they compose accounts of their own lives and their original work. The male voice-overs of television science documentaries announce the miracles of science and rehearse the clichés about objectivity and reason, geniuses and nerds. In the sections of their biographies in which they discuss their ideas, many of these scientists resort to the same disembodied voice-over style.

Roland Barthes (1977, 1981, 1982) has written about the ways images of Einstein's head have come to represent genius in popular cultures. There have been many plays and films in which a character called Albert Einstein has had a crucial role. Just now I am more interested in the visual images of Einstein circulating among physicists. There are two posters of Einstein that I have seen on the walls of many physicists' offices and laboratories. One is of Einstein's head, back-lit so that his rumpled curly hair takes on the appearance of a saint's illuminated halo. This is a common iconographic motif in the Christian art of the Catholic Church, signifying that the individual represented radiated divine grace. In cartoons a charged light bulb over someone's head suggests that the character is having an idea. In US popular culture, long, rumpled hair on men signifies rebellion, just as unkempt hair, mismatched clothes, and a befuddled expression typify a nerd.

The other Einstein image popular in laboratories and department offices is the poster in which he is seen awkwardly riding a bicycle and appears to be leaning or falling. Here again Einstein wears ill-fitting, rumpled clothes. His hair is blowing in the wind, and his expression is quite cheerful. Physicists tell me they really enjoy this image of Einstein because of the implied physical forces at play; they also like that childlike expression.

Richard Feynman's life in physics almost exactly parallels the fifty years of the hot and cold wars of the twentieth century: he began his career at Los Alamos during World War II and died in 1988. He is held in awe by physicists for the ease with which he could grasp and visually represent very complex ideas. The favored image of Feynman, which appeared on the jacket of one of his extremely well regarded undergraduate physics textbooks, shows him happily playing bongo drums. In the 1950s and early 1960s bongos represented a certain bohemianism and flamboyance to the European-American middle classes. Legends tell of Feynman routinely doing physics at his regular table in a small bar that featured topless dancers. Feynman's stoic suffering as his young first wife died after a long illness is often recounted, too. These images, memoirs, and legends juxtapose great intellectual subtlety and simplicity with childlike pleasures and the flaunting of social conventions. The message is that genius, like grace, comes only to the unworldly (Feynman 1985, 1988; Gleick 1992).

Epidemic Epistemes

At this point I have only explored the range and variation in some of the images commonly employed by particle physicists, finding some themes and some fault lines in their aesthetic pleasures. What has interested me here is the everyday world of images and image play, particularly in the images that everyone knows, likes, believes, and takes for granted. The first component of a paradigm, according to Thomas Kuhn, is the "symbolic generalizations" that everyone in a field accepts as given and that have not been disputed for decades. In the formal knowledge of particle physics those symbolic generalizations would be classical mechanics, as modified by quantum electrodynamics (QED) and quantum chromodynamics (QCD), and those fundamental assumptions would include their accepted, unquestioned ambiguities (Kuhn 1970). I am suggesting here that the commonplace images and image play in physics laboratories, classrooms, and corridors are part of these symbolic generalizations, so I am interested in the ways physics images are made, consumed, revised, maintained, and deployed.

I argued in my first book, *Beamtimes and Lifetimes* (Traweek 1988), that the presence of machines in the laboratory is what makes the work there scientific. These machines are not black boxes purchased from a catalog and remembered only as laboratory furniture; anthropologically speaking, they are the material culture of the people I study. Discussing the research equipment called detectors, I argued that they are simultaneously icons of the community, emblems of the community's desire for knowledge, and the means of production of that knowledge. To put the same point more plumly, making research equipment is making scientific knowledge and vice versa; making scientific knowledge is a mechanical and electronic craft; science is engineering and engineering is science; science is technologically and electronically and mechanically constructed. It is also the case that computing, like designing and operating detectors, can be, simultaneously, icon, emblem of the scientist's desire, and the means of production.

The teams needed to build these subtle detectors have grown larger and larger. Each part is now designed, constructed, and analyzed by a separate team, and those teams come from different institutions, often located in different countries. The detector's complex physical and social structure is reenacted in the electronic data collected from it. No analysis can be done on the data unless each team, and each smaller part of it, is willing to cooperate. Each group must disclose to the others its "local knowledge," its "tacit knowledge," its "rules of thumb" for making sense of its own artifact. Each of these constituent groups may have different research styles or research priorities, different levels of funding, different time constraints, or different kinds of careers. The detector has become a borderland where all these forces implode.

In the last decade, anthropologists have become interested in borderlands. High energy physicists, too, are interested in regions of intersection; they have been building borderland accelerators, borderland laboratories, and borderland detectors for some time. They are now building borderland simulations.

Evans-Pritchard (1976, 1978) wrote that if we understood the diverse senses that the Nuer made of cattle, we could understand the process by which the Nuer made sense. I am suggesting that if we understood what sense particle physicists made of detectors and computers, we would begin to understand the way physicists make sense. It is this "ground state," as the physicists might call it, that I tried to characterize in my first book. In Clifford Geertz's language I described their common sense, "recognizable by the maddening air of certainty with which it is always expressed" (Geertz 1973). I argued that, just like the Nuer, as physicists probe the world they try to structure their material and social worlds like their cosmology, and vice versa. I take all their worlds as simultaneously received and occasionally under revision and debate. I argued, for example, that they puzzle about where to make cuts among the novices, just as they puzzle about what signals to cut from their data. They also work hard to prevent revision and debate about what they take to be fundamental. In this paper I argue that this is what they do with charts, graphs, timelines and with images of their heroes: what is known is etched on the chart, preventing revision and debate; new knowledge is expected to fill the blank boxes. The form, the genre, and the coordinates, recorded in black and white, are not open for discussion.

In my recent work I have attended to another aspect of paradigms: the templates or formulaic modes of thought used in making problems and solutions. I take these templates of craft knowledge to be just as embodied, just as performative, as the "common sense" I described about detectors. There is a generational shift in aesthetic pleasures in formulaic modes of thought, just as there is a shift among those who are beginning to enjoy working with unconventional resources in unconventional sites on unconventional projects. Physicists over the age of forty-five tend to enjoy the aesthetic pleasures of taxonomies, classifications, stabilities, hierarchies, and uniformities. They enjoy making cuts and exclusions. They like law and order.

Younger physicists are more likely to share the aesthetic pleasures of complexity, instabilities, variation, transformations, and irregularities. They like interesting differences and enjoy kaleidoscopically changing patterns. They find pleasure in alternating foregrounds and backgrounds. They could use these different aesthetic approaches for problem formulation, research design, equipment design, data analysis, or education. The choice is between log books and simulations, between taxonomies and complexity. It is important to note that both of these formulaic modes of thought can easily accommodate certain conventional tools and techniques in the community, such as Lorenz transformations and renormalization, just as both can use the ordinary tools of their craft: colliders, detectors, charts, and timelines.

None of these differences are timeless. The material manifestation of these styles in specific detectors, accelerators, computers, images, charts, and timelines, and the money that pays for them, do not remain in place forever. The people who like one way of doing physics or another do not last forever, just as we know that the current debates are ephemeral and the current topics of experimental research, such as the search for the top quark, will be seen as boring, if not misguided, by the young. As younger physicists began to turn away from an aesthetic of cuts and to search for anomalous data against a controlled background, anomalous physicists, usually working at some distance from the centers of power, also came to share the same ground.

I am not claiming that the formerly anomalous physicists are somehow uniquely creative or prescient; the interesting aesthetic differences between people who have access to all the resources they want and those who must patch together their resources have always been with us. The generational aesthetic fault line I have suggested here certainly has been felt in many fields during the past twenty years. Something similar has happened in women's studies, history, and anthropology. Current work by anthropologists of science and technology has been profoundly influenced by our participation in the "virtual communities" created in electronic mail systems. Computing has changed our fieldwork and our writing in profound ways. For some people the coordinates of standard visual arrays are being challenged; for others the charts are even more deeply etched in their minds' eyes.

Profound and subtle gaps have emerged between the practices of those who participate in these new ways of making knowledge, knowledge producers, and communities, and those who are uninterested or unable to participate. These gaps are most obvious to those of us who are engaged or trying to become engaged in these new landscapes with their new pleasures; those who are not so engaged only notice some sort of deviance, strangeness, or incompetence in the rest of us as we begin to need these linked computing prostheses to think, to be, to teach, to do our work. We do not want to do it alone anymore. Like the physicists I study, many researchers today embody such a shift in their emergent aesthetic pleasures, with the absorbing ambivalence and ambiguity that attend to the change.

ENGINEERING SELVES / Hiring In to a Contested Field of Education / **Gary Lee Downey and Juan C. Lucena**

O NE DAY IN AN ENGINEERING THERMODYNAMICS CLASS, the professor pulled out an Ann Landers column, titled "A Different Breed," from that morning's newspaper. He asked the seventy or so third-year students, "How many of you think this is reasonably accurate?" The column contained excerpts from letters, mostly from frustrated spouses, that portrayed engineers as technical, inflexible, and socially inept. The first provided a summary statement:

> Dear Ann: This letter, my first ever to a columnist, was sparked by your column about the engineer's wife who asked, "Are engineers really different?" The answer is ABSOLUTELY! My father was an engineer. My three brothers and four uncles are engineers. Engineers ARE a different breed. They are precise, logical, and great at problem solving, but they know very little about human interaction. My engineer husband makes a fine living, but when it comes to expressing emotions, on a scale of 10, he's about a 4. *A wife in Houston*

Three other letters added credibility to this characterization by attributing it even to the best engineers, describing the difficulties of living with such a person, and suggesting that things could be otherwise:

FROM TUCSON, ARIZ: Engineers ARE different. My engineer husband (graduate of MIT) tells me when my skirt is ⅛ inch shorter in the back. If the floor in the bathroom looks uneven, he gets out a tape measure for "proof." A crooked window must be adjusted at once. If, however, I am crawling around the house with a killer migraine, he doesn't notice.

SANTA BARBARA: My husband the engineer has no tolerance for the gray areas of life. He sees everything in absolutes. It's black or white, right or wrong, yes or no. Never a maybe. He feels no joy, but he is never depressed either. Everything is in perfect order, or there is hell to pay. It is not easy to live with such a man.

NO CITY, PLEASE: Thirty years ago, I married an engineer. Our marriage was an emotional wasteland. He would have been a better father if our children had been robots he could program. Engineers can figure out everything except how to be human and caring.

The sole letter from a man removed gender as the determining feature:

CHICAGO: My wife is an engineer. She is precise, analytical, and definite in her views, and she always thinks before she speaks. She's as cold as ice and so sure of herself, she makes me sick. My next wife will be an empty-headed, bubbly moron, and it will be a relief.

Finally, letters from an optician and an interior designer established that the problems with engineers go beyond family relations, for engineers "are murder to work with" as well.

The only two letters expressing disagreement contested not that engineers were somehow a different breed but the judgment that this assessment is negative:

CARBONDALE, ILL: You're damned right engineers are different. I am still happily married to mine after 35 years. They tend to look before they leap and have stable marriages. By nature, they are problem solvers, sensitive, and caring. The woman who wrote to complain ended up with the wrong man, not the wrong profession.

INDIANAPOLIS: My engineer husband doesn't send me flowers. In fact, some days we don't even have a decent conversation, but I'll take this nerdy looking guy with his assortment of pens and eyeglass cases in his shirt pocket, his dull tie and wrinkled trousers over any of the men I've ever known. He is loyal, decent, dependable, and real. He'll never cheat or lie to me. That's worth a lot these days.

Prodded by the instructor to offer their own judgments, roughly a third of the students raised their hands to say they agreed that engineers were a different breed, nearly half said they disagreed, and the remainder did not respond. In sharp contrast with the usual practice when asking about today's homework or next week's test, not a single hand rose above shoulder level. The quick, low wave was the rule, as if students felt vulnerable and hoped to avoid calling attention to themselves. After all, why risk being wrong about something that would not be on the test? Despite the fact that the professor was a gentle man who was not afraid to discuss matters of the heart, he made no move to offer his own assessment. Did it even cross his mind to do so? To hear how he felt about things would have been interesting but unusual, and no one dared ask. Class began.

The two of us are interested in the making of engineers, the mechanisms of self-fashioning that take place in undergraduate engineering education. The professor and students in this class were grappling with what they understood to be the stereotype that engineers regularly criticize as a preconceived and oversimplified idea of the characteristics of the typical engineer trained in the United States. Questioning its accuracy involves asking whether or not people called engineers actually conform to the stereotype. During the course of our fieldwork among engineering students and teachers at Virginia Tech, a land-grant university with roughly four thousand undergraduate engineering students, we heard many discussions about the stereotypical engineer. In every case, someone disputed the image by claiming that it did not characterize accurately many engineers and hence was not true, only partly true, or at least too narrow. Yet the image persists, and students seemed to take for granted its existence and its power.

We were initially drawn to study engineering education by our own experiences as engineering students. Both of us completed undergraduate degrees but felt somehow that the sort of people we were being asked to become did not fit with the sort of people that we already were or wanted to be. Above all, we felt constrained. Over the years we have observed that people who left engineering, including ourselves, seemed to feel a need to explain why. It was not that they couldn't "hack it"—the engineering student's term for not having what it takes—but that somehow there was a lack of fit. Understanding learning solely as the transmission of knowledge from the heads of faculty to the heads of students did not begin to account for the bodily experiences of constraint so many of us experienced. How might we make these and similar feelings more visible? And might doing so suggest ways to shift engineering toward a place where people like us would want to be?

As a cultural anthropologist teaching in a graduate program in science and technology studies (STS) and a Ph.D. student in STS interested in cultural and anthropological studies, we organized a research project that would follow engineering students through their curricula to explore the changing demands they experienced as persons.[1] Do stereotypes count? How was the knowledge content of engineering related to the social dimensions of engineering personhood? Long-term participant-observation and extensive interviewing and document collection became strategies to explore students' experiences.

We quickly learned, however, that this work involved more than just studying students. As we elaborate below, we had organized our project in the midst of great debate among engineers over the contents of engineering education. Our study was even funded by the National Science Foundation, which had emerged as one of the leaders in this debate. Unless our written work appeared entirely irrelevant or uninteresting to engineers, it would likely be captured and positioned by this debate (see Rapp on abortion and Hess on capturing, this volume). Not only did running away or hiding seem pointless; we also wanted very much to participate in the process of retheorizing engineering education. We sorted out a pathway we call "hiring in."

Hiring In

As a metaphor of employment, hiring in indicates a willingness on the part of social researchers to allow their work to be assessed and evaluated in the theoretical terms current in the field of analysis and intervention. It means becoming employees in a sense, whether paid or unpaid. Although one need not accept at face value what people say about themselves and what they consider desirable or worthwhile, hiring in involves, at minimum, acknowledging that established modes of theorizing constitute established power relations and that contributing new theorizing captures one within those relations. Maximally, hiring in involves following all the pathways to critical participation that one can identify and attending to all the details and doing all the work necessary to position, assess, and, if warranted, try to achieve a theoretical shift.[2]

Accepting the responsibility of hiring in, as many researchers studying science and technology, including anthropologists, have already done, may provide an opportunity to contribute directly and genuinely to the theorizing that takes place in a contested field. It offers the possibility of convincing people to shift their modes of theorizing from here to there by acknowledging and emphasizing that one can only start where one is. As a practice for researchers combining cultural perspectives and ethnographic fieldwork, hiring in can involve making visible modes of theorizing that are otherwise hidden, thus possibly legitimizing alternate perspectives that are rooted in the field itself.

However, complementary risks of cooptation and social engineering are substantial, each leading in its own way to marginalized ineffectiveness and self-delusion. The cooptation of a project involves its transformation into something indistinguishable from that which it studies. More than gaining participation or otherwise becoming located as part of the field, cooptation dissolves the identity of the researcher(s) entirely into the field. A coopted project not only goes native; it is nothing else. Social engineering involves presuming that one's expertise warrants the authority to legislate change through a research project. The arrogance of social engineering keeps a project permanently outside the door, preventing it from participating critically in that which it studies. A social engineer in the field never leaves the hotel.

We locate hiring in as one approach to a more general academic practice that one of us has elsewhere called "partner theorizing" (Downey and Rogers 1995), limiting it to those situations in which one is not already located in the field of intervention but seeks to gain entry. Hiring in contrasts with debates, for example, whose interlocutors are presumably located within the field of intervention. The main goal of partner theorizing is to encourage the growth of collaborative relations in academic work and relocate the agonistic politics of rebuttal from a necessity to an option in the everyday practices of academic researchers. It asserts that the practice of theorizing neither can nor should be a proprietary feature of academic work, for much theorizing, in fact the major proportion of theorizing, takes place outside the institutionalized Western academy. Alongside the anthropological perspectives represented in this volume and elsewhere, partner theorizing thus reconceptualizes relationships within and between academic disciplines, as well as between modes of academic and popular theorizing, as flows of metaphors in all directions rather than the necessary diffusion of truthful knowledge and power from the inside out.

Our project on engineering education illustrates three different moments of partner theorizing. First, it envisions all acts of theorizing as undertaken in partner relations with their interlocutors in collective, but temporary, negotiations of knowledge production. We are thinking of partner theorizing not as a market activity, a business partnership, but as a variety of activities of exchange among committed cohabitants, married or otherwise. One's work always intervenes in the context of other theoretical agendas. Competing theorists, both academic and popular, live together.[3]

The dominant theoretical agenda with which our project has to contend is the doctrine of "competitiveness." Since the early 1980s official United States dogma has redefined international struggle from a political and military to an economic idiom, transforming understandings of the nation from a site within which individual interests compete into a single economic actor maximizing a collective interest. The power of patriotic commitment to this economic call to arms became concentrated in the slogan of competitiveness and its logic of productivity, which locates humans alongside technology and capital as resources for the production of consumer goods. Popular theorizing about competitiveness seems to reach into everyday lives and selves much more than the military logic of the Cold War did, because it turns every action into an economic defense of the nation. Something is good if it enhances competitiveness and bad if it does not. Engineering education has gained particular salience in these developments because engineers figure as key participants in virtually every image of increased national productivity (see Business-Higher Education Forum 1983; National Academy of Engineering 1986; President's Commission on Industrial Competitiveness 1985).

National visibility is new for engineering education (Lucena 1996). In the years after World War II and before Sputnik (October 1957), engineering education stood alongside other forms of technical and scientific education as an integral component of a military struggle against communism. "Our schools are strong points in our national defense," said President Eisenhower early in 1957, "more important than our Nike batteries, more necessary than our radar warning nets, and more powerful even than the energy of the atom" (US Congress 1957). Sputnik shifted concerns somewhat because it was read as a shocking accomplishment of science rather than of engineering, and nationalist interest in education during the 1960s narrowed to an exclusive focus on science and the production of scientists for basic and applied research. The National Science Foundation took Edward Teller's advice that emphasizing engineering education would miss the point:

> It is my belief that it [engineering] should not be considered a weak link in our scientific and technological effort [and therefore has sufficient funding]. We should put the greatest possible emphasis on higher education in applied science. (National Academy of Sciences 1965:259)

The 1970s maintained an emphasis on basic and applied science but expanded the range of legitimate problems to include energy, transportation, pollution control, and other nonmilitary arenas, through such programs as NSF's Research Applied to National Needs.

The nationalist reinterpretation of economic competition as national struggle and strategic risk in the 1980s was sudden and dramatic, embodied and epitomized in the Reagan election. Engineers gained the opportunity to become leaders on the battlefield, as when President Reagan called upon the National Academy of Engineering in 1985 to

> marshal the nation's technical engineering-based expertise in a campaign that will ensure America's scientific, technological and engineering leadership into the 21st century. . . . These efforts . . . are essential to the goal of helping American businesses and workers to modernize and compete. (National Academy of Engineering 1986:3)

The NAE began by mapping engineering "education and utilization" visually in a computer-printed model that looked like a dense piping diagram. Tracing flows in from secondary school on the left to flows out through "death," "disability," "emigration," and so forth on the right, the model linked education and employment together in infrastructural movement with lots of connections and feedback loops. Education became a flow of bodies through engineering schools whose primary problems were, accordingly, "recruitment" and "retention."

The piping image stuck. Tracking flows of engineers made it possible to reimagine education and utilization in the economic terms of supply and demand and establish the goal of enabling supply to respond more flexibly to demand. The National Science Board, which sets NSF policy, said in 1988:

> If compelled to single out one determinant of US competitiveness in the era of the global, technology-based economy, we would have to choose education, for in the end people are the ultimate asset in global competition. . . . Economic performance and competitiveness will be particularly affected by undergraduate engineering education. (National Science Board 1988)

By 1988, NSF had not only explicitly adopted the pipeline image but also transformed it into a linear image that defined continuity of flow as the goal and leaks as the problem.

Within American industry, becoming more competitive meant becoming more flexible. The flexible accumulation of capital in a struggling nation needs flexible people, lots of them (see Martin 1994). NSF is now at work reengineering engineers, supported by the National Research Council:

> We have to be thinking now what we want to see 10 and 15 years from now in terms of what is coming out of the pipeline with respect to science and engineering . . . human resources that will be flexible enough in terms of their training so that if they don't quite match what is at that time the need for their skills, they can be retooled very quickly. (Task Force on Science Policy 1985:43, 64–65)

NSF has pumped more than $200 million into research and innovations in engineering education to build more flexibility into engineering curricula and produce more graduates, mainly under its flagship program, the Engineering Coalitions.

Hiring in to engineering education in the 1990s thus involves recognizing that engineering educators in the United States can fairly easily construe what they do as being centrally in the national interest. Engineering education is widely understood not only as a place where good students prepare themselves for career tracks that promise financial stability and upward mobility but also as a test site for the refiguring of patriotism. Research that seeks participation in the fashioning of engineering selves risks simply contributing to this nationalistic fervor by improving students' abilities to pursue the goals of competitiveness without critically examining its contents. At the same time, it risks giving the impression that one seeks to prescribe change, to fix people who are presumably broken. One could easily come across as antiknowledge, antiengineering, and anti-American. Images count.

A second moment in partner theorizing involves viewing all theorizing as totalizing in content but not necessarily totalitarian in effect, in the sense that theorizing depends for its insights on a metanarrative, or background story, that builds a world within which its interpretations have meaning and power. From this perspective, the metaphor "temporary" may serve better than "partial" to describe the limited claims of totalizing theories participating in exchange relations. That is, the value of any form of theorizing is temporary, connected to changing circumstances.[4] If all forms of theorizing are temporary, then hiring in involves a historically and culturally specific encounter between distinct modes of theorizing.

The point of contact between our project and the ongoing retheorizing of engineering education lies in images of personhood. With the background story of competitiveness, we have to deal in particular with theorizing about "underrepresentation" and "flexibility." Because the pipeline model is an aggregate mathematical image, it counts people as individuals grouped according to distinct statistical categories. Sorting out persons biologically by sex and race, for example, the pipeline called attention to categories that were statistically underrepresented in engineering, namely women and minorities.

"If we want to supply our industries and government and our universities with the human power that we need in the future," NSF director Erich Bloch told Congress in 1987, "we need to concentrate on the groups which are underrepresented today in the scientific engineering areas—women and minorities" (US Congress 1987:9). Echoed the National Science Board, "From the perspective of economic competitiveness (as well as other perspectives), NSF programs and management efforts designed to help bring women, minorities, and the economically, socially, and educationally disadvantaged into the mainstream of science and engineering deserve continued focus" (National Science Board 1988).

By the late 1980s the dominant argument was that the country needed more engineers, but the pool of college-age people was declining, compounded by a declining interest in engineering among first-year college students (roughly 8% of college degrees). Since approximately three-quarters of the engineering bachelor's degrees are granted to white males, "greater participation on the part of women and underrepresented minorities in engineering studies would be one way of addressing the supply-side problem" (Bowen 1988:734). While constituting 51 percent of the population, women make up roughly 15 percent of first-year engineering students. Blacks make up 12 percent of the population but only 6 percent of first-year students, Hispanics 10 percent and 4 percent, and American Indians 0.7 percent and 0.5 percent, accordingly. In addition, while retention rates for white males hovered around 70 percent, retention rates for women were roughly 40 to 50 percent, 30 percent for blacks, 48 percent for Hispanics, and 33 percent for American Indians. Students of Asian descent were not considered a problem in this demographic profile, for with 3 percent of the population, they accounted for 6 percent of first-year students. Also, their retention rate exceeded 100 percent as more students subsequently transferred into engineering programs than departed (National Science Foundation 1990).

The nagging problem for engineers in theorizing underrepresentation concerns how to explain it in the first place. If students enroll in engineering programs because of innate capabilities and dispositions, then are nonwhites and females less capable or otherwise naturally predisposed away from this career track? From our perspective, the problem of underrepresentation is a citadel effect, an effect of theorizing learning entirely in terms of a diffusion model of knowledge. Colleges of engineering have been able to respond only by establishing new recruiting strategies and support systems for minorities and women students to increase enrollments and retention rates. These must struggle to maintain legitimacy since providing support programs for students does not fit the dominant model of challenging students to prove they belong.

Statistical calculations turn biological groupings into socially significant objects, making each woman stand for all women and each African American stand for all blacks. An anthropologist might be inclined to challenge this essentialist view of the person because it seems to reduce people to biology by showing these biological categories themselves to be historically and culturally specific constructions. Maybe the problem of underrepresentation would simply dissolve away as a misleading construct. Although this pathway to critique and opposition is fairly straightforward conceptually, it is likely incomplete as an approach to intervening, at least in the near term. One might be able to alter the importance engineers attribute to biological categories, but the categories themselves are probably here to stay.

Furthermore, the allocation of resources through the pipeline image, which draws on biological categories, actually provided new access to the corridors of power for interest groups that explicitly defined themselves as representing women and minorities. One newly funded female professor gleefully told us, "I can't believe I'm so deeply involved in this. I'm making connections all over the place. We're building an old girls' network." Similarly, the National Association of Minority Engineering Programs Administrators (NAMEPA) proudly announced the theme, "Partners in the Pipeline," for its 1994 meeting in Washington, which included scheduled lobbying trips to Capitol Hill. Congress itself authorized formation of the Task Force on Women, Minorities, and Disabled in Science and Engineering. In other words, classifying students by race and sex made some people visible who were otherwise hidden (Lucena 1996:180–81). As Donna Haraway (1989, 1991a) has shown us, biology can be useful.

In order for our work to hire in to the problem of underrepresentation, or have any role at all, must we force our data into artificial, predefined groups of women, minorities, whites, nonwhites, etc.? For us, these are interesting as cultural categories of persons that people apply to themselves rather than as distinct types or categories of humans that should be taken as real because based in biology. What sorts of categories might following students' experiences produce, and would these help to account for underrepresentation as an outcome without dividing up the world by race and sex at the outset? Applied in this case, partner theorizing thus involves going beyond showing that students' experiences cannot easily be parceled up by race and sex and, hence, making the problem of underrepresentation seem illegitimate. We must also try to account for how students themselves identify people by race and sex and ask if this process has any implications for statistical underrepresentation.

The problem of "flexibility" raises different issues. At present, engineering policy makers are at a loss to establish what flexibility means in curricular terms, although many different groups, from NSF officials to Boeing engineers, are vying to control its definition. Generally, flexibility seems to mean "malleable," as in making the bodies of engineers sufficiently malleable to fit changing job definitions. Because no single mode of theorizing flexibility has become established, we have the opportunity to contest what flexibility could mean rather than having to limit ourselves to accounting for the effects of a given model.

A third moment in partner theorizing is that it focuses attention on the power relations between alternate modes of theorizing by accepting that knowledge is never simply knowledge *of something* but is also knowledge *for someone*. Accordingly, the practice of partner theorizing encourages one to look for ways of factoring into one's own thinking the views of others in the field of intervention without necessarily seeking the consensus that is often unrealizable. Rather, by looking for reasons to accept the legitimacy of others, even if one finds limitations in their perspectives, partner theorizing shifts the goal from simply advancing one position in a debate to advancing or replacing the debate as a whole. One theorizes in terms of both one's siblings and ancestors—the traditional mode of defining a theoretical stance—and one's interlocutors or the positions one is trying to engage. Together these locate one's theoretical position at the start of an analysis. Advancing a new mode of theorizing in a contested field through partner theorizing thus involves greater entanglement in existing power relations than either mastery through truth or opposition through resistance. Furthermore, when the goal is to hire in from an outside position, paying attention to the power dimensions of theorizing is crucial to make sure that one even gains the legitimacy to participate.

Our first step in hiring in to engineering education is to make visible the experiences of students as they move through their curricula, thus confining our intervention to what happens within engineering education. Were our own experiences idiosyncratic? Do the students themselves offer alternative ways of thinking about learning? The next section briefly summarizes a handful of these experiences, drawing material especially from three students labeled minority in order to engage theorizing about underrepresentation and flexibility. We then search for pathways for intervention that take account of the current structure of engineering education and do not demand the resources that would be necessary to redesign curricula from scratch.

Weed-out

Jen Lopez is a Hispanic woman; Glenn Phillips and Rick Williams are African-American men. Although these labels identifying race and sex[5] were non-negotiable for students, the significance they played in students' lives is more problematic. Despite the expectations associated with her status as both minority and woman, Jen appeared in many respects to be the prototypical engineering student who sought the appropriate goals and adjusted herself properly to fit the curricular demands. Although Glenn and Rick had difficulty finding ways of fitting themselves to engineering, in neither case was race the problem or issue.

The core knowledge content of engineering curricula is what engineers call "engineering problem solving." Learning how to draw a boundary around a problem, abstract out the mathematical contents and solve it in mathematical terms, and then plug the solution back into the original problem is central to the fashioning of engineers and a major challenge to the bodies and minds of students. Students regularly asserted that the goal of certain courses was to "weed out" students from engineering curricula.[6] For students who stayed, these and other courses also appeared to weed out a part of themselves as persons.

Establishing disciplined habits and attention to detail are typically a student's first adjustments to the practice of engineering problem solving.[7] Rick found the content of this discipline different from his experience in the Navy:

> Oh, it's a different type of discipline. When I was in the Navy, they would say you have to stand this guard pose, and you gotta do this, and you gotta do that. They would lay it out for you. Here, it takes more self-discipline because you gotta figure it out for yourself.

An associate dean of engineering articulated a key feature of this self-discipline when he told incoming would-be engineers and their parents at freshman orientation that "engineers have to learn how to have fun . . . efficiently." We later repeated this to a friend who was completing his Ph.D. in mechanical engineering. After laughing heartily for a minute or so, the student stopped suddenly and said, "You know, he's right." The discipline in engineering problem solving is a total body experience that extends across all engineering "disciplines," a term that seems particularly appropriate in engineering (see Foucault 1979). As a dean told graduating seniors in engineering:

> What you've all learned in engineering is how to attack and solve problems. It doesn't matter what discipline you go into. You all learn the same thing. Solving problems is what engineering is all about.

The first two engineering courses that students take at Virginia Tech, appropriately called Engineering Fundamentals, explicitly describe disciplining the body and the mind as essential prerequisites to later success. One instructor in the first course described his course as "probably as much training as it is education." On the first day of class, for example, this professor announced, "The wooden pencil is dead for you." Holding one up, he said, "This is gone. You don't use it anymore. You are an engineer in training. You are on your way to the top. This looks crummy." Then, holding up a mechanical pencil: "This looks great. . . . Zero point five millimeter HD lead." In the second class he frightened students with a pop quiz, had them grade it themselves on the honor system, and then informed them that it would not count, to their great relief: "I just wanted to get across to you [that] you are always responsible for knowledge contained in a previous class." In the third class he outlined the connections between good habits, including a regular bedtime, and success in engineering:

> One of the biggest mistakes students make is they have an irregular bedtime because tomorrow their classes don't start 'til eleven, the next day they start at eight . . . and your body, instead of developing a habit, you just change that every day. Folks . . . it won't work. I haven't met any student that's been successful that way. . . . The ones that oscillate back and forth, in the long run do not succeed.

New engineering students find out quickly that engineering problem solving revolves around homework exercises and that engineering courses have more daily homework than any other curriculum. To get a good grade, every correct homework assignment must include the student's name, course number, and date lettered properly at the top of a sheet of engineering paper, which is a cross between unlined paper and graph paper. One writes on the unlined side, guided by faint lines that show through from the other side.

It is significant that every problem undergraduates encounter begins with the word "Given." The boundary has already been drawn. Engineering problem solving confines itself to the ideal world of mathematics. All the nonmathematical features of a problem, such as its politics, its connections to other sorts of problems, its power implications for those who solve it, and so forth, are taken as given. This contrasts sharply with physics problem solving, in which the main challenge is to learn to "think like a physicist" (White 1996) so that one can bring that unique genius to bear in a process of discovery. In engineering, students learn they must keep reactions, intuitions, or any feelings they might have about the problem out of the process of drawing a boundary around it and solving it. These are irrelevant and can only get in the way.

Good problem solving follows a strict five-step sequence: Given, Find, Equations, Diagram, Solution. Students start by abstracting from a narrative description of the problem mathematical forms for both given data and what they must find in order to solve the problem. Then, invoking established equations and drawing an idealized visual diagram of the various forces or other mechanisms theoretically at work in the problem, they systematically calculate the solution in mathematical terms. They must write down each mathematical translation or risk losing credit. Also, if they write a numerical solution without including units of measurement, for example, as feet, meters per second, or pounds per square inch, the answer has no meaning and frequently receives no credit. The students we followed still learned the computer language FORTRAN, which translates engineering problem solving into computer code, even though far simpler programming strategies were available. It seemed to us that instructors continued to insist on FORTRAN because it penalizes small mistakes, thereby forcing attention to detail. During the undergraduate years, students are expected to solve thousands of problems either on paper or in programs.

All engineers learn something about "engineering ethics," but not to enable them to critique the ethical dimensions of the problems they solve. Rather, in order to be good problem solvers, engineers need to behave ethically. Being ethical is about controlling one's passions and impulses, making it more akin to a habit than a commitment. An administrative memo distributed to new students, for example, introduced the subject of ethics in the context of good problem solving. "Engineers in professional practice," it stated, "are required to perform their work in a neat, orderly, timely, and efficient manner. . . . Since they design and construct facilities upon which the safety and health of the public depends, professional engineers must conform to a strict code of ethics in their work." An introductory reader collated by faculty abstracted different types of ethical principles from the writings of philosophers. Students learned that "utilitarian ethics" alone could be dangerous because self-interest can lead to self-destruction, "duty ethics" alone could lead one to neglect the needs of individuals, and "rights ethics" alone could lead one to overlook the general welfare. "Value ethics," attributed to Aristotle, were perfect for engineers because these encouraged "tendencies, acquired through habit formation, to reach a proper balance between extremes in conduct, emotion, desire, and attitude."

As students become transformed into engineering problem solvers, what gets weeded out is everything else. That is, engineering students experience a compelling demand to separate the work part of their lives from the nonwork parts. Work is about rigorously applying the engineering method to gain control over technology and is simply not about any other stuff. Budding engineers can have the other stuff in their lives, but not in their practices as engineers.

We found that the constraints of engineering problem solving fit well in bodies where "other stuff" meant forces of identity or challenges to personhood whose meanings did not make competing demands of personhood. Jen, for example, was convinced of a deep connection with engineering because she had been interested as a kid in fixing cars and other things and had attended a science and technology high school. For her, the habits and attention to detail in engineering problem solving were simply an advanced form of tinkering, and high school had added the appropriate mathematics. Moving from tinkering to engineering problem solving was simply a shift from a private activity to a collective one. Jen thus found the demands she encountered entirely reasonable and appropriate. "It's just to make everything easier to read," she said, "like for instance if you need to present something to your boss, you're not going to just slap the answer on a piece of paper. You have to say, 'This is what I was given, this is what you asked me to do, and this is what I found out.'" Jen had made engineering a personal goal at a fairly early age. "I always knew I wanted to do engineering," she told us.

Accepting the rigors of engineering problem solving also fit very well Jen's desires for upward class mobility. Raised in a working-class family, she clearly felt class identity as a challenge and source of force: "I think eventually I will leave engineering because I want to work with management. I think I'm ambitious." She even arrived with some understanding of how being an engineer could be positioned as a job in a corporate environment. In high school she had worked ten hours a week as an intern at an engineering firm: "I got to see what it was like, and I liked it. I liked the atmosphere, the job atmosphere." Granting authority outside the self was no problem because Jen expected to work in a job with a boss. It was also important that, in Jen's image of class mobility, success in life was the outcome of sacrifice and individual determination. At freshman orientation, she recalled, "They were saying that, if you look around the room, half the people won't be graduating." Asked if that intimidated her, she replied, "A little bit, but I think if you really want something you can get it, if you try hard enough."

Finally, the expectation in engineering problem solving that one will control one's emotions fits the stereotypic definition of a mature man, one who is strong and in control and who exercises considered judgment. For Glenn, who had trouble keeping his emotions out of his problem solving, this part of the challenge in engineering courses posed a significant problem. He told us, for example, that he had a tendency to say "I feel" in expressing his views, "I don't think it was accepted too much," he said. "Maybe 'I think' was accepted more." As Rick explained in another interview: "I've never seen someone say, 'Well, sir, the answer is because I feel that.' The professor will just cut him to pieces." Referring to his feelings got Glenn in trouble in job interviews, where he appeared to be wishy-washy and lacking in self-confidence. His strategy for getting through engineering courses was to link himself to people whom he believed were in a similar situation, namely women. That is, rather than contest the challenge from engineering or the stereotypic man, Glenn made use of the stereotypic woman by seeking out emotional support from a network of female engineering students and avoiding men, who he believed would not talk about such things.

Because the stereotypic woman is emotional by nature, she is not the ideal engineering problem solver. For Jen, the continued currency of the stereotypic woman gave her a chance to stand out as an individual: "[Being a woman in engineering] feels kinda neat, because not many women are engineers. I kinda like that I stand out in a way because I don't want to be one of the nameless engineers. Girls aren't supposed to be [engineers], so I'm glad that I am one." Jen had indeed encountered overt discrimination from at least one male professor who drew on the stereotype in treating women, but she discounted this experience as exceptional. "You've always got those chauvinistic types," she said somewhat casually, such as the one who put on a worksheet a question "something like 'an engineer goes home to the housewife da da da.'" When she went in to ask for help, this professor "treated me like I was, you know, like a baby or something." But she insisted he was unlike the others: "Most professors are really nice."

Standing out as a woman did call for special strategies if Jen was to make sure that faculty were blind to the stereotype in judging her merit. One strategy she used to demonstrate her capabilities and commitment to male professors was to go and talk with each one, making sure they would have to treat her as a full person. "Most of the time professors like me a lot," Jen said. "I'll go talk to a professor just so they know who I am, not necessarily so they'll know my name or anything, but just so they know my face and they recognize me." She also had to compensate for a lack of study partners. Because she did not identify herself as a "woman engineer," she did not seek out other female students: "Not like real good friends or something." Also, there were few engineering students in her dorm:

> Jen: It's especially hard. I live in an all-girls dorm. How many girls are engineers? Most of the guys, you know, had at least two or three on their hall. So you just walk down the hall and there's a guy doing his engineering homework too. But there weren't any other girls on my hall that were engineers. So it's kind of like, "Well, what do I do?"
> So, what do you do?
> Jen: Well, I kinda figure it out by myself.

The dominant image of the graduate engineer is one who controls technology, who creates by translating internal knowledge into object form; in short, one who designs. One department brochure summarized its curriculum simply with the words: "Engineering teaches students to design." The College of Engineering routinely advertised itself with photographs of solar cars designed and built each year by students. First-year students meeting in small group interviews regularly described how they long "to design something." Said one, "I want to be the person that draws it . . . that kind of designs it in a way, and then hands it to somebody else and they go do it."

However, the images of design that incoming students carry with them often differ greatly from those that discipline their work. Most students we encountered started out viewing engineering design along the lines of the stereotypic architect, whose designs are a deep, personal expression of some distinctive perspective, subjective orientation, or emotional reaction. Indeed, the romantic fantasies that drew students toward engineering in the first place, such as helping society through new designs, helping one's people improve economically, or advancing civilization through space, generally included a heavy measure of agency or even autonomy for the individual engineer. But it was not easy to hang on to these visions.

In engineering problem solving, design is the timely, disciplined application of the engineering method to real-life problems. In this image, the genius of design shifts from the person to the method itself, and authority shifts to the curriculum. Because seniors were most likely to have mastered the method, they were usually the ones who got to practice design. The solar-powered car, for example, was designed and built each year by a group of senior students. But for newer students, becoming successful engineers meant giving up on the fantasies that made them the geniuses behind great designs. Instead, they had to accept the constraints imposed by the curriculum and learn to solve problems properly. As we followed students through these challenges, images of creative invention tended to dissolve away, and engineering design lost its romance.

Rick's main fantasy was to solve problems "relevant to ordinary life," but the commitment to mathematical problem solving in engineering course work soon left him with a strong sense of personal loss:

> Drawing a cylinder with a hole in the middle and at two different angles had to be perfect. You had to have it just right. That's what I hated the

most. I really did. Took me at least fifteen hours a week sitting at the computer to get it right. I hated it.

In Rick's view, the heart had to be connected somehow with the job:

My opinion is that anyone can do engineering but not everyone can love engineering. I didn't love it, so I couldn't do it. My ex-girlfriend loves it. Every other word out of her mouth is in touch with systems engineering. She loved it.

Rick united his work with his heart by switching to biochemistry, whose activities felt more relevant to ordinary life:

I had gone from an engineering major that morning to a biochemistry major by dinner, and I already felt the difference. I could really feel the difference. Now, my biochemistry class, I love it. I mean, I sit down to do my biochemistry problems, and I can't believe the Navy is paying me to do these things. Just yesterday we were analyzing protein structures and how they interact with each other due to spatial orientation and how they function. I was sitting there, you know, "Wow, I can't believe I'm getting paid for this."

Class identity played a significant role in Rick's career path, as it did with Jen. A naturalized US citizen born in Ethiopia, Rick was graduated twenty-second from his high school class in upstate New York. He had enlisted in the Navy because his mother could not afford to send him to college. Moving to biochemistry rather than say, anthropology, preserved his goal of upward mobility, because the Navy would pay for his education and he could anticipate a well-paying job later.

In contrast with both Rick and Jen, Glenn's main fantasy had to do with race. Raised in urban Washington, D.C., he wanted to stand out as a black man in a white world, thereby challenging the racial stereotype of black people as mentally slow, physically lazy or undisciplined, and, hence, inappropriate:

My high school was all black. I chose to come here over North Carolina A&T [a predominantly African-American school] 'cause I thought I needed to experience the world. It's not just going to school with white people; it's living with them.

Glenn elaborated on how he wanted to become a role model for other African-American students: "I want to inspire others to get into engineering. I guess me becoming [an engineer] encourages others to do that also. That's a contributing factor to why I stuck with it." A sense of loss would come later.

As we have already seen, Jen's fantasy embraced the identity of an engineer as a pathway to a job and income, but all this was alongside another passion: ballet. Jen had studied ballet since childhood and remained deeply committed to it, hoping also to learn choreography. However, doing both engineering and dance brought great stress:

Every single night we had practices from after dinner 'til at least 12:00. My classes ended at 3:00 so there was no time to do homework. Sophomore year there's like three problems to do in every class and it takes you a long time. The people in charge of the dance were all like, "Go talk to your professor to get an extension on your homework." I was like, "You've got to be kidding me. No engineering professor is gonna do that."

Jen talked to an engineering professor after receiving a poor grade on an exam. "Worst choice," she said. "He's like, 'Why did you come to Virginia Tech? You've gotta decide for yourself, you've got to set your priorities straight. Did you come here to dance or to become an engineer?' He told me that directly to my face." Jen was angry about the image that "if you're an engineer at Virginia Tech, that's all you can do." She wanted to contest the view that mathematical problem solving was everything for a prospective engineer, "because you have to be well-rounded. . . . If you just sit at your desk or work at your computer all day long, you can't survive in society," she insisted, "because work is not just what you can do on your computer, it's how you associate with people." She reacted to the challenge with even greater determination: "It pissed me off and I was like, 'Well, no, I'm gonna do them both.'" Yet ultimately she gave in, adjusted, and reduced her involvement in dance:

> I just did the best I could. I did less dancing. I went to another dance company but I didn't do as much. I didn't choreograph because that takes a lot of time. I knew I wanted to be an engineer and I was gonna do it. I was only in one dance in that group whereas the year before I was in four. It worked out fine. I did well with my grades.

Students have no room or opportunity to challenge the priority given to mathematical problem solving in engineering lives and work. The curriculum is just there, demanding that they make all the adjustments and informing them through grades of the extent to which they were succeeding. Homework after homework, test after test, and course after course rank each student on a linear scale. Glenn struggled for a long time just to achieve the C average he needed as a minimum demonstration of membership; in other words, to graduate. When asked if he ever felt engineering was not for him, he said:

> Probably every other day. Freshman year, I kind of just said, "Well, I have to adjust and do better next semester." The grades were the biggest thing that kind of told me. They made me tell myself to struggle to stay here and stay in engineering. I always believed that I could do better but I didn't know how. I messed up.

Fortunately for him, Glenn was able to locate one essential, that is, natural, connection to engineering: "I am more technically minded than most people." So he persevered. He had come to Tech with three black friends from D.C. Before the end of second year, all three of Glenn's friends had left school.

By the time students reached their junior year, the vast majority appeared to have found strategies for accommodating their bodies and minds to engineering problem solving. As courses became specialized within majors, the mathematical challenges in engineering problem solving became more complex, and strategies accumulated for commanding greater control of the world in mathematical terms. Back in engineering statics, for example, which many called the "first real engineering course," students had learned how to apply a single mathematical equation to a range of different circumstances. However, in the engineering thermodynamics course we attended, students found that any problem could have several pathways to a solution. They faced a whole menu of equations that may be appropriate in any given case and had to decide which assumptions to make in choosing which particular configuration of pathways to follow. "It's more intense," Jen said. "It's harder stuff, so it's [all about] how much you can handle . . . how much work you can handle."

One cost was a sense that the rain never stops. The experience of isolated struggle in the early years of engineering education had been replaced by a more shared struggle just to get through whatever came next. When we asked upper-division students how they were doing, we often heard the simple mantra, "Eat-sleep-study." By this point Rick had already left. Glenn was clinging to his female groupmates, still struggling to survive. Having failed thermodynamics, he said, "When I got into my major, it was like starting over for me. It really did feel like starting over freshman year."

Like many students we encountered, Jen coped not only by disciplining herself for engineering work but also by making sure her life had other things in it as well. The key lay both in maintaining a sharp boundary between the work part and the other parts and in making sure that their meanings for one's identity as a person did not conflict. Consider Jen's images of bounded, efficient play in a group interview of advanced students who had just outlined the pitfalls of dating nonengineers:

> *Deepak*: I think it's engineers. They don't want you to have a social life. The ME [mechanical engineering] department, or any department.
> *Jen*: You *can* have a social life.
> *You can? Do you guys all say that?*
> *Thuy*: We get as much as we have a chance to get.
> *Dan*: Budgeted.
> *Deepak*: Budgeted, yeah.
> *Did you hear what they said in freshman orientation, that engineering students are not like the other people? You have to learn how to have fun efficiently.*
> *Jen*: That's true.
> *Dan*: That makes sense, I think.
> *Thuy*: Very true, yes.
> *Dan*: Some people's idea of having fun is just sitting around and just yakking. Just really nonproductive.
> *Jen*: I think that's why I don't watch TV. I'd rather be out with my friends having a good time instead of sitting in front of the TV.
> *Dan*: Do a good quality two hours of fun time.
> *Jen*: Yeah.

Establishing an identity as an engineer can mean allowing engineering values to diffuse into other areas of one's life, even while holding these separate.

As the classroom encounter with Ann Landers illustrated, students by this point had come to treat instructors narrowly as functionaries who simply transmitted the knowledge students needed to pass tests rather than as independent sources of reflection and interpretation. Although professors might bring human characteristics to their work (one "is a lot of fun" while another "will slam you"), such factors were relevant only around the margins of standard pedagogy—presenting and testing mathematical knowledge. Students knew that the curriculum had been established by some past authorities and that the truth or validity of its contents was not subject to question:

> *Jen*: I took a class in family and child development. It was like the biggest trip. The teacher's up there talking, people aren't paying attention. People are fighting over a multiple-choice test, saying the teacher didn't have the right answer. If they walked into an engineering class, everyone would be like, "Who the hell is she saying that professor is wrong in their answer?"

The value of engineering knowledge in the world persists over time. We were amused but not surprised when the thermodynamics class used the same textbook Gary Downey's own class had used in 1971.

Having survived the solitary struggles of the first two years, students had adopted a range of strategies for getting through their courses efficiently. Said one: "I now know that the homeworks and tests are what's important, so I'm a lot more efficient. I hate textbooks and never read them. I just listen to the lectures and work the problems." Also, a student no longer stood alone as an engineer but had become part of a larger group, an engineering major. "I see a lot more familiar faces in my classes now," said one student in a group interview. We heard all sorts of strategies for doing group work, including what qualities make good study partners and when group work helps the most or gets in the way. We followed one organized trio of students who divided up their three toughest courses in order to conquer them together. Each did the homework for one class and prepared the others for the tests.

In sharp contrast with the entering student, the engineering graduate who emerges from the curriculum is understood to be a disciplined, knowledgeable, and powerful person, at least in terms of engineering problem solving. Knowledgeable students have gained control over technology in a way that is unavailable to other persons, whether human or corporate. Through the logical, precise actions of identifying and solving problems in mathematical terms, each student has, by definition, succeeded in extending the realm of human control to include technology, transforming technology into a tool for human use. The stereotypic engineer is this and nothing else.

However, unilinear ranking also insures that students do not all attain the same level of control or receive the same credit. As students reached their last year of school and began looking for work, they focused on their grade-point averages to an extent they could only have imagined earlier. In a profession that does not make graduate school a prerequisite of employment, one's value as a potential employee depends in the first instance on the grade-point average. GPA is the key line in every résumé because it is read as the main indicator of accomplishment. We once listened to an African-American recruiter for a corporation advise African-American students to keep their résumés from looking "too black," or employers might become suspicious that a student was putting racial identity before engineering identity. Glenn was one of several students we encountered who tried to resist this system of evaluation by leaving his GPA of 1.95 (C-) off his résumé at the annual job fair:

> At the first booth I went to, he pretty much told me, "Keep on walking." I remember one lady gave me a lecture. She was like, "Isn't your GPA any better?" She went on telling me all this stuff and she said, "Well, you need to work on your problem." I was like, "I don't have a problem." I felt my potential is much greater than what my grades say about me.

Glenn had been very active in student organizations and had accumulated considerable work experience, but he was unsuccessful in championing these as indicators of talent and motivation apart from his grades. Six months after graduation, Glenn had not yet found a job.

Interviewing for jobs also brings a new set of challenges to personhood as students emerge from the cocoon of mathematical problem solving and begin to face the vagaries of economic and social life. Jen's B average surpassed the minimum standard of acceptance, which students consider to be a B– average, yet she received far more interviews than students with much higher grade-point averages. She attributed this primarily to her status as a minority woman. In contrast with her experiences as a student, this time she felt the challenge deeply:

> I've never had any type of conflict [within engineering from being] Hispanic at all. I mean, me personally, I don't like the word "minority." That bugs me because the definition of minority means "less." So I really don't associate myself as being a minority. I'm Hispanic, not a minority.

Of course it was easier for Jen to avoid the label minority or woman as a student taking tests than as a potential employee trying to sell her labor in the marketplace. How could she maintain a sense of accomplishment for what she had achieved?:

> *So does it bother you, for example, policies like affirmative action?*
> Jen: Well, it helps me, so I don't mind, but I wouldn't like people saying "She only got the job because she's a girl," or "She only got the job because she's Hispanic." Because I don't think just because of that they should pick me. They should pick me on my merit, not from where my parents are from or what sex I am. It is an advantage because I get a lot more opportunities because I'm a woman and a minority.

This was a problem that mathematics could not solve. Jen's work life suddenly threatened to become Hispanic and/or female in content, undercutting her long-standing efforts to separate her work from her race and sex. She wanted a job, the best possible job with the most money and highest potential for advancement into management. Affirmative action had explicitly mixed race and sex with merit in a way that made these inseparable, yet without affirmative action would the currency of stereotypes by race or sex in corporations have prevented Jen's many opportunities from appearing in the first place? For her, this sort of problem could only be resolved through long-term job performance, demonstrating her worth as a person rather than as a Hispanic woman. She accepted a position and got started.

In sum, while the challenges to personhood that engineering students experienced overlapped challenges from the stereotypic man and a desire for upward class mobility, these conflicted with challenges from the stereotypic woman, the stereotypic black person, the stereotypic Hispanic, and a desire to link work with nonwork in some sort of organic whole or total self. In none of the three cases discussed here did stereotypic expectations drive students away. Wanting to challenge the stereotypic black person had actually motivated Glenn both to enter an engineering program and to stay, despite poor grades. The curricular insistence on keeping emotions out of problem solving added to his sense of being marginalized, however, for in seeking out female friends in engineering Glenn kept the distinction between work and nonwork permanently blurred. For Jen, being Hispanic was largely irrelevant, while being a woman provided great motivation. Her bodily adjustments involved trying to keep instructors or prospective colleagues from using stereotypes to classify or judge her. Finally, Rick left not because of a racial stereotype but because he could not live with a work self that not only had to be kept separate from everything else but that lived entirely in an ideal, esoteric world of mathematical problem solving.

Pathways

But what about other students? How might this account hire in to theorizing and accounting for the problems of underrepresentation in engineering education and the flexibility of engineers? Because it will take far more than three cases to map students' experiences sufficiently to make the investigation plausible, we expect a book manuscript to be a necessary vehicle. Still, as we elaborate below, an argument or hypothesis that emerges from this account of students' experiences is that statistical underrepresentation is a special type of citadel effect, an effect of conflicting challenges to personhood. That is, the challenge from engineering problem solving to bifurcate the person between work and self, where work is dedicated wholly to solving bounded mathematical problems, might be driving away people who experience this as a denial of self rather than simply a novel encounter with discipline. Also, helping students understand and grapple with this bifurcation may have the effect of shifting the meaning of flexibility in engineering education from malleability to sophistication in critical reflection.

Stereotypes do count. For people who already feel challenged by stereotypes that somehow make them or the people they care about invisible or subordinate, might separating a raceless, sexless work self from a nonwork self feel like a double whammy, yet another demand for invisibility? Might one feel a need to leave or avoid engineering to demonstrate one's wholeness as a person? We have collected a great many stories that appear to fit such an account.

Wholeness in personhood is a Western problematic, built on a stereotypic image of society as a collectivity of autonomous individuals. This powerful image values coherence in personhood and challenges people to pursue coherent selves as a means to fulfilling lives. Indeed, any self in an adult body that is less than a coherent whole is considered downright pathological. Rick clearly sought such coherence in his person and life, as have the two of us. For Rick, the challenge from engineering education to live half his life in mathematical problem solving separated sharply from everything else he did or wanted was plainly intolerable. We propose there are many more who, like Rick, felt they had to weed out their persons in order to become engineers. One does not, for example, become an engineer-woman, engineer-African American, or engineer-world-helper, but only an engineer who happens to be a woman, an African American, or a person interested in helping the world. Unlike professional training in law and medicine, which understand themselves as adding expertise to an already educated, mature, and complete person, engineering education seeks out the high school graduate or, as one professor described it, the "blank slate." The disciplining in engineering education must be all or nothing.

Although our proposed generalizations are still tentative, Jen and Glenn may illustrate the strategies of many women and minorities who remain in engineering, resisting stereotypes by standing out as individuals. Knowing that "girls aren't supposed to be engineers" motivated Jen to prove that the stereotype did not apply to her. Similarly, Glenn wanted not only to prove to himself that he could "live with" whites but also to encourage other African Americans to follow him. Yet for the student who wants to "be around people like me," as several told us, standing out as an individual among white male engineers could feel like a lonely sacrifice of selfhood.

In sum, statistically significant differences of race or sex between students who stay and students who leave may depend upon a larger, more pervasive source of difference—the challenge made to diverse students to parcel off a mathematical work self from a nonwork self. In emphasizing such a separation, could engineering problem solving be driving away many passionate, motivated people who might be most likely to challenge the boundaries of engineering and/or want to spend their lives improving or changing the world? Based on following the experiences of students, we think that is the case. But must it be so? Must engineering be structured to fit best those who accept established authority and believe the world is pretty good right now? Might these same student experiences make visible pathways for linking engineering work to other dimensions of selfhood?

Attracting and retaining more members of underrepresented groups through support programs appears to be one such pathway. That is, making visible the experiences of engineering students may help affirm the value of recently developed programs for women and minorities that try to increase retention rates by reducing the extent to which such people might feel isolated or alone. One can argue that even white males need some nurturing and support, for the early disciplining in engineering education forces everyone to survive solitary struggles, and later courses keep the burden on students to prove they belong. However, because women and minorities also have to deal with stereotypes that label them invisible or inappropriate as they grapple with the challenge from engineering problem solving to ignore everything else about themselves, offering such groups extra opportunities to identify with other students makes good sense. For example, being able to frequent a "safe space" that is populated with other stereotypically "inappropriate" people may be one way to reduce the extent to which the stereotypes feel relevant or significant. Choosing to leave engineering should be acceptable as a legitimate decision by informed adults, but not if leaving is solely the product of having to cope with the stress of conflicting challenges to personhood.

A second possible pathway to hiring in is to reimagine the role of mathematical problem solving in engineering education and, accordingly, relocate the sharp separation between work and self. What if, for example, disciplined attention to mathematical detail and the avoidance of extreme emotion and desire were no longer celebrated as more important than anything else in an engineering problem solver? What if engineers located mathematical problem solving as simply one valuable resource among many they might use? Perhaps the stereotypic images of women and minorities would become irrelevant for prospective engineering students rather than congruent with images of incompetence, and the statistical problem of underrepresentation could potentially dissolve away.

This pathway could be elaborated in several ways, of which we have so far tried only one. We designed and teach a course called Engineering Cultures, which tries to demonstrate that placing the highest value on mathematical problem solving in the lives and decision making of engineers is both historically and culturally specific. That is, things could have been otherwise. Although this is only a single course offered as a humanities elective, it does stimulate students to reflect critically on the curricula that shape their lives.

The course begins in the present by illustrating how mathematical problem solving is never the totality, and frequently constitutes only a small part, of the activities of practicing engineers. In other words, the work of practicing engineers extends well beyond the limits of mathematical problem solving. Then, working backward in time to excavate a series of genealogical layers, the course helps students understand how the great emphasis on mathematical problem solving in engineering curricula was molded in response to the perceived threat of Sputnik to American science. Exploring a long-term tension between "design" and "manufacturing" in industry illustrates how competing perspectives with equal value can exist simultaneously and that the very act of drawing a boundary around a problem establishes a claim of authority over it. Tracing several disciplines through mechanical engineering and onto the factory floor establishes a novel connection between the identities of engineering management and labor.

The course then pushes hard on the boundary around engineering knowledge by briefly visiting a number of different systems through readings and guest lecturers trained in different countries and demonstrating that a range of ways of locating engineers is already available. Exploring connections through colonial traditions and the contemporary organization of multinational capitalism offers images of global relations that contrast with the doctrine of competitiveness, which pictures a population of autonomous nations competing with one another on a level playing field. The course concludes by encouraging students to hold on to their romantic images of life as engineers by treating engineering problem solving as but one resource among many in careers that could make a difference.

Other steps we might take to reimagine the role of problem solving in engineering education involve using our narrative of students' experiences as a test bed for assessing existing proposals to modify engineering education and for formulating some new ones. These proposals range from the structure of engineering curricula to the daily organization of classroom pedagogy. Proposals to increase the flexibility of engineering curricula, for example, often involve introducing design activities prior to the senior year while still conceiving design entirely as application of the engineering method to real-world problems. Perhaps modifying such activities to examine how engineering problems are connected to other sorts of problems and how the boundary around a given design problem can be drawn differently from different perspectives could help engineering students both to develop greater tolerance for different perspectives and to better assess how and when to apply the engineering method to the problems they encounter. Engineers trained to reflect critically on their own practices are not likely to be malleable but may indeed be more likely to figure out ways of making their disciplines and workplaces adapt to their dreams and fantasies of helping society or otherwise making a difference through engineering work.

Furthermore, must engineering education require a full body transformation? Might it be possible, for example, to formulate and structure a graduate degree in engineering for students with other undergraduate backgrounds, helping them to understand how engineering problem solving works and enabling them to map differences in engineering problem solving across fields and disciplines without necessarily having to master esoteric fields of engineering science? Couldn't one be just a little bit engineer, and might such people be capable of linking engineering problem solving to other social and personal agendas?

Lastly, what if engineering pedagogy admitted nurturing as necessary to its success? Could engineering instructors help individual students learn how to link engineering problem solving to their long-term fantasies and desires by offering their own personal stories and experiences? What if the thermodynamics professor offered his own reactions to the Ann Landers column or outlined what he was trying to achieve through research and teaching in thermodynamics? Could establishing legitimacy for welcoming and nurturing students improve their abilities to imagine new and different ways for using engineering problem solving as a resource in their later lives and work?

In sum, a theoretical approach that draws on the experiences of engineering students to intervene in engineering education need not demand a total transformation of engineering curricula to do so in potentially significant ways. Fully accepting the responsibility of hiring in to this contested field of education will require us to publish in places engineers respect, visit educators in their own spaces, and use reactions and critiques as opportunities to advance the discussion as a whole. This might involve as little as making presentations to engineering audiences and publishing specific recommendations in engineering publications, or as much as joining key organizations, committees, and ongoing policy debates. In either case, hiring in through ethnographic field work and applying a cultural perspective on the fashioning of selves might encourage engineers to foreground practices that constantly address and critically rethink the question: What is engineering for?

Notes

1. For lengthy reviews of research on engineers and engineering, see Downey, Donovan, and Elliott (1989) and Downey and Lucena (1994). Some other relevant work in the anthropology of education in science and engineering includes Chaiklin and Lave (1993); Eisenhart (1994); Eisenhart and Marion (1996); Lave (1990); Lave and Wenger (1991); Nespor (1994); Seymour and Hewitt (1994); and Tonso (1996a, 1996b).

2. This mode of critical participation joins with and draws from feminist critiques of science and literary cultural studies, both of which are challenging citadel effects of the academy from positions within it. It is lived most intensely by those people and perspectives that work to make a difference while accepting the risks of life in the so-called private, public, and/or nonprofit sectors, away from the guaranteed paychecks of tenured professorships.

3. Because the image of committed cohabitation is a white, middle-class ideal that can work to hide inequalities and abuses of power, we see the danger of narrowness. But even in everyday language partners can be more than two, be of various sexual orientations, develop a variety of styles of commitment, extend partnering to some areas of life but not to others, etc. Because the key assumption in partnering is that each position presumes the legitimacy (but not necessarily the value) of others in principle, we mean the practice to be multidimensional, constantly involving work, and having varying power dimensions at the capillary level. We worry that the image of commitment may be limited by its individualistic overtones, but it does emphasize the mutual dependence of alternate modes of theorizing.

4. In anthropology, for example, the linear evolutionism that thrived during the nineteenth century could not account for a twentieth-century world of nation-states. Likewise, the structural-functionalism of the 1940s and '50s lost its relevance in a '60s world of rapid change, and a '70s symbolic anthropology could not maintain its holistic image of cultures in a contemporary world where the production of hybrids appears more the rule than the exception. Today, theories of postmodernism vie with theories of late capitalism for control over how to interpret the contemporary scene; eventually both will fall silent in the face of changing circumstances. Arguably, each step has been a historically specific improvement, but to label the whole as progress is to suggest that later forms of theorizing would have been appropriate, indeed desirable, in earlier periods. In fact it seems unlikely that, say, structural-functionalism would have played well in the nineteenth century, or postmodernism in the '40s.

5. We use the word "sex" here rather than "gender" to call attention to the fact that, in popular theorizing, the words "woman" and "man" are uttered and heard as labels rooted in biology. The word "gender" loses this dimension by displacing the label into academic theorizing that treats "woman" and "man" as cultural terms. We must maintain a focus on how our everyday theorizing uses biology to naturalize our categories.

6. Whether or not faculty actually tried or wanted to weed out students does not matter, for weed-out was a student category.

7. In this account we do not address the problem of variation in curricula among the three hundred or so schools that offer engineering degrees in the United States.

Sometimes these differences are indeed significant. Often they are not, owing to the fact that almost all schools seek accreditation from ABET, the Accreditation Board for Engineering and Technology, which offers strict guidelines for engineering curricula. One instructor told us, "I tell parents . . . that a student who goes to Clemson, Ohio State, Michigan, Penn State, or Georgia Tech will come out looking pretty much like one who comes through here."

IF YOU'RE THINKING OF LIVING IN STS / A Guide for the Perplexed / David J. Hess

C ULTURAL AND ANTHROPOLOGICAL STUDIES of science and technology in the United States have become something of a growth industry in the 1990s. The list of North American anthropologists interested in science, technology, and computing issues now includes more than two hundred names.[1] The topic is covered in growing numbers of panels on the programs of the American Anthropological Association and the Society for Social Studies of Science as well as in a burgeoning number of publications. Yet anthropologists and their siblings in cultural studies who move into this area sometimes assume that they will be living in a remote village that no one else has ever studied. It does not take long before they begin bumping into others who claim authority as students of science and technology and who may also expect anthropologists to prove that they have something new or interesting to say. In this essay I provide in somewhat idiosyncratic terms a partial map of STS (science and technology studies) that focuses on researchers and research likely to be of interest to readers of this book.

The discussion takes the form of a critical literature review, but it is rooted in several years of fieldwork. As an anthropologist I have done fieldwork among Spiritist intellectuals in Brazil and various alternative medical and scientific groups in the United States, and in the process I have negotiated theories and frameworks from the social studies of science and cultural anthropology. I have also lived for the better half of a decade in one of the leading departments of science and technology studies in the United States, where I have negotiated the interdisciplinary intersection of anthropology with STS. As a result, I can offer a perspective of both "insider" and "stranger."

STS and SSK

"STS" is usually taken to mean science, technology, and society studies, although on occasion it is glossed as science and technology studies. At Rensselaer and some other schools the faculty tend to think of STS as an interdisciplinary field with constituent disciplines in the anthropology, cultural studies, feminist studies, history, philosophy, political science, rhetoric, social psychology, and sociology of science and technology. In North America STS is organized at a professional level around a number of disciplinary societies, each with its own acronym and affiliated journal. Among the major organizations are the History of Science Society (HSS, *Isis*), Philosophy of Science Association (PSA, *Philosophy of Science*), Society for the History of Technology (SHOT, *Technology and Culture*), Society for Literature and Science (SLS, *Configurations*), and Society for Social Studies of Science (4S, *Science, Technology, and Human Values*). Usually the societies hold their annual meetings separately, but occasionally two or more convene for joint meetings. There is also a Society for Philosophy and Technology, with an annual volume titled *Research in Philosophy and Technology*, and in 1993 yet another organization was formed, the American Association for the Rhetoric of Science and Technology (AARST). That list covers only the major North American organizations. Probably the most relevant institutions for social scientists outside North America are the European Association for the Study of Science and Technology (EASST) and the European (but not EASST) journal *Social Studies of Science*.

My forthcoming book (Hess 1997b) provides an overview of some of the key concepts in the major constituent disciplines of STS, including the philosophy of science, the institutional sociology of science, the sociology of scientific knowledge, critical/feminist STS, and cultural/historical studies of science and technology. There are also several other reviews of various aspects of the interdisciplinary field (Fuller 1993; Rouse 1991, 1996b; Webster 1991; Woolgar 1988b; the review articles in Jasanoff et al. 1994). Traweek (1993) has provided perhaps the most comprehensive overview of the field for those interested in anthropological, feminist, and cultural studies in the United States.

In this essay I will focus on the particular branch of STS known as the "sociology of scientific knowledge" (SSK), its relations to anthropology and ethnography, and the role of anthropology and cultural studies in shaping the future of the interdisciplinary STS dialogue. Given the prominence of SSK in this dialogue, it usually is not long before a newcomer encounters its texts and members. Furthermore, because there is a tradition of "anthropological" or "ethnographic" studies within SSK, it should be of particular interest to anthropologists.

The "core set" of SSK members, according to Malcolm Ashmore's (1989:16–19) reflexive sociological study of SSK, includes Ashmore, Barry Barnes, David Bloor, Harry Collins, Nigel Gilbert, William Harvey, Jon Harwood, Karin Knorr-Cetina, Bruno Latour, Michael Lynch, Donald MacKenzie, Michael Mulkay, Andrew Pickering, Trevor Pinch, Jonathan Potter, David Travis, Steve Woolgar, and Steven Yearley. Of course, conjuring up a network or school and naming its main members is problematic. As Ashmore himself recognized, other names could be added to his list. Candidates would include Wiebe Bijker, Michel Callon, David Edge, John Law, and Brian Wynne. Conversely, some of the people on the list might not classify themselves as part of SSK. For example, in an article published after Ashmore's study, Lynch (1992) distinguishes between SSK and his own program of ESW (ethnomethodological studies of work in sciences and mathematics).

Furthermore, the term "SSK" is now somewhat out of date. Given the subsequent "turn to technology" and "practices" in what was originally known as "science studies" (Pickering 1992; Woolgar 1991a), the subfield might better be called SSKP or SSKT. Many outsiders also refer to the group not as SSK but as "constructivists"; however, the term "constructivism" or "social constructivism" is not universally accepted within the group and there are many people not affiliated with SSK who accept some version of the social construction of knowledge and technology or the co-construction of technoscience and society. Within the SSK point of reference, constructivism or social constructivism may refer more narrowly to the programs associated with Michael Mulkay and his students as well as with continental Europeans such as Knorr-Cetina and Latour.

As the attentive reader has already noticed, almost all the SSK members are men. Most are British; a few are from other countries, mostly in western Europe. Corridor talk of the interdiscipline suggests that many of them have scientific or technical backgrounds, and several passed through the British polytechnics rather than the elite Oxbridge system. I have heard that their apparent proclivity toward theory, programs, and acronyms was influenced by their socialization in the polytechnics, but it is also similar to the use of jargon in philosophical circles. Their non-elite background has sometimes been used to explain their hostility to the traditional philosophy and history of science of the elite universities. I have heard the suggestion that the entire debate between SSK and the traditional philosophies of science is shaped by the cultures of the British class system; a similar dynamic may be at work in the US in the opposition between STS programs, which are often housed in technical universities, and the more traditional history and philosophy of science programs.

Certainly the SSK social scientists view themselves as radicals, if only epistemological ones, and in the 1970s and early 1980s they were the Young Turks of the sociology, philosophy, and history of science. Over time, it seems, the Young Turks have become silverbacks (to mix metaphors) and they now find themselves occupying what is in some ways a conservative position with respect to the increasingly international, diverse, and politicized field of science and technology studies.

Corridor talk or folk sociological theorizing on SSK can only go so far; it soon runs into the problem of internal diversity that undermines generalizations of the type made in the previous paragraphs. Perhaps a better way of generalizing about SSK is to say that its members share a belief that knowledge and artifacts are socially shaped or "socially constructed," a central rubric that, as a kind of Burkean God term, might best be left undefined. In addition to the belief in some version of the social shaping or construction of knowledge and technology, one often encounters a shared origin narrative that positions the SSKers against several Others, usually positivist/Popperian philosophers, internalist historians, and institutional sociologists of science (sometimes erroneously lumped together as "Mertonian" and sometimes with overtones suggesting the vulgarity of ugly American empiricism). These Others all would and do contest the SSK narrative. Furthermore, the SSK origin narrative varies from person to person and from context to context, and those variations constitute significant rhetorical resources that mark internal identities. For the purposes and space limitations of this essay, however, I construct one narrative that gives an overall flavor of SSK. If pressed, I could locate shreds and patches of this narrative throughout the SSK literature.

An SSK Narrative

In the 1920s and 1930s Karl Mannheim (1966) extended the project of a sociology of knowledge as it had been handed down from ancestors such as Marx, Durkheim, and Weber. However, Mannheim suffered a loss of nerve and ruled out social studies of the content of science (in other words, its theories, facts, methods, and so on). In subsequent decades Robert Merton (1973) built a sociology of science that focused on institutions and social structure but left the content in a black box. Merton assumed that the knowledge-production process was governed by the institutional norms of universalism, communality, organized skepticism, and disinterestedness, and by technical norms such as a concern with evidence and simplicity. In effect, he saw the content of knowledge production as objective and asocial, and he left theorizing about content to the philosophers.

Thomas Kuhn's *The Structure of Scientific Revolutions* (1970) helped pave the way for the new sociology of science in the form of SSK by stirring up waves in the philosophy and history of science. However, Kuhn soon backed away from the radical philosophical implications of his research, and today many regard him as something of a traitor to his own cause who may have even impeded the development of a thoroughly sociological approach to the study of scientific knowledge. Several researchers (e.g., Restivo 1983) also argued that Kuhn's work was similar to that of Merton in fundamental ways and not nearly as revolutionary as some had claimed. Nevertheless the black box of content had been opened, and soon the new sociologists of science were finding other, more reliable precedents. For example, Ludwik Fleck's *Genesis and Development of a Scientific Fact* (1979) is now seen as a precursor to Kuhn, and SSK researchers often point to a tradition of conventionalist accounts of knowledge within the philosophy of science. Most frequently mentioned is the controversial Duhem-Quine thesis of underdetermination, which holds that theories can be maintained in the face of contradictory evidence provided that sufficient adjustments are made elsewhere in the whole theoretical system (e.g., Knorr-Cetina and Mulkay 1983:3).

In the 1970s a group of primarily British sociologists completed the dismantling of the legacies of Mertonian sociology and positivist/Popperian philosophy. For example, Barnes and Dolby (1970), Mulkay (1976), and others showed the nonnormative nature of Mertonian norms; Collins (1975) showed how replication rested on social negotiation; and in *Knowledge and Social Imagery*, first published in 1976, Bloor (1991) articulated the "strong program" in the sociology of scientific knowledge. Thus, by the mid-1970s sociology of science had witnessed a dramatic shift from the Mertonian paradigm to the SSK paradigm.

The narrative of a dramatic rupture or paradigm shift has been hotly contested. Institutional sociologists of science have pointed out that the dismantling of Mertonian norms began with a paper by Merton (1957) that marked the transition to reward and stratification studies in the American sociology of science. The May 1982 issue of *Social Studies of Science* was devoted to a debate between Merton's student Thomas Gieryn and the SSKers over the extent to which the strong program was new or worth pursuing. Likewise, in "The Other Merton Thesis," Zuckerman (1989) argued that Merton's early work on Protestantism and science anticipated constructivism in his discussion of shifts of foci of inquiry and problems within and among sciences. Philosophers of science were even more contentious: many argued that the new sociology of scientific knowledge did not have the revolutionary philosophical implications sometimes claimed for it; rather, SSK led to a radical relativism and philosophical incoherence (e.g., Hull 1988; Laudan 1990).

At the heart of the strong program were four controversial principles: (1) causality: social studies of science would explain beliefs or states of knowledge; (2) impartiality: SSK would be impartial with respect to truth or falsity, rationality or irrationality, or success or failure of knowledge (and, presumably, technology); (3) symmetry: the same types of cause would explain true and false beliefs, and so on (in other words, one would not explain "true" science by referring it to nature and "false" science by referring it to society); and (4) reflexivity: the same explanations that apply to science would also apply to the social studies of science.

The symmetry principle is probably the most important tenet of the strong program. Bijker (1993), following Woolgar (1992), has characterized the intellectual history of the sociology of science in terms of progressive extensions of the symmetry principle: from Merton's symmetry between science and other social institutions to Bloor's symmetry between types of content to later developments that argue for symmetry between science and technology, the analyst and analyzed, humans and machines, and the social (context) and technical (content).

An early version of empirical research related to the strong program was interests analysis, associated with Barnes, MacKenzie, and (at that time) Pickering and Shapin. They, like Bloor, were at Edinburgh and are sometimes referred to collectively as the Edinburgh school. The interests studies explained historical controversies in science by reference to interests ranging from the Habermasian to the more identifiably Marxist conflict of classes. In several of the more notable studies, the scholars explained two rival theories by referring them to two conflicting social networks that in turn were related to class antagonisms (see Barnes and Shapin 1979; Barnes and MacKenzie 1979).

The interests approach soon encountered a number of criticisms even from within networks that were broadly friendly to the SSK project. From the perspective of laboratory- or interview-oriented methods, the historical studies of the Edinburgh school suffered from problems of interpretation. In Chubin and Restivo's (1983:54) phrase, interest theory seemed to explain "everything and nothing—and [did] so retrospectively." Perhaps even more damaging was a detailed criticism from Woolgar (1981b:375), which included the memorable complaint that science studies had almost returned to the original sin of

Mertonianism except that "instead of norms we have interests." A debate erupted in the STS journals, after which discussions of class interests took on a decidedly retro flavor (Barnes 1981; Callon and Law 1982; MacKenzie 1981, 1983, 1984; Woolgar 1981a, 1981b; Yearley 1982). The debate is significant because today the analysis of how class or macrostructural interests shape the technical content of science and technology has largely disappeared from the SSK agenda. Instead, the concept of interests survives in a slightly different form via the actor-network analysis of how scientists and other actors can create interest in their work, to be discussed below.

Another of the early empirical research programs is sometimes called the Bath school. In effect, the Bath school is Harry Collins, but it is also associated with his collaborator Pinch and his student Travis. Collins accepted the symmetry principle of the strong program but was less enthusiastic about some of the other principles (Ashmore 1989). His "empirical program of relativism" (EPOR) postulated three stages for the analysis of controversies: (1) documenting the "interpretive flexibility" of experimental results, that is, showing how a number of positions were possible among the "core set" of actors in a scientific controversy; (2) analyzing the mechanisms of "closure," or showing how the core set came to an agreement, such as through a social process of negotiation of replication; (3) relating the mechanisms of closure to the wider social and political structure, a problem that Collins (1983) tended not to tackle and instead relegated to Edinburgh-style interests analysis. Subsequently, Pinch and Bijker (Bijker, Hughes, and Pinch 1987; cf. Bijker 1993) extended EPOR to technology via the "social construction of technology" (SCOT) program that posited a similar series of stages moving from "relevant social groups" to "stabilization."

A third area of research in SSK involved field studies of laboratories, sometimes called "laboratory ethnographies" and usually associated with constructivism proper. Latour and Woolgar's *Laboratory Life*, first published in 1979, introduced a number of significant new concepts. Perhaps most influential was the analysis of fact construction as a rhetorical process that involves increasing deletions of markers of the social origins of the fact. In other words, the idea of a fact can be interpreted as a deletion of "modalities" that qualify a given statement. Facts were then viewed as historical outcomes of a process of movement across "types" of facts ranging from conjectures that are connected to specific people and contexts to anonymous, taken-for-granted knowledge of the sort that is found in textbooks or that everyone merely assumes to be true. As facts move from the former to the latter, the connection with their producers and social contexts is progressively deleted. The study also developed the related "splitting and inversion" model of the discovery process, in which "discoveries" were invented, then split from their inventors, and finally inverted to be seen as products of a real, natural world rather than the social world of their inventors. Furthermore, the study presented a modification of economic models of scientific behavior that saw scientists as investors of credibility and reapers of credit.

Knorr-Cetina's *Manufacture of Knowledge* (1981) developed the idea of the construction of knowledge in somewhat different terms. She used the metaphors of fabrication and manufacture to portray the constructed nature of the "discovery" process in the laboratory. She pointed to the locally situated nature of knowledge production, in which inquiry and products were "impregnated" with indexical and contingent decisions. She also presented a critique of the concept of scientific communities as well as of market models (for which the "market" involved similar assumptions about a community) and posited the idea of trans-scientific fields.

Mulkay and students such as Gilbert, Potter, and Yearley developed a related area of SSK known as "discourse analysis" (e.g., Mulkay, Potter, and Yearley 1983). Their studies demonstrated how scientists' accounts of their actions varied considerably over time and across genres or registers (such as conversations, letters, and reports). As a result discourse analysts could destabilize accounts of science that rested on one type of informant's account, such as reports or biographies. They also argued that by failing to study the full range of variability of participants' accounts, social scientists would naively take over native accounts and make them their own. At least some of their destabilizing studies were directed at fellow SSK accounts from Edinburgh and Bath.

Other students of Mulkay, most notably Woolgar and Ashmore (1988), developed the "reflexive" tenet of the strong program. Essays in the reflexivist vein attempt to inscribe the constructed nature of constructivist accounts in their texts. The more theoretically interesting reflexive studies have turned constructivism back on itself to explore philosophical and theoretical paradoxes. Ashmore (1989), for example, did meta-analyses of attempts to replicate Collins's replication finding as well as variable accounts of discourse analysts regarding the variability of scientists' discourse. Woolgar (1983, 1988b) explored the paradoxes of what he called the "reflective" or naive view of the relationship between scientists' accounts and the out-thereness of reality, which SSK researchers rejected only to have it reappear in their practice. As in some discussions of reflexivity in anthropological fieldwork, the theorization of reflexivity in SSK tended not to consider reflexivity in broader social terms that include the relations between discursive communities (Hess 1991, 1992).

The actor-network approach of Callon and Latour returns, in a sense, to the naturalistic flavor of the earlier Bath and Edinburgh studies (see Callon 1980, 1986; Callon and Law 1982; Latour 1983, 1987, 1988). Actor-network analysis views the truthfulness of knowledge and the success of technology as the outcome of processes of social negotiation and conflict that involve marshaling resources via sociotechnical networks that in turn produce changes in society. Thus, unlike social constructivism, in which the context of society (either macro or micro) shapes the content of science and technology, the actor-network analyses point to the "seamless web" or co-construction of technoscience and society. (This form of analysis may therefore be better termed "constructivism" in contrast to "social constructivism.")

The political process of knowledge/technology construction is conceptualized through yet a new set of terms, which in a very rough and preliminary way can be glossed as follows: the *problematization* of the issues that forces others to go through one's own network as an "obligatory passage point"; the *interessement* of other actors that locks them into roles defined by one's own program; *enrollment* strategies that interrelate the roles that one has allocated to others; and the *mobilization* of the spokespersons of the relevant social groups to make sure that they continue to represent or control their constituencies (Callon 1986). Networks are heterogeneous conglomerations of "actants": people, institutions, and things, all of which have agency in the sense that they generate effects on the world. In general, the concept of heterogeneous networks has been highly influential, although American social historians of technology are more likely to refer to a similar theorization by Thomas Hughes. Hughes's work brings yet another concept to the study of networks: the concept of "reverse salients," or bottlenecks that stall the expansion process (see, for example, Bijker, Hughes, and Pinch 1987).

A much misunderstood point, which Callon clarified in a conversation with me, is that his framework does not ascribe agency to things but instead focuses on the ways in which agency is attributed or delegated to things. In this way he provides a counterargument to the criticism I raised with him that his theory involves a version of reification, commodity fetishism, or even animism (see also Latour 1992). To the extent that actor-network analyses do indeed examine attributions of agency, the framework provides a point of contact with a cultural perspective more familiar to anthropologists, because the analysis of attributions in case studies could be tilted in the direction of a methodology that enters into the cultural world of the people involved. By studying the historical processes by which people grant nonhumans a degree of agency, such as conferring the legal status of the person on a corporation, it is possible to bring out the critical potential of Callon's approach to agency.

Through the actor-network approach, the SCOT program, and other developments, SSK has diversified in recent years toward the study of technology and of science in society. As a result, SSK has come closer to issues that are of concern in "post-Mertonian" American sociology of science (e.g., Cozzens and Gieryn 1990; Nelkin 1992) as well as the "social worlds" approach of the American sociology of Anselm Strauss. Students of the latter approach have creatively blended their own sociological tradition with SSK frameworks (see Clarke and Fujimura 1992; Fujimura 1992). Likewise, American ethnomethodologists have produced careful analyses of conversation and texts that have led to collaboration and dialogue with the discourse analysis/reflexivist tradition within SSK (Lynch 1985; Lynch and Woolgar 1985). Some philosophers, such as Steve Fuller (1993), who edits the journal *Social Epistemology*, have also developed a dialogue with SSK. The expansion of SSK and fuzziness of the boundaries is evident in Pickering's edited volume *Science as Practice and Culture* (1992), which even includes an American feminist and anthropologist, Sharon Traweek (1992). It is to the question of anthropology and ethnography, and its construction within SSK, that I now turn.

Theorizing Knowledge: The Anthropologist as Resource

In a book review in *Current Anthropology* of an "anthropological" study of science, the sociologist Steve Woolgar (1991b:79) asks, "What is 'anthropological' about the anthropology of science?" Although he admits that the ethnography under review repairs some of the "descriptive inadequacies" of the laboratory studies, he finds that it lacks "theoretical purchase." Woolgar then defines his own version of an approach that is recognizably "anthropological," which I shall outline later in the essay. Although I am not entirely comfortable with Woolgar's definition, I am here interested less in disputing his argument than in the phenomenon of a British sociologist writing in an American anthropol-ogy journal and telling us what anthropology is, using as text or touchstone a book in the "anthropology" of science that was written by Australian re-searchers who may not be anthropologists themselves.

To understand the phenomenon, it is necessary to begin with the point that anthropologists are latecomers to STS conversations. Of course there is a long and rich history within anthropology of studies of material culture, ethno-knowledges, culture and medicine, technology and evolution, magical and ra-tional thought, and the social impact of technology in the development context. However, it was not until the 1980s and 1990s that anthropologists in signifi-cant numbers began to study contemporary, cosmopolitan science and tech-nology and to take part in the interdisciplinary STS conversation. In contrast, in SSK there is a relationship with anthropology and ethnography that dates back to the 1970s. The role of anthropology/ethnography in the construction of SSK is another important aspect of STS that anthropologists and cultural studies scholars will soon encounter, and it warrants further inspection because the pos-sibilities for misunderstanding are very high.

One early example of anthropology as a resource in SSK appears in "Homo Phrenologicus: Anthropological Perspectives on an Historical Problem," by Steven Shapin (1979a). As occurred with the Annales school, the (then) "Edin-burgh-school" historian borrowed anthropology to write better history. (I put "Edinburgh-school" in quotation marks because Shapin did his graduate work at Penn, where, according to my colleague and Penn graduate Tom Carroll, fac-ulty and graduate students were combining anthropology and history indepen-dently of the Edinburgh school.) Why use anthropology to do a better history or sociology of science? Shapin answered as follows: "Cultural anthropologists have not been so frequently or so deeply committed to the forms of culture they have studied as have historians of science." Anthropologists might question the attribution of a lack of commitment; many of us in some way have shown deep commitment to political issues in the countries where we have lived. However, Shapin seems to be thinking less of anthropology's politically engaged side than of the image of the cultural relativist as a neutral, outside participant observer who, like an extraterrestrial, tries to make sense of radically different ways of life. Shapin therefore draws on a version of anthropology that could help his-torians to escape from their hagiographic tendencies; it could help them to think about science and technology as profane—that is, as not set apart from society.

What was Shapin's anthropology? He turned to the British school of Horton, Firth, Beattie, and Douglas, and he examined their different positions on the relationship between social structure and ideas, including both neo-Frazerian intellectualism and versions of functionalism. He then articulated those posi-tions with a framework informed by Barnes's (1977) development of interest theory. The result was a sensitive portrait of the relationship between nine-teenth-century Scottish phrenology and Scottish society. SSK researchers today would probably fault the essay for the unproblematic use of interest theory or the unproblematized division between knowledge and society. Anthropologists reading the essay today might fault Shapin for remaining within the narrow confines of British social anthropology without exploring alternatives posed by

American cultural anthropology, French structuralisms, or other anthropological research traditions. Nevertheless, the essay remains a competent application of anthropological theory to a history of science problem, especially for the time when it was written. Furthermore, because Shapin located the heterodox science in a historical context of changing class relations, he made it possible to put on the agenda macrosociological issues involving class and power. Those questions have been largely lost in a number of subsequent strands of SSK research.

Another way in which anthropology entered into the construction of SSK involved more explicit uses of the principle of cultural relativism. Collins's empirical program of relativism, for example, used the term "relativism" as a heuristic to signal his stance of neutrality in the face of opposing native (scientific) views of true and false knowledge. That usage certainly was similar to cultural relativism, although when applied to science it can be interpreted as endorsing epistemological relativism. In general, the impartiality and symmetry principles of the strong program came to be associated with anthropology's principle of cultural relativism. As Woolgar and Ashmore (1988:18) noted, "The espousal of a relativism traditionally associated with cultural anthropology enabled the social study of science to treat the achievements, beliefs, knowledge claims, and artifacts of subjects as socially/culturally contingent."

As a resource, then, not only did anthropology provide a theory of knowledge/society relationships (as in the Shapin paper), it also provided a metaphor of cultural relativism to aid in the application of the principles of the strong program. By imagining sciences as foreign cultures and themselves as anthropologists, sociologists and historians were able to describe their relativist position—epistemological, cultural, moral, or otherwise—regarding the content of scientific knowledge. At the same time, however, SSK researchers tended to be fuzzy on distinctions among the types of relativism, and consequently they became vulnerable to criticisms from philosophers who insisted that at least some variants of SSK self-destruct in the contradictions of social idealism and epistemological skepticism. (On the types of relativism and their relationship to constructivism, see Hess 1995:chap. 1; 1997b:chap. 2.)

Anthropology also served as a resource for SSK in the more general sense of providing a metaphor for the excitement that the SSK researchers felt as intellectual pioneers in the study of the content of science and technology. They became heroic explorers of test-tube jungles. For example, Latour and Woolgar began their classic *Laboratory Life* (1986:17) with an anthropological metaphor that is found throughout the SSK literature:

> Since the turn of the century, scores of men and women have penetrated deep forests, lived in hostile climates, and weathered hostility, boredom, and disease in order to gather the remnants of primitive society. By contrast to the frequency of these anthropological excursions, relatively few attempts have been made to penetrate the intimacy of life among tribes which are much nearer at hand.

Armed with their colonialist and masculinist metaphors, much in the tradition of Carolyn Merchant's (1980) portrait of Francis Bacon, the SSK researchers were ready to "penetrate" the secret of the content of science that Merton, like a good Puritan, had left modestly covered.

Anthropology also provided a method or, more accurately, a metaphor of method. Indeed, this use of anthropology came to displace the theoretical use as seen in Shapin's essay (1979a), and "anthropology" came to be synonymous with "ethnography." For example, in *Laboratory Life*, Latour and Woolgar developed the argument that historical studies (such as the Edinburgh school interests research) suffered from the limitation of having to rely on scientists' own statements about their work. An anthropology of science as a form of "ethnographic" observation provided a better alternative:

> Not only do scientists' statements create problems for historical elucidation; they also systematically conceal the nature of the activity which typically gives rise to their research reports. In other words, the fact that scientists often change the manner and content of their statements when talking to outsiders causes problems both for outsiders' reconstruction of scientific events and for an appreciation of how science is done. It is therefore necessary to retrieve some of the craft character of scientific activity through in situ observations of scientific practice. (Latour and Woolgar 1986:28–29)

Thus, whereas historical studies suffered from the problem of having to rely heavily on scientists' own retrospective accounts, in the laboratory studies sociologists were able to observe for themselves the unmasked and unclothed content of science.

In *Science: The Very Idea*, Woolgar (1988b:84) explained in more detail what the ethnography of science involved as a method. Usually, the ethnographer takes a menial position in the laboratory and works there for eighteen months until becoming "part of the day-to-day work of the laboratory." In other words, in the Malinowskian tradition, one comes in off the library veranda of archives or surveys and instead lives with a people for a sustained period of time. Woolgar described the ethnographer's task as one of note-taking, interviewing, and collecting documents. Those descriptions of ethnographic method are likely to be familiar to anthropologists; however, another aspect of Latour and Woolgar's construction of ethnographic method, the stranger device, is apt to be less so.

In *Laboratory Life* as well as in Woolgar's *Science: The Very Idea*, Latour and Woolgar were concerned that laboratory culture was too familiar, a problem that anthropologists who work in cultures unlike their own are less likely to face. Because of the cultural proximity of scientists and SSK researchers, Latour and Woolgar became preoccupied with going native and accepting uncritically the accounts of scientists about their work. In order to demonstrate the social construction of knowledge, Latour and Woolgar wanted to achieve distance from the sciences and scientists under study, and they appealed to the idea of "anthropological strangeness" for that sense of distance. They cited as their theoretical inspiration a 1944 essay by the phenomenological sociologist Alfred Schutz (1971). Although in *Science: The Very Idea*, Woolgar (1988b:84) noted that "'ethno-graphy' means literally description from the natives' point of view," he added that the scientists' point of view "must be perceived as strange." "Just as in any good anthropological inquiry," Woolgar wrote, "the ethnographer of science must bracket her familiarity with the mundane objects of study and resist at all times the temptation to go native" (1988b:86). The hoped-for result was a demystification of science. As Latour and Woolgar (1986:29) wrote, "Paradoxically, our utilization of the notion of anthropological strangeness is intended to dissolve rather than reaffirm the exoticism with which science is sometimes associated."

For anthropologists who study non- or semi-Western cultures and who have been, like me, confronted with practices such as spirits who perform surgery, achieving a sense of strangeness or distance is not a problem. Rather, the trajectory tends to be in the opposite direction: to take ideas and practices that educated Westerners would describe as irrational and show how they form a coherent system once the different set of assumptions is understood. However, that trajectory is only half the journey. As Marcus and Fischer emphasize in *Anthropology as Cultural Critique* (1986), understanding other cultures provides a vantage point for critical inspection of the values and assumptions of Western culture (including modern science). In contrast, *Laboratory Life* starts with an assumed rationality for Western science, then exoticizes it through the stranger device, and finally reveals a gap between the assumed rationality of the scientists' self-representations and a nonrational or other-rational practice that is revealed through observation. Rather than showing the hidden rationality of the scientific Other, Latour and Woolgar show the hidden irrationality of the scientific Self.

In the other laboratory studies, different aspects of anthropology as ethnography served as a resource. Knorr-Cetina (1981) used anthropology to help pose an alternative to the "frigid" methodologies of data collection in sociology and psychology (a metaphor that, like her use of "impregnated" above, I flag in contrast to "penetration" to suggest possible feminist resonances in her work). The frigid methods, Knorr-Cetina argued, rely on the questionable assumption that the meanings of scientists' language can be taken at face value. In their place she called for a more sensitive sociology that would achieve "an intersubjectivity which does not as yet exist." She suggested that this more sensitive sociology could "be found in a return to the anthropological method of participant-observation," and she described the history of anthropology as involving "progressive attempts to establish intersubjectivity at the core of the ethnographic encounter" (Knorr-Cetina 1981:17).

Collins and Pinch (1982) articulated a similar view in their "ethnography" of science, *Frames of Meaning*. They began the introduction to their book with a discussion of the rationality debate. Framed in terms of a "relationship between different cultures" that are likened to Kuhnian paradigms, the distance between the social scientist and the scientist is likened to a divergence between two cultures. Like Knorr-Cetina and unlike Latour and Woolgar, Collins and Pinch viewed the problem as one of achieving understanding across different scientific cultures rather than going native by taking scientists' statements at face value. Collins (1994:383) also showed concern with the stranger device and the means by which "ethnographers" of science may obtain an "estranged viewpoint."

For Collins and Pinch, the problem was not achieving strangeness and distance but instead achieving competence in another scientific culture. Achieving competence in turn involved both practical and theoretical difficulties. As a practical problem, the jobs of both the sociologist and the scientist are full-time positions that require years of socialization. As a theoretical problem, the sociologist never becomes an entirely native member of the other scientific discipline and consequently may be prevented "from understanding native members both by virtue of his untypical array of competences and by virtue of his position as sociologist/outsider with regard to the native community" (Collins and Pinch 1982:20). Sustained fieldwork in the culture of the scientific Other was the solution proposed by the Bath school, which espoused an interpretive sociology that in some ways was reminiscent of Geertzian cultural interpretation (Collins 1981). As in Geertzian cultural interpretation, the Bath school's position did not imply that the goal was to accept uncritically scientists' accounts as their own; understanding the Other's world was instead a prerequisite to a more theorized account of that world (cf. Mulkay, Potter, and Yearley 1983; reply by Pinch and Collins in Collins 1983).

To summarize, the understanding of anthropology, ethnography, the ethnographer-informant relationship, and related concepts was by no means uniform across the various members and texts of the SSK school. Their understandings also changed over time. For example, Latour (1990a:146) admitted that the first laboratory ethnographies, including his own work, "used the most outdated version of anthropology." Likewise Woolgar (1982, 1988a, 1988b:91–95; Woolgar and Ashmore 1988) drew on subsequent discussions related to the "new ethnography" in anthropology, including the SAR seminar that produced *Writing Culture* (Clifford and Marcus 1986), to advance his own version of reflexive ethnography as the "second generation" of the ethnography of science that would replace the older "instrumental" ethnography. (Our current SAR seminar may someday be seen as an exemplar of yet another generation of ethnographic studies of science and technology.)

In *Leviathan and the Air Pump*, Shapin and Schaffer (1985) also showed some significant developments in comparison with Shapin's (1979a) essay on phrenology. They opened the historical study of Boyle and Hobbes with a distinction between the accounts of "members" of a culture and those of "strangers." In order to move away from self-evidence, they followed Latour and Woolgar in contrast to Collins and Pinch. They noted that in *Laboratory Life* Latour and Woolgar were "wary of the methodological dangers of identifying with the scientists they study." Their position contrasted with that of Collins (1981:6), who argued "that only by becoming a competent member of the community under study can one reliably test one's understanding." Shapin and Schaffer (1985) argued that "we need to *play* the stranger," because the stranger to the experimental culture is in the position of "knowing" that there are alternatives. Finally, after noting that "of course we are not anthropologists but historians," Shapin and Schaffer provided a method for playing stranger to the experimental culture.

At a theoretical level, *Leviathan and the Air Pump* deconstructs the laboratory/society division in ways similar to Latour's post–*Laboratory Life* work on Pasteur (Latour 1983, 1988). Shapin and Schaffer show that Boyle was building not only a laboratory and an experimental method but also a new type of society that recognized a boundary between science and society. The argument is consistent with actor-network theory in general and with Latour's emphasis on the laboratory as a site for the coproduction of science and society. Latour (1990a) subsequently returned the favor to Shapin and Schaffer in a book review of *Leviathan* that called for an anthropology of science "without anthropologists." The review marks what is perhaps the final step in the SSK construction of anthropology and ethnography. In the review, Latour leaves the impression that SSK has done such a good job of appropriating anthropology that, as Modleski (1991) argues is the case for constructions of feminism without women, anthropologists are no longer necessary or interesting. Anthropology without anthropologists.

I will close this section with a simple question: Did they get it? Notwithstanding all the internal differences and the changes over time, there is a way in which the SSK laboratory studies and some of the related historical studies can be seen as a unity. This unity or *doxa* has to do with how those studies are all likely to appear "strange" to anthropologists who read them for the first time. As several other anthropologists have commented to me, when we read SSK laboratory "ethnographies" or the "anthropology of science" we have a sense that we are not reading ethnography or anthropology at all. For example, for me the question of whether one is a stranger or insider is less interesting than how the fieldwork begins to unravel connections among various cultural domains: exchange structures, funding flows, institutional positions, theoretical allegiances and divergences, methodological preferences, and so on. In the SSK "ethnographies" there is little if any thick description or semiotic analysis of local categories, contradictions, and complexities; there is little sense of cultivating informants, talking to people, finding out what they think, understanding their social relations, and analyzing the play of similarity and difference across domains of discourse and practice. In short, there is little if any culture. What tends to happen instead is that the sociological theories and (anti)philosophical arguments upstage the stories and worlds of the informants.

By explicating this difference I do not mean to put down the achievements of the SSK laboratory studies, nor do I wish to engage in gratuitous boundary-work. The laboratory studies have produced theoretical arguments that merit consideration, even if they are ultimately rejected or reconstructed. However, the value that I place on those studies does not change my perception that the books do not read like anthropology. Anthropological ethnography is often more like a historical case study than a treatise in empirical philosophy or a social theory with fieldwork-based examples. The difference between anthropological and SSK ethnography could be a productive tension, but in order for that to be the case both sides would have to recognize first that the difference exists. When there is no mutual understanding and respect, anthropologists can experience SSKers as arrogant, dismissive, and imperialistic because they want to tell us what anthropology and ethnography are. The result can mean that anthropologists become just another of the excluded voices in the SSK conversation. I and other anthropologists have experienced this misunderstanding, and at the cost of slowed, blocked, or misunderstood publication and review (for examples in print, see Fleck 1994; Forsythe 1993a, 1994). Of course, as anthropologists become more integrated into STS networks, the process can go the other way (e.g., Gusterson 1992). It is to the question of looking at SSK from the other side of the mirror that I now turn.

Other Voices: Toward Counternarratives

In the essay "The Critique of Science Becomes Academic," the radical Australian STS analyst Brian Martin (1993) examines a footnote in Harry Collins's book *Artificial Experts: Social Knowledge and the Intelligent Machine* (1990). The footnote reviews case studies in the sociology of scientific knowledge, and the usual suspects are rounded up: Collins, Harvey, Knorr-Cetina, Latour and Woolgar, Pickering, Pinch, Shapin and Schaffer, and Travis. Martin takes Collins, and SSK in general, to task for citation practices that exclude radical voices. In their place, Martin provides a counternarrative that roots STS research in radical social movements: radical science, feminism, women's health, civil rights, environmental justice, peace, and so on. In providing another narrative for the history of STS, he also urges STS to forsake its current tendency toward professionalization and to return to its roots in progressive social movements.

Brian Martin is one of the prominent voices in what I will call, for lack of a better name, "critical STS." The term seems least offensive to the largest number of people, and it has appeared in the literature in ways that explicitly link conventional radical agendas with feminist ones (see, for example, Restivo 1988; Restivo and Loughlin 1987). Critical STS—which, again, is only one of the many neighborhoods of STS—is much less coherent than SSK; I would characterize it as a series of interwoven sociointellectual networks and countertraditions. There is no closely integrated cocitation cluster, no single counternarrative, and no dialogue of clearly articulated programs with neat acronyms. Instead of appearing as a London men's club, in which vigorous but carefully chosen debates end with a good smoke being had by all, this branch of STS might better be likened to a querulous New York neighborhood in which there are many disciplinary transients and where many people do not know—or even want to know—their neighbors.

I think of the diversity of this wing of STS as a positive rather than negative feature, for diversity and anarchy may be one way to insure the vitality of dissent that is at the core of democratic research. Here I echo Traweek (1992:433, 440) in her discussions of the positive aspects of diversity and diversification. Furthermore, by invoking critical STS as a counterpoint to SSK, I do not mean to imply that the relations between these two groups are entirely polarized. Still, there is considerable evidence that supports a conflictual characterization of the relationship: conflicts over naming a new 4S prize after a man or a woman, holding the 4S meetings at the same time as the meetings of the American anthropologists (a group that includes several feminists and profeminist men[2]), celebrating or condemning the supposed politicization of the 4S, and deciding who controls the 4S board and the review process for journals and book series, not to mention what actually gets said in the book reviews, conference talks, essays, and books. I might also point to citation practices and reviewer comments, which indicate mutual ignorance and at times mutual hostility.

What is this other neighborhood of STS like? I order this heterogeneity (and to some extent others do as well) in terms of clusters of people with related interests. Examples include, but are by no means limited to, the technology-and-society critics from Jacques Ellul (1964) to Richard Sclove (1995) and Langdon Winner (1986), and from feminist perspectives work on topics like reproductive

157

technologies, such as Judith Wajcman (1991); feminist/critical philosophers of science such as Sandra Harding (1992), Helen Longino (1994), and Joseph Rouse (1996b); radical science studies from Hilary and Steven Rose (1976a, 1976b) to David Dickson (1984), Brian Martin (Martin et al. 1986), and Robert Young (1972, 1977); antiracist studies such as those by Robert Bullard (1990), Richard Lewontin, Steven Rose, and Leon Kamin (1984), and others gathered in Harding's *The Racial Economy of Science* (1993); radical work studies from Harry Braverman (1975) to David Noble (1984) for the workplace and Ruth Schwartz Cowan (1983) for domestic work; environmental and appropriate technology studies that followed in the decades after Rachel Carson's *Silent Spring* ([1962] 1987) and E. F. Schumacher's *Small is Beautiful* (1973); Third World and global perspectives such as Antonio Botelho (1993), Shiv Visvanathan (1991), and Richard Worthington (1993); and critical feminist and profeminist sociologists such as Adele Clarke and Theresa Montini (1993), Sal Restivo and Julia Loughlin (1987), and Susan Leigh Star (1991).

If I were to construct a narrative for this branch of STS, the ancestors would not be Mannheim, Duhem, or Fleck but instead—to name a few other dead white males who are frequently cited—Bernal (1939), Hessen ([1931] 1971) and Mumford ([1934] 1964); or, better, the intellectual precursors of antiracist and feminist science studies such as W. E. B. Dubois's *Health and Physique of the Negro American* (1906) and Simone de Beauvoir's *Second Sex* ([1949] 1989). Both Dubois and Beauvoir were studying biological ideas as constructions long before the idea became fashionable. Likewise, the "events" of the 1970s and early 1980s might be displaced from building a strong program to creating movements and related journals such as the British Society for Social Responsibility in Science (*Science for People*), Scientists and Engineers for Social and Political Action (*Science for the People*), and *Radical Science Journal* (now *Science as Culture*), as well as developing organizational sites such as the Radical Science Collective, the Rensselaer STS Department, and movement organizations such as the women's health movement organizations (see Clarke and Montini 1993). By the late 1980s and early 1990s, instead of a turn to technology I would posit a turn to race and gender or, more generally, culture-and-power perspectives that move away from foundational analyses rooted in a single dimension (such as class) to the interactions of race, class, gender, age, nation, sexual orientation, and other markers of difference, power, and hierarchy.

How are the two hundred–plus anthropologists and their siblings now working in cultural studies contributing to the STS dialogue, without reduplicating work already done in critical STS or SSK, not to mention any of the other disciplines and schools associated with STS? I suggest five interwoven strands that mark a distinctive anthropological/cultural studies contribution to STS. Perhaps the most obvious contribution of anthropologists has been our redefinition of ethnography as a research method and a way of knowing in general. The SSK "ethnographies" focused on the laboratory, addressed questions about theoretical issues in the sociology and philosophy of knowledge, and were the product largely of Europeans with training in sociology and philosophy. The anthropological ethnographies work with larger field sites such as transnational disciplines or geographic regions, address questions defined largely by a concern with various social problems (e.g., sexism, racism, colonialism, national/ethnic difference, class conflict, ecology) that are framed by hybrid feminist/cultural/ social theories, and are much more the product of Americans with graduate training in anthropology. Traweek's ethnographic studies of physicists (this volume), based on over a decade of ethnographic fieldwork and substantial graduate training in anthropology (even if, as she has said modestly, she only has a green card), are often regarded as a landmark for the beginning of the second wave of ethnography.

A second contribution of the anthropology of science and technology has been to reframe research on the public understanding of science. Models based on how scientists protect their legitimacy through boundary-work, or on how expert knowledge can be most efficiently conveyed to a public that is sliding down the slippery slope toward antiscience and New Age occultism, have been modified by culturally rich accounts that show how nonexpert lay groups and geographically localized communities actively reconstruct science and technology, often with high levels of sophistication (for a review, see Hess 1995 : chap. 6). Samples of this work include the reconstruction of medical knowledge (Martin 1994; Treichler 1991), workplace technologies (Hakken 1993), religiously relevant scientific theories (Hess 1991; Toumey 1994), theories of development (Escobar 1995), and environmental knowledge (Laughlin 1995).

Feminist anthropologists and cultural studies analysts have made a third contribution to STS by expanding feminist STS from the critique of reproductive technologies, the theorization of standpoint epistemologies, and the analysis of career attainment patterns for women to a much more general study of the culture of science as female and the institution of science as patriarchal. For example, studies by anthropologists Davis-Floyd (1992), Layne (1992), and Rapp (this volume) of reproductive/birth technologies provide a richness based on patient/user perspectives that was not evident in the first waves of feminist/ STS critiques of reproductive technologies. Likewise, studies by Haraway (1989), Keller (1985), E. Martin (1987), Merchant (1980), and other feminists have brought semiotic, cultural, and related frameworks into STS accounts of the content of science as not merely constructed but gendered.

Closely interwoven with the third strand is the shift in the understanding of what the word "construction" means. Although SSK prided itself on opening the black box of the "content" of science and technology, the stories of content that emerge from SSK are themselves highly technical ones. Stories of content are often told in a causal sequence, in which contingent social factors "S" are variables that cause technical content "C": $S \rightarrow C$. When content is conceived of in this way, it becomes difficult to discuss it in anything other than local, microsociological terms. From this perspective, broader markers of social difference such as class, race, and gender become a problematic background set (BS) of social factors that only tenuously shape microsocial factors (MS): $BS \rightarrow MS \rightarrow C$.

However, content can also be understood in a more anthropological sense. Consider an anthropological lineage of theories of cultural difference and meaning that runs from Boas, Benedict, Mauss, Peirce, and Saussure through Douglas, Dumont, Geertz, Lévi-Strauss, Sahlins, and Turner, and on to the feminist, subaltern, and variously engaged "critics" of later generations. Rather than ask how class, gender, race, and so on serve as variables that shape science and technology, this tradition would ask what science and technology mean to

159

different groups of people as marked by culturally significant categories of gender, class, race, and so on. Instead of opening only black boxes, one opens red boxes, pink boxes, purple boxes, brown boxes, and a rainbow of other boxes. The fundamental SSK insight that the technical is the social/political (like the old feminist adage that the personal is the political) is retained but recast in a different light. Divisions among facts, methods, theories, machines, and so on are seen as culturally meaningful and as interpretable in terms of locally constituted social divisions. In short, they are "technototems."

This relationship, unlike that of totemism as discussed in the SSK literature (Bloor 1982; Latour 1990a), opens the door to interpretive methods. Anthropology's culture concept via semiotic theories provides a new approach to the analysis of construction, one based on the interpretation of meaning rather than a sociological explanation of the content of science with reference to social factors or variables. To be clear, the two approaches—what I call cultural and social constructivism—are complementary and work best when used together. The point is worth emphasizing because SSKers are already misinterpreting me to be advocating an acausal analysis; instead, I am showing how anthropology and related fields bring a symbolic/semiotic level to STS that complements the accounts of social constructivism.

Finally, anthropology and cultural studies have contributed to STS by shifting discussions of the position(ality) of the researcher from reflexivity and policy (in the sense of how to manage science and technology) to issues linked to intervention, activism, and popular movements for social justice. This shift is taking place in a variety of ways illustrated in this volume, such as through theorizing intervention (Downey's partner theorizing, Heath's modest interventions), through studying scientists (Haraway's women primatologists [1989], or her comparison of Crouch and Hinchee), by analyzing technoscientific activism (as in Emily Martin's studies of AIDS activists and my own research on the alternative cancer therapy movement [1997a]), or by intervening in scientific controversies by helping one side get a hearing (Brian Martin 1996). The question of intervention and the problem of thinking about it in a rigorous way deserve more attention and, as I will argue, can benefit greatly from the resources of STS as a transdiscipline.

Theorizing Intervention: The Sociologist as Resource

The tendency of many associated with critical STS is to make a blanket rejection of the ideas and arguments of SSK. The alternative considered here is to appropriate and reconstruct SSK as a resource in much the same ways that it appropriated and reconstructed anthropology. Two examples will suffice: the impartiality principle of the strong program and the analysis of networks.

A tempting move would be simply to reject the impartiality principle as a reinscription of the very positivism, value neutrality, and objectivity that at another level it attempts to put into question. The impartiality tenet is perhaps the most vulnerable of the strong program principles, and some critics have interpreted it as a continuation of the value-neutral social science tradition that most practitioners of critical STS, not to mention many in anthropology and cultural studies, have long left behind. The obvious question is: If the social studies of science and technology are supposed to be neutral or impartial regarding what counts as truthful knowledge or successful technology, how does one adopt an engaged position as a proponent of one side of a scientific or technological controversy? If the technical is the social/political, then this form of impartiality seems to imply political impartiality. But why should one buy in to impartiality, when science and technology often embody and legitimate social relationships that the researcher finds unjust? As Winner (1993:374) argues, "One must move on to offer coherent arguments about which ends, principles, and conditions deserve not only our attention but also our commitment."

Although I am sympathetic with this line of argument and agree with Winner and others who have challenged value neutrality as political indifference, I think there may be a way in which the impartiality tenet might be preserved under some conditions for use as a rhetorical resource in attempts at intervention. To understand those conditions, it is useful to refer to the literature on capturing in relation to neutrality. It has been noted that in many cases of polarized controversies, epistemologically balanced or "impartial" treatments of scientific debates are rarely interpreted as such. Woolgar (1983:254) also notes that when social scientists offer alternative accounts even in a rhetoric of neutrality, "the proffered alternative account will be heard as a comment on the adequacy of the original account." Moreover, neutral accounts will often lead to capturing by the proponents of controversies, usually by the ones with less authority (e.g., Hess 1993; Martin, Richards, and Scott 1991; Scott, Richards, and Martin 1990). In other words, the party with the lower credibility may seize a neutral account because it implicitly levels the playing field.

The theorizing on capturing suggests that in some circumstances a neutral account may be a more effective form of intervention than an engaged or positioned account. As a resource, then, neutrality or impartiality can be used strategically for more effective intervention. Although this argument by no means implies endorsing the impartiality tenet of the strong program, it suggests a way in which the strong program brings up ideas that can be useful for those concerned with intervention. Certainly this argument may have a more general application to considerations of the role of the social scientist in movement organizations.

For a second general example of the possibility for a fruitful dialogue between SSK and various projects of intervention and activism, consider an obvious and fairly frequent charge leveled against the actor-network approach: It tends not to ask why certain people are able to build successful networks and others are not. Structural issues regarding glass ceilings and the politics of exclusion are backgrounded or forgotten in a theoretical model that assumes a level playing field on which competing networks duke it out in a masculinist game that is somewhere between market competition and all-out warfare. As a result, actor-network theory seems largely irrelevant for those who are concerned with issues of fairness and justice.

However, as I read actor-network theory, I also think about my experiences in coalition politics, especially when I worked with the diverse progressive groups and complex identity politics of the San Francisco Bay Area in the late

1970s. Coalition politics are based on heterogeneous networks that seek to expand and make their truth flow through their networks and into the larger society. Likewise, as I have studied various groups of heterodox scientific and medical researchers—many of whom have good ideas that merit more inspection from the broader scientific community—I am struck by the naiveté of their sociology of science. They should all read Callon, not to mention both Collinses (Collins 1985; Collins and Restivo 1983). Concepts such as enrollment and obligatory passage points can be useful as part of the package of tactics, strategies, and rules for radicals who go about organizing successful coalition politics inside and outside the citadel. If science is politics by other means, then coalition politics can be actor-networks with other ends. In other words, although actor-network theory has problems because of what is excluded from its analytical frame, some of its concepts can be of use for interventionist projects.

In short, critical STS analysts who are attuned to issues of power and culture (a general rubric that I prefer for issues such as gender, class, race, age, and so on) need to go beyond the strong program, but they should not reject it in a facile way. In my book on STS and its application to the evaluation of a medical controversy (Hess 1997a), I suggest that rather than explanation, impartiality, symmetry, and reflexivity, a set of rubrics that better describes a more viable program of critical/cultural studies of science and technology is power, culture, evaluation, and intervention.

First, the analysis is political; it explores the operation of power in the history of a field of knowledge that is constituted by a consensus and by attendant heterodoxies. For example, I (Hess 1997a) study several research trajectories on bacteria and cancer from a political perspective to show that a substantial body of research was systematically excluded—intellectually suppressed, to use Brian Martin's phrase (Martin et al. 1986)—from what became mainstream cancer research.

Second, the analysis is cultural in the sense that it develops a sophisticated, noninstrumentalist explanation and explication of the dynamics of power that have been described in the first step. Although some researchers may prefer the term "sociological" or "social," the term "cultural" is used instead to flag a kind of analysis that does not reduce the explanation of consensus knowledge and heterodoxy to sociological variables and the explanation of power to what Marshall Sahlins (1976b) calls practical reason. In other words, it is far too easy to explain the history of repression and suppression as the result of a coalition of interested parties who act in a mechanical way to attain status, enhance symbolic capital, protect their interests, or simply gain and maintain power. Instead, instrumental explanations are encompassed by a more complex interpretation of the growth of the autonomy of research cultures that respond with some internal integrity to theoretical developments and new research findings, ecological changes in the political economy, general cultural values involving standardization and gender, and cross-cultural flows of patients and clinicians who support alternative approaches.

Third, the analysis is evaluative; it weighs the accuracy, consistency, pragmatic value, and potential social biases of the knowledge claims of the consensus and alternative research traditions. This step or principle assumes that a fully interdisciplinary STS analysis moves out of the traditional plane of social scientific analysis/critique (here formulated around the two strands of culture and power) to a prescriptive level. This level involves two stages: the evaluation of knowledge claims and the evaluation of proposed policy or political changes. The evaluation of knowledge claims is necessary because of the capturing problem; it is accomplished in a heterogeneous manner that takes into account the cultural politics analyzed in the previous stages. The evaluation is based on the standards of the best scientific knowledge available at the time of the evaluator's analysis, but it also assumes that those standards may themselves be biased against the research under analysis due to the same political and cultural processes already analyzed.

Rather than provide an impartial or symmetrical analysis, I evaluate the content of the science itself from the philosophical perspective of constructive realism, that is, a position that recognizes both the constructed and the representational aspects of knowledge. The view of knowledge is neither relativist (as for the ideal typical radical constructivist, who does not allow for the power of the world to constrain evidence) nor algorithmic (as for the ideal typical naive realist, who believes that the crucial experiment can generally resolve disputes over evidence). Rather, the nature of knowledge is assumed to be more like that of the legal profession and the qualitative social sciences, in which evidence can be established but always within a social situation that recognizes the power of cross-examination and interpretation. To establish criteria for evaluating the alternative research program, a wide range of sources in the philosophy of science are used, including the work of feminists such as Longino (1994).

Finally, the analysis is positioned; it provides an evaluation of alternative policy and political goals that could result in beneficial institutional and research program changes. As a social scientist I therefore assume that I will be positioned inside the controversy, as the capturing literature demonstrates is inescapable, and that I am better off positioning myself rather than letting someone else do it for me. In the terminology of the STS field, this level of analysis can be described as a type of reflexivity, but one that is more profoundly sociological or anthropological than previously discussed forms.

In short, an alternative to the strong program should move beyond a social scientific analysis of science to the evaluation of competing knowledge claims: What alternative research traditions or theories are available or possible? Are they any good? If so, what kinds of institutional changes are necessary to move toward the alternatives? Yet moving beyond the strong program does not mean forgetting what it and SSK achieved; my argument is for a both-and rather than an either-or view of SSK and its Others in the neighborhoods of critical STS, cultural studies, anthropology, feminist studies, and so on. In arguing for this view, I hope I can make the interdisciplinary turf somewhat more inviting to readers who are thinking of living in STS or at least spending some time here. In constructing a map and countermap of SSK and making some articulations with anthropology, I have also been constructing a vision of a field that not only theorizes but also *does* more about exclusion, marginalization, hierarchy, and difference, including our own tendencies to reproduce those processes. That is the kind of community in which I would like to live.

163

Notes

I wish to thank Rayna Rapp for many helpful comments and for giving me the title, one for which I had a great affinity as a fellow New Yorker. I also wish to thank all the participants of the SAR seminar, as well as Brian Martin, Sal Restivo, and Stewart Russell, for their criticisms and suggestions. I owe the use of the term "silverbacks" in this essay to my colleague Roxanne Mountford, who introduced the term into feminist circles at Rensselaer. According to Donna Haraway, it is also used in similar ways among primatologists.

1. The estimate is based on the current number of subscribers to the list moderated by Joe Dumit and run for CASTAC, the Committee for the Anthropology of Science, Technology, and Computing of the American Anthropological Association. To subscribe to this low-traffic list, send a message to LISTSERV@MITVMA.MIT.EDU with the following text: SUBSCRIBE CASTAC-L your name. In the past I edited a newsletter and list of publications by anthropologists interested in science and technology (*The Anthropology of Science and Technology*), but I discontinued the project after the list became available.

2. The term "profeminist" is sometimes preferred in the wake of male attempts at appropriation of feminism that results in a possible "feminism without women" (Modleski 1991).

123456789 1011

ETHNOGRAPHIC FETISHISM OR CYBORG ANTHROPOLOGY? / Human Scientists, Rebellious Rats, and Their Mazes at El Delirio and in the Land of the Long White Cloud / **Sarah Williams** in collaboration with **Frederick Klemmer**

This space is where I would have liked to present a complete ethnography of the SAR advanced seminar on cyborg anthropology, because it seems to me next to impossible to convey the points I want to make about anthropology as science without a thorough account of this fascinating event. But some of my colleagues told me that studying them would be problematic—which vindicates the power not only of ethnographic authority but also of the prohibitions against its reflexive application, strenuously noted but often ignored by the participants themselves (including myself).

Rubbing Against Biotechnopower

B Y BEGINNING WITH THIS NOT-SO-EMPTY SPACE, my intention is not only to cite its representational force as described in Michael Taussig's (1992) "Maleficium: State Fetishism" (in which his unrepresentable object is the sacred Churinga of the Arunta) but to introduce the reader to the ethnographiclike writing that does follow. For just as Taussig would have liked to present an image of the frog totem that was central to Durkheim's study of the Arunta, I would have liked to produce an ethnography of the politics and pleasures that danced throughout all aspects of the event that produced this book. And just as Taussig declined because he was told that such a presentation would offend the Arunta, so have I refrained because I was told that it might trouble people's careers, complicate their access to field sites, and disrupt the collegiality of the seminar itself.

Perhaps there will be readers of this chapter who do see ethnography in/as such an empty space. What could be more ethnographic than a writing that fails to represent that which it was prohibited from representing?

I was invited to participate in the SAR seminar as an anthropologist of anthropology, and I arrived with the understanding that my role for the week was that of resident ethnographer of the seminar itself. However, as I will detail later, the organizers did not inform the other participants of this plan early enough for them to take part in its development, and when I arrived, some participants expressed concerns that preempted my study. In short, they did not want the entire event to be "on the record." I was not, however, asked to leave. Rather, at the end of the week, I would be able to lead one conversation "on the record." Some also suggested that I produce an ethnographic text about the seminar that mimed the form of a scientific report. That text—the text that follows—has a long, difficult, surreal history.[1]

Still present are some traces of the text's original vision: I once believed an anthropology of anthropology could produce an ethnographic writing not circumscribed by the fetishistic reembodiment of reason and the legitimation of violence. Yet, while the anthropologists I attempted to study have explicitly, self-consciously had to negotiate violent and embodied relationships between knowledge and power, these negotiations, like those that help sustain the sacredness of the Churinga, are consistently devalued relative to the "data" such negotiations help produce and the system of proprieties such negotiations presuppose. As is evident in the transcript selections that follow, the roles of the observer and the observed, the "native" and the "anthropologist," the Self and the Other were not merely maintained despite the postcolonial, postmodern politicization of these roles, but were reified anew in ways that actually contribute to their intellectual force. Thus, contrary to the original vision with which I began my anthropology of anthropology, the conscious negotiation of relations of power and knowledge has not, after all, overturned the system of proprieties and fetishizing desires through which anthropology-as-science reproduces culture. As Taussig puts it in relation to anthropology's evident inability to reflexively recognize the social taboos it serves to reproduce, "There is no anthropology of the ruling class that rules over us. . . . And the time is long past for that project to have been initiated. There are institutional reasons for it not having happened" (Taussig 1992:134).[2]

While not an ending point, the empty space that begins this paper does express the acute vulnerability (others' as well as my own) that results when the ethnographic encounter mirrored by an ethnography of anthropologists makes tangible the complicities of knowledge and power that cannot be spoken yet empower the force of research itself. The empty space foreshadows the vulnerability of the subject (both observer and observed) in relation to the force of writing. The SAR seminar was structured around the production and reproduction of texts. Yet although all participants recognized that who gets to write about whom determines, to a large degree, the symbolic capital of ethnography, this recognition only heightened the vulnerabilities of us all. As the case studies presented here demonstrate, there are institutional reasons, both empowered and obscured through science's own formal mechanisms of fetishization, that over-determine individuals' personal negotiations of speaking positions.[3] By juxtaposing a case study from Aotearoa/New Zealand with the SAR seminar, what becomes all too visible are the differences race, ethnicity, sex, and class as well as, and perhaps more importantly, nationality and the logic of global capitalism make in the negotiation of the complicities inherent in the relationships between knowledge and politics.

If the comparison between the Arunta culture and the culture of (cyborg) anthropologists—their systems of proprieties and their fetishizing desires—is pushed a bit more, it becomes obvious why the power of the fetish objects of the anthropologists of science and technology, like the Churinga of the Arunta, can be strenuously noted yet not reflexively recognized. As several of the strangely "Durkheimian" chapters in this volume demonstrate, because thinking and objectification reciprocate each other, understanding "factoids" (facts-in-the-world) is crucial to understanding ourselves. *Enactments of biotechnopower*, not Churinga, are what contemporary "natives" (and their "ethnographers") rub themselves on to gain power and knowledge.

The practice of cyborg anthropology does not escape being a mode of knowledge production. Rather, such a practice remains (perhaps necessarily) involved in (1) the violent proliferation of fetish objects and fetishizing desires and (2) the reproduction of proprieties and governmentalities through which some channels of proliferation are rendered legitimate and others illegitimate. Moreover, cyborg anthropology must also admit its participation in the specifically anthropological mode of knowledge production through which other people's fetish objects and fetishizing desires are examined and exhibited, their potency being thus appropriated and made to serve the fetishizing desires and object-effective investments of the anthropologists themselves. The main difference between cyborg anthropology and more traditional anthropology is not that the practice of cyborg anthropology in any way evades the anthropological usurpation of totemic potencies but rather that, insofar as cyborg anthropology reflexively recognizes its participation within the same processes of "cultural" enunciation and signification it serves to objectify (e.g., the culture of technoscience or the culture of epistemic colonialism), it is led to be more rather than less explicit in this usurpation.

This book is in many ways the material object that expresses well the conscious, collective sentiment of the SAR seminar. Every member of the group was committed not only to developing and maintaining a supportive and collegial seminar experience, but also to producing a text that represented a collectivist ethos. Whereas the original gesture that informed my participation was to facilitate the interactive fashioning of a self-study that "would actually perform cyborg anthropology in the process of advocating it" (see fig. 9.1), this gesture was disabled before it could begin to be enacted. I find that what I am now enabled to do is rather to advocate cyborg anthropology in the process of not performing it.

Where my nonethnography succeeds, or by traditional standards fails, is precisely in its effort to name or at least mirror the sacredness it refuses to objectify. A weak smile seems the only response to a reviewer's comments, on an earlier draft of this paper, asking me to map the power relations and lines of affinity among seminar participants. But what of a response to these, another reviewer's comments?

> I'm uncomfortable that the seminar participants aren't quoted by name, or even identified by age, sex, ethnicity—as all other "informants" in this volume are! It puts into stark relief the whole problem of doing ethnography and representing the "other." When it's ourselves (the SAR, the seminar participants), we don't like it. But shouldn't this question be explored and problematized in greater depth in this chapter, and in the volume Introduction as well? You argue for a politics of transparency, of involvement and "intervention"; you get scientists and others to reveal themselves on the record, but you won't. What's going on here?

What is going on here, indeed?! An appropriate response is now embodied, I hope, in the very content and form of the following mimetic enactment of a scientific report. "Get with it!" exclaims Taussig (1992:122). "Get in touch with the fetish." Let the abstract, the introduction, the materials and methods, the results, and the discussion of this chapter put you in touch with the fetishes of technoscience and its ethnographers.

Abstract

Experiments in participant-observation were conducted among anthropologists in order to ethnographically render the dilemmas associated with the contemporary politics and pleasures of anthropological practice. Analysis of the preliminary data from two field sites suggests two conclusions: (1) that anthropological practices can, though only with great difficulty, be at odds with the discipline's original symbolic structure as a *human science*, and (2) that the very doing of an anthropology of anthropology makes participant-observation itself into an experiment that mirrors, through distortion, magnification, and refraction, the precise ways in which anthropology has functioned as a mirror for those entities known as man, science, and humanity.

Data from the first experiment—excerpts from a May 1993 interview with a senior anthropologist in Aotearoa/New Zealand—describe an "irreconcilable tension" between his emotional commitment to his part-Maori cultural identity and his intellectual commitment to archaeology as a science. Data from the second experiment—an on-the-record, taped discussion among anthropologists about cyborg anthropology during an October 1993 seminar at the School of American Research in Santa Fe, New Mexico—describe an irreconcilable tension between reflexivity in theory and reflexivity in practice. Especially when it comes to reflexive engagement, anthropologists (including this author) tend to like theoretical projections but are often haunted by their practical manifestations. Conducting and writing up these experiments sets up "culture" both as laboratory and as a field of experimental objects. When essayed as a human prosthetic device, "culture" is recognized both as a process of enunciating difference and as an object/field for examinations of categorical diversity.

Introduction

Although this report describes only two case studies, these experiments are part of a larger study concerned with the anthropology of anthropology. First conceived in 1984, this larger study is the direct result of data obtained while I did fieldwork in the company of anthropologists who were studying a "premodern" culture in a place of nature renowned to be "remote and isolated": the semi-arid grasslands of the sub-Sahara. The US National Science Foundation and several universities supported the study by cultural anthropologists, biological anthropologists, and ecologists of the "traditional, sustainable" interaction among people, plants, and animals in Turkana District, northwest Kenya. My research was an ethnoarchaeological study of the material culture of Turkana, which was to supplement anthropological understanding of the interface between the Turkana people and their nonhuman environments. However, I became fascinated with Turkana women's and children's beads. And this fascination had consequences.

An average woman's bead necklace weighs between twenty and forty pounds, is worn around the neck twenty-four hours a day, seven days a week, and is worth several goats, sheep, cows, and/or camels. The beads are given to females by their dominant male figures: before-marriage beads are a gift from the father, after-marriage beads are given by the husband. The quantity of beads is a direct reflection of the wealth—the quantity of animals—owned by a father or husband. Wives who do not produce children can be returned to their fathers, and bride wealth (animals and beads) reclaimed. Beads placed around the ankles, wrists, waist, and neck of a Turkana child who survives the high incidence of infant mortality simultaneously marks and blurs the boundaries between human and nonhuman. Neither beads nor humanness are attributed to children before approximately two years of age (Williams 1987).

My preoccupation with these beads forced me outside the constitutional parameters of modernist science, including the categorical distinctiveness of objectivity relative to subjectivity, the West relative to the non-West, the human relative to the nonhuman, cultural diversity relative to cultural difference, politics relative to truth, and intellect relative to body and emotion. In short, while I was trying to understand how these beads could be perceived by some as mere receptacles of human intentions and by others as having intrinsic properties, a scholar at Nairobi University said to me, "But surely you know about the relationship between beads and the underdevelopment of Africa?" I didn't. But learning that many cultures maintained a bead-based exchange economy irrespective of the gold standard enabled me also to realize that cultures, irrespective of their relative adequacy in anthropological terms (i.e., cultural relativism), have differently valued computational (from the Latin *computare*, to think together) values. Consider the correlation Bruno Latour (1993:51–52) makes between different computational values and the different meaning that human scientists, as opposed to "ordinary people," ascribe to material objects:

> Ordinary people imagine that the power of gods, the objectivity of money, the attraction of fashion, the beauty of art, come from some objective properties intrinsic to the nature of things. Fortunately, social

scientists know better and they show that the arrow goes in fact in the other direction, from society to the objects. Gods, money, fashion, and art offer only a surface for the projection of our social needs and interests. At least since Emile Durkheim, such has been the price of entry into the sociology profession. To become a social scientist is to realize that the inner properties of objects do not count, that they are mere receptacles for human categories.

Meta-anthropology began for me with the study of Turkana beads not because, as Latour would have it, anthropologists had failed to recognize their intrinsic properties, but precisely because the projections through which the beads were made meaningful were articulated through irreconcilable symbolic economies, thus making cultural interpretation itself suspect. Once considered a trivial interest in ornamentation, the study of Turkana beads eventually made visible not only the political economy of tribal culture within a modern nation-state but also the symbolic economy of anthropology's own globalizing culture. To give a twist to Clifford Geertz's (1973) classic formulation of human distinctiveness, those beads were not simply suspended by webs of meaning but were themselves prosthetic extensions of meaning. Although the beads were part of Turkana culture, the nature of their contemporary fabrication in Czechoslovakia, their highly restricted importation into Kenya, and their monetary value in light of their historical symbolic value revealed much about the culture of anthropology, the function of cultural relativism, and the cultural limitations of the ethnographic encounter.

During the process of realizing this about those beads, my own position shifted from that of an apprentice—being initiated into anthropology through the ritual of fieldwork among a tribal people—to a meta- or comparative anthropologist of anthropology itself. (This failure of my ritual initiation into the profession continues, as discussed below, to have unforeseen, occult consequences.) In learning about the protocyborgic relationship between the Turkana and beads, I became aware of the protocyborgic relationship between "the Turkana" and anthropology. That is, the Other has been as necessary for the symbolic system of human science as the beads were for that of the Turkana. Beads and the Other embody the machinery of culture. But just how exactly do they work as meaning-making technologies? Why did some meanings of beads become devalued, and why did the value of the Other become naturalized?

My interest in how the facts or, to use Dumit's term, the "factoids" of human science are produced drew my attention to the apparatus of anthropology's production. I began keeping records regarding the petrol and food expenditures of anthropologists in the field, the cost of notebooks, pencils, calculators, tires, antimalarial medicines, water gauges, skin-fold calipers. I noted how much tobacco, *ghee*, tea, sugar, and *posho* we were giving to informants. I found I could track the history of anthropological study among the Turkana through the material culture anthropologists had left behind. Children who had been named after a particular anthropologist all wore plastic crosses attached to their neck beads. Another anthropologist's gift of a camel for his primary informant crossed some line regarding ethnographic protocol; he had to reimburse the cost of the camel from his private funds.

Locating culture within the everyday practices of anthropology changes the cultural interpretation of those practices. Suspicion, fear, and defensiveness often have characterized the default position of my Others for interpreting my shift from doing anthropology to doing an anthropology of anthropology. For example, in 1983 my desire to study Turkana beads and then to study "Turkana anthropologists" was considered, by some, a threat to the scientific status of the research project. And according to one informant, the perception in 1993 of a lack of "mutual trust and goodwill" in my introductory account of my work in Africa presented at the School of American Research "led to more than one corridor chat speculating about [my] motives in the SAR project."

The discussion section of this report elaborates on this dilemma whereby an anthropology of anthropology becomes an overdetermined site of anxious speculation. Rather than have the leader repeat my painful experiments with setting up culture as a laboratory, I would like here to describe some interpretive technology useful for essaying "culture" as a distinctly human prosthetic device.

There are many conceptual programs available for such an essaying, but what I will describe as Homi Bhabha's "ambivalence" is recommended, due to the computational values it makes explicit. Bhabha's interpretive technology rests on the following distinction between the nounlike qualities of cultural diversity and the verblike qualities of cultural difference:

> Cultural diversity is an epistemological object—culture as an object of empirical knowledge—whereas cultural difference is the process of the *enunciation* of culture as "knowledge*able*", authoritative, adequate to the construction of systems of cultural identification. If cultural diversity is a category of comparative ethics, aesthetics or ethnology, cultural difference is a process of signification through which statements *of* culture or *on* culture differentiate, discriminate and authorize the production of fields of force, reference, applicability and capacity. Cultural diversity is the recognition of pre-given cultural contents and customs; held in a time-frame of relativism it gives rise to liberal notions of multiculturalism, cultural exchange or the culture of humanity. Cultural diversity is also the representation of a radical rhetoric of the separation of totalized cultures that live unsullied by the intertextuality of their historical locations, safe in the Utopianism of a mythic memory of a unique collective identity. (Bhabha 1994:34)

Bhabha's analysis of the distinction between culture as an object of knowledge and the *process* of the enunciation of culture as knowledgeable relies on the experiments of those postcolonial intellectuals—the granddaughters and grandsons of anthropologists' native informants—who have written about the instabilities of relationship between selves and cultures. These experiments often occur in what (before *The Satanic Verses*) could have been considered merely works of fiction. But as recent anthropological interest in magic realism attests, such fictions, which are "alive to the ambivalent structure of subjectivity and sociality," demonstrate computational values favorable to the production of cultural understandings that are ethnographically interesting (Bhabha 1994: 32). Bhabha's theorization of the inherent ambivalence of human subjectivity relative to sociality draws on Foucault (and Lacan) to articulate a position from which the limits of human science and its culture can be productively enunciated:

> Foucault's archaeology of the emergence of modern, Western man as a problem of finitude, inextricable from its afterbirth, its Other, enables the linear, progressivist claims of the social sciences—the major imperializing discourses—to be confronted by their own historicist limitations. (Bhabha 1994:32)

Bhabha's project is precisely that of causing the progressivist claims of the social sciences, particularly as they are embodied in the anthropological imaginary of literary texts, to be confronted by their own historicist limitations. And I find this project is also useful for considering the contemporary confrontations between anthropology and culture's own historicist limitations.

The constitutional foundation of anthropology rests on the conceptual utility of culture, a universal and universalizing interpretive technology that has progressively produced a collective human identity—the unity of humanity—from the encounter between Western man and his Others. But there are quite profound historicist limitations to this encounter. Rey Chow's (1993) query, "Where have all the natives gone?," indicates just how fundamental the limitations of this encounter have proven. Indeed, for whom can culture be an object of knowledge rather than an embattled and embodied process of political negotiation?

If it has been the work of the culture professional (the human scientist, the anthropologist) to produce a cultural narrative that represents the essential human unity of the ethnographic encounter, then the work of the meta-anthropologist is to render ethnographically the cultural practices that enunciate culture as a human universal. If successful, an anthropology of anthropology might enable us to confront the historicist limitations of culture itself. However, if the originary symbolic structure of human science was characterized by the ambivalence of Western man studying, and, in his study, disavowing the difference of Others, the recognition of this ambivalence produces yet more ambivalence. For how does studying the studier not produce an even more problematic discourse?

The encounter between meta-anthropologist and anthropologist, which is without question characterized by eccentric ambivalence and considerable subjective and cultural uncertainties, might be an example of Victor Turner's liminality. It might also be theorized as a site of what Bhabha (1994), drawing on Fanon, calls an "occult instability." Ambivalence conceived of as instability beyond human understanding points us in the direction of cyborg anthropology as I imagine it. And enunciating the ambivalence of the cultural difference that is set up in the ethnographic encounter might be an alternative to disavowal, to moral discourse, and to the all-too-human pain of being the Other who is studied, critiqued, and objectified (or the Self who is accused of studying, critiquing, and objectifying).

But, I discovered, the productive energy of such profound and occult ambivalence relative to the ethnographic encounter is often not easily or willingly cultivated. I have learned that being told my project is "too political" means that too many conflicting forces must be negotiated. It means that the ethical dilemmas of the ethnographic encounter exceed the available iterations of the limit-texts of man, humanity, science, and culture itself. The following critique by Bhabha of the politics of negotiation expressed in contemporary critical theory speaks to the politics of negotiation problematized in my meta-anthropology of anthropology's culture:

> Culture only emerges as a problem, or a problematic, at the point at which there is a loss of meaning in the contestation and articulation of everyday life, between classes, genders, races, nations. Yet the reality of the limit or limit-text of culture is rarely theorized outside of well-intentioned moralist polemics against prejudice and stereotype, or the blanket assertion of individual or institutional racism—that describe the effect rather than the structure of the problem. The need to think the limit of culture as a problem of enunciation of cultural difference is disavowed.
>
> The concept of cultural difference focuses on the problem of the ambivalence of cultural authority: the attempt to dominate in the name of a cultural supremacy which is itself produced only in the moment of differentiation. And it is the very authority of culture as a knowledge of referential truth which is at issue in the concept and moment of enunciation. The process introduces a split between the traditional culturalist demand for a model, a tradition, a community, a stable system of reference, and the necessary negation of the certitude in the articulation of new cultural demands, meanings, strategies in the political present, as a practice of domination, or resistance. The struggle is often between the historicist teleological or mythical time and narrative of traditionalism—of the right or the left—and the shifting, strategically displaced time of the articulation of a historical politics of negotiation. The time of liberation is, as Fanon powerfully evokes, a time of cultural uncertainty, and, most crucially, of significatory or representational undecidability. (Bhabha 1994:34–35)

In short, if what one wants to understand is the cultural problematic embedded in cultural interpretation, then one's project and its politics must be made capable of negotiating the uncertain, ambivalent, and excessive flows of cultural enunciation that form what Fanon calls "the zone of occult instability wherein people dwell" (Bhabha 1994:35). And insofar as one's project is made capable of this negotiation, it will itself become excessive in relation to the governmentality of cultural diversity implemented through employments of "culture" as a prosthetic device.

Materials and Methods

Experiment One
The first case study is the result of research among anthropologists in Aotearoa/ New Zealand.[4] This location, at this time, can be seen as a "cultural studies" petri dish. For those whom the media present as the dominant population, New

Zealand is still very much part of the British Commonwealth. For proponents of Maori sovereignty, however, it is still the land of an indigenous people: Aotea-roa, "the land of the long white cloud." In contrast to the recent challenges else-where to policies of multiculturalism, Aotearoa/New Zealand's official policy of biculturalism still rests, perhaps with increasing unease, on successful monetary practices, ongoing negotiations regarding the Treaty of Waitangi, and restrictive immigration policies (see McLoughlin 1994).

The determination, identification, and declaration of ethnicity in Aotearoa/ New Zealand is complex, fluid, and extremely political; immigration battles, like internecine struggles between Maori *iwi* (tribes), are a historical and on-going contemporary reality. Official census data break the population into the following categories: Pacific Island, Chinese, Indian, Fijian, Maori, New Zea-land European (Pakeha), other European, and other (New Zealand Official Yearbook 1994). Approximately 13 percent of the country's three million people are indigenous Maori. One million people, or one-third of the country's entire population, live in Auckland. And 85 percent of the Maori population live on the north island. The experiment discussed below, however, was conducted in the heart of Pakeha "settler country" at Otago University, which is located at the south tip of the south island where the local Maori iwi, the Kai Tahu, Kati Ma-mae, and Waitaha, are still waiting for the resolution of their land and fishing claims under the Waitangi Treaty.

Otago University is the world's southernmost university and New Zealand's oldest and most conservative. Founded in 1869, Otago currently has some eight hundred teaching staff and thirteen thousand students. It provides a primary fo-cus of cultural life in Dunedin, population 114,000. Dunedin (the name means "little Edinburgh") was founded in 1848 by Scottish Free Kirk Presbyterians, and the spirit of their militant Protestantism still thrives.

In January 1993 I began a lectureship in women's studies at Otago Univer-sity with a contract that stated that this appointment was based in the depart-ment of anthropology. In July 1993 I applied to the Otago University Ethics Committee for approval to conduct interviews with members of the department. The creation, submission, and approval of this application in consultation with members of the department of anthropology essentially made my place of employment into a field site for my participant-observation among anthropol-ogists. Although the Ethics Committee gave unanimous approval to my appli-cation, the approval letter from the deputy academic registrar contained the following statement, which itself provides a great deal of ethnographic infor-mation regarding my experiment:

> I am pleased to be able to inform you that no member of the Commit-tee has raised any objection to the project on ethical grounds.
>
> However, I have been asked to forward the following comment to you: "because of the specific nature of individuals' research fields, and the small number of individuals concerned, those contributing infor-mation will be readily recognizable. The subsequent citation of their views etc. could have an impact on their own careers. With such a so-phisticated group of people, it is up to them to be aware of this. How-

174

ever, it is suggested that if you have not already done so, you might like to see whether the Head of the Department of Anthropology feels that this is likely to be a problem." (Otago Ethics Committee, n.d.)

The word that I would like to consider in this statement in order to demonstrate the method of my research and analysis is "sophisticated." But first I must juxtapose it with the word "backward." Backward is the adjective that was used by one of the United States' most prominent cultural anthropologists to describe to me the anthropologists I would be working among if I accepted a job at Otago. I must be explicit: What is at issue for me is *not* whether the tribe of anthropologists in my laboratory is sophisticated or backward. Rather, my juxtaposition of these antonyms makes blatant the modernist dilemma of anthropology introduced by my Turkana fieldwork.

"Sophisticated" and "backward" are markers of cultural diversity, some form of which is necessary for classic anthropological fieldwork. Because, from one perspective, the anthropologists I was studying were "sophisticated," my research among them was considered ethical despite being potentially threatening. Because, from another perspective, the anthropologists I was going to work among were "backward," my research would constitute real fieldwork. The symbolic structure of the discipline is clear: The objects of the anthropologist's study, however subjective their construction, must be symbolically robust. The objects of anthropological study must be capable of retaining the boundaries of their distinction; they must provide the anthropologist a particular kind of computational value.

The symbolic economy of this sophisticated/backward split exemplifies the ambivalence inherent in cultural diversity. The sophisticated/backward distinction legitimized my research even as it made problematical my position relative to Otago anthropologists. Could anthropologists be an object of anthropological study? Was I one of "them"? Although cultural difference is a prerequisite, a constitutional foundation, of the ethnographic encounter, the political negotiations foregrounded by the occult instabilities of cultural difference are resolved into cultural categories, stereotypes that must be both assumed and disavowed by the progressive historicism of human science.

Gyan Prakash (1992:168) presents a similar analysis of the cultural limit of the science of man. "The project of science," he argues, "had begun by targeting the subaltern as the object to be transformed by the exposure to new forms of knowledge. But those defined as ignorant and superstitious could never be fully understood or completely appropriated—for if they ever became fully intelligible and completely assimilable, the project of educating them would have come to an end." It follows that the ultimate anthropological taboo in postcolonial British colonies is anthropologists "going native" not in the field but in their cultural narratives about encounters in the field. The discipline has a rich tradition for dealing with the inherent intellectual colonialism of this taboo. Anthropologists split their subjectivity (they even assume new names) and supplement their ethnographic reports with diaries, works of fiction, and poetry.[5] But what happens to the foundations of human science when diaries are put into conversation with ethnographies? What would happen if anthropologists' split subjectivities were made into objective social facts that reflected the culture of their profession? Would it then no longer make sense to ask whether native anthropologists represent the study of culture or merely(!) their own identity politics?

How will the discipline deal with the postcolonial circumstances of the nativity taboo gone awry? Anthropology has been remarkably resilient in its ability to ignore what Derrida (1970) nearly thirty years ago saw would be a monstrous offspring of the liaison between anthropology and postmodernity. It might be sufficient to suggest that anthropology's incestuous histories with psychoanalysis will continue to narrate an ego-centered, rationalized practice whereby cultural diversity, infused with sexual and racial difference, remains sacred.[6] But granting anthropological membership to the cyborg might promote a post-Oedipal turn whereby anthropology sees its own image and embraces not "woman," as the father of modernist anthropology, Malinowski, observed her, but its own culture.[7]

Historically the symbolic structure of the discipline has been such that when cultural difference is not objectified and assimilated as cultural diversity, anthropology becomes unethical, impossible, or *not* anthropology. For example, fears of informant-readers of early drafts of this report were both (1) that I had "gone native" and was representing only what my informants wanted represented, and (2) that I was not empathetic enough and would offend my informants and the institutions constituting my field sites. For the meta-anthropologist, the universal and universalizing proclivities of culture, which have so successfully produced cultural diversity from the consumption of difference, now mirror the distortions of this interpretive technology. Indeed, if the culture of human science reflects the social technology of anthropology as a mirror for man, then meta-anthropology refracts this reflection and renders human culture alien. My experiment is precisely an attempt to denaturalize and make positively monstrous the "sophistication" encountered when anthropologists study up, the "backwardness" encountered when anthropologists study down, and the "humanity" encountered when anthropologists study man.

Experiment Two

The second case study is the result of research that took place during a School of American Research advanced seminar on cyborg anthropology in October 1993. In the early days following its founding in 1907, SAR was known as "the Smithsonian of the West" due to its focus on the prehistory of the southwestern United States. But twenty-five years ago SAR's mission changed to that of "a premier center for advanced study," which supports, in the words of its president, Douglas Schwartz, "the most creative scholarship in anthropology and related disciplines" (Schwartz 1992:4). The editors of *Lingua franca* concur in their recent article "Ivory Bowers," subtitled "The finger bowls and olive groves of the world's cushiest scholars-in-residence programs." Despite the name of the Santa Fe estate it occupies—El Delirio ("the madness")—SAR "is now considered one of the country's most respectable anthropological institutions" (Kittay and Shulevitz 1993:52). It supports a resident scholar program, the Indian Arts Research Center, the SAR Press, a library, the J. I. Staley Prize, archaeological research, and the advanced seminar program.

Several times a year the advanced seminar program brings ten scholars to-gether for five days of intensive discussion. According to SAR's advanced semi-nar brochure, the program promotes in-depth communication among scholars who are working at a critical stage on a research topic and whose interaction will help move the human sciences forward with new insights into patterns of human culture, behavior, or evolution. Participants may "appraise ongoing re-search, assess recent innovations in theory and methods, and share data and knowledge relevant to broad anthropological problems." The program favors seminars that promise significant results for the field of anthropology as a whole. In addition to symbolic rewards, including the production of a seminar manuscript to which the SAR Press has first publication rights, advanced semi-nar participants receive round-trip transportation to Santa Fe as well as food and lodging in the SAR Seminar House.

The proposal for a seminar on cyborg anthropology was prepared and sub-mitted to SAR by Gary Downey, Joseph Dumit, and Sharon Traweek. Following the proposal's selection, these organizers invited seven additional scholars to participate in the seminar. The document reproduced here as figure 9.1 de-scribes the original vision of my participation as the resident ethnographer. This letter, with accompanying materials, was sent by E-mail to all participants by the seminar's coordinating organizer.

I arrived at the Seminar House late at night on 22 October, having per-formed very poorly in the cyborgian obstacle course of conference travel: stop-overs, delays, cancellations, plastic food, chemical air, sleep deprivation, and media-controlled environments. And I promptly burst into marginally human tears when told by the two seminar participants still awake that there were widespread and potentially preemptive concerns regarding my participation. These included confusion about what I actually intended to write; concern re-garding the source or sources of my data; confusion regarding the stipulation of, and conditions for, anonymity; desire for the establishment of explicit ground rules; concern regarding the possible loss of the informal nature of discussion; and a suspicion that an important topic—the elitist nature of SAR—would re-quire very sensitive and delicate negotiations.

My project and participants' concerns about it were discussed during the in-troductory session on Sunday, 23 October. An additional concern discussed was whether my project could have negative consequences for, or even jeopardize, participants' relations with informants and their continued access to field sites. Although all these concerns were by no means shared by everyone, or shared equally, it did feel to me that my "rats"—a term of endearment suggested by one participant—had ostensibly rebelled. It was agreed that the first morning session on the final day, Thursday, 27 October, would be devoted to assessing what the outcome of my proposed project could be. This final session, unlike all the others, was to be on record, and a tape recording of it was made available for my use.

Although the format of the seminar and the physical environment of SAR were in some ways curiously reminiscent of my field site in Turkana, there were important differences as well. The Turkana are known simply to move camp in the middle of the night when they want no more interaction with an anthro-pologist. And I knew anthropologists who were able to spend days cruising the

From: Gary Downey
Subject: Resident Participant-Observer

Dear Donna, Deborah, David, Emily, Paul, and Rayna,

On behalf of Joe and Sharon, I am writing to inform you about Sarah Williams's role in the SAR seminar as a resident participant-observer. At the end of this message is a letter and proposed statement of informed consent from her, which present more details.

As some of you already know, Sarah has considerable experience doing anthropological research on anthropology and anthropologists. As co-author with Joe and Gary of the short piece, "Granting Membership to the Cyborg Image," she also has strong interest in the idea of cyborg anthropology.

Last spring, she suggested that she write about the seminar itself, exploring the politics and pleasures of cyborg anthropology. We were enthusiastic about the idea for two reasons: (1) Including a self-study in the book would actually perform cyborg anthropology in the process of advocating it. By displaying our willingness to describe the political dimensions of our relationships as an explicit component of our own academic theorizing, we could show readers how such strategies can help shift academic theorizing away from oppositional politics and enhance the possibilities for positive exchange relations among competing perspectives. (2) Because of reason #1, we thought a self-study might provide a perfect conclusion to the volume.

However, we agreed to Sarah's proposal without consulting all of you. I took responsibility for informing everyone and initiating a conversation so that everyone would be clear about Sarah's project by the time we got to Santa Fe, but I have not done this in a timely fashion. Because Sarah is strongly committed to fulfilling her ethical responsibilities in this project while also producing an interesting account, she has occasionally and gently pushed me to get going and contact you. I apologize for the late notice.

Sarah's statement of informed consent is worded very carefully to offer each of us (1) the choice of anonymity and (2) the opportunity to edit any transcripts she might plan to use in a publication. She is committing herself not to use statements we might make other than those that are recorded and transcribed. At the same time, Sarah is not offering us authorial control over her writing. She will have to submit her paper for editorial review, but we have assured her that we will confine our editing to the same intellectual criteria that all contributors will have to face. This could get tricky in a volume that explicitly seeks to blur the boundary between conceptual and political contents, but we are all counting on relationships built on trust and mutual respect.

As I read in a message from her yesterday, Sarah says it best herself: "Trust is a big issue. As you know, I'm not out to get anybody or anything. My hope is that my work encourages an awareness of the politics and pleasures of what we do as anthropologists. I hope that by calling what I'm doing fieldwork, it gives others a context for critical self-reflection on their own cultural locations within the profession."

For the next week, I will serve as a clearinghouse of comments on Sarah's project. I will forward all messages to everyone, unless you ask me otherwise. Since our group is small and its membership published, choosing anonymity probably means choosing to be excluded completely, unless Sarah wants to do a major job of fictionalization. I hope you will choose to participate in what I think could be a significant expression of the world imagined by this volume.

Again, I'm sorry for not giving you more time to consider this.

Best to you all,
Gary

Figure 9.1. E-mail sent to SAR seminar participants.

desert trying to find "their Turkana." For the seminar participants, leaving each other and the SAR campus was not so easy. One cannot easily leave one's own people, either as "native" or as "ethnographer," and this relationship becomes especially complicated when an ethnographer's "people" (a category of cultural diversity) is also her people (a medium of cultural difference).

There is another comparison. The seminar schedule, like pastoral nomadic life, had the appearance of leisure. But like the elite among the Turkana, the elderly herd owners who spend their days sitting under acacia out of the desert sun, exchanging information and making decisions, seminar participants found that talking all day, trying to make sense of the world, and determining the direction and consequences of one's individual and collective behaviors were as tiresome as they were stimulating, and as burdensome as they were pleasurable.

A final note is necessary on methods and materials in relation to my results and their presentation in this volume. Given the multiple complexities and sensitivities of these experiments and their outcomes, my data are explicitly not attributed to specific anthropologists. I agreed to ascribe neither words nor identities to individual informants. Cross-cultural comparison suggests why such strategies regarding the negotiation of human collective behavior should be respected. Among the Turkana, counting animals not only is bad manners, it is culturally inappropriate. I was warned that I would bring bad luck if I persisted in counting animals. But this did not mean that the Turkana did not account for their animals. An interface beyond anthropologists' human perception enabled the Turkana to know exactly which animal among herds of hundreds was whose. Likewise, among a tribe of anthropologist-authors, herders of words and identities, it is culturally inappropriate to account for words and identities without conveying authorship. So please, read with care and respect; identifying animals or humans inappropriately jeopardizes anthropology and its future experiments.

Results

Experiment One
The cracks in anthropology's mirror make visible the all-too-human limitations of "humanity" and the excessively cultural limitations of "culture." It is through these cracks that the meta-anthropologist must travel in order to appreciate the relationship between the mapping of cultural diversity and the enunciations of cultural difference inherent in anthropological practice. One such crack/ opening to meta-anthropology is formed when the people an anthropologist studies are also the people to whom he or she belongs. What follows is an excerpt from an interview with an anthropologist for whom the personal and public recognition of his Maori ancestry has both facilitated and problematized his work as an archaeologist of Maori culture.

SW: Maori have told you that as you grow older the Maori part of you will become stronger. Do you feel that will be a tension? What are the pleasures of going the critical-thought route versus the rewards of not having that kind of critical consciousness which most cultures, as you put it, have been defined as not having?

A: Yeah, I wish I could answer that question. It's something which in various forms I often think about and I'm never really very sure about or come to any conclusions. In some ways I've managed to stave it off by going to University X.

SW: Which was going to be my next question . . .

A: I'm not saying that's why I'm going to University X, but in some respects it's a way of avoiding precisely that contradiction that I ran into by coming back to New Zealand in general, I suppose, but coming back to Otago/Southland in particular. There is still for me a lot of tension between my feelings and people's expectations of me, their perceptions of what it means to be a Maori, or part-Maori, or part of a tribe. On the one hand, there are people who say, "You're one of us, you're a Maori." And there are my feelings about that. Then, on the other hand, there are my intellectual thoughts about archaeology. I really can't resolve them. I mean they're two different categories that just can't overlap. You can take one into the other for specific purposes. You can take your critical faculty into Maori things but to do that then you have to create as it were a category of Maori intellectual things in order to attack them. I mean you can't take what you think, the training in Western science and so on, this doesn't have any bearing on your feelings about Maori things. You have to elevate those Maori things or depress them, whatever way you look at it, into intellectual things in order to talk about them, or to interface with them in that way. In other words, the two categories of feeling about Maori things and thinking about archaeological and anthropological things, there's no interface between them unless you change one or the other in order to create it. And that always is a problem, and as I say I've never been able quite to resolve that problem in my own mind.

In this first case study, the anthropologist's location within different cultures, his simultaneous suspension within the webs of significance spun by human science and Maori culture, presents an "irreconcilable tension." The anthropologist was able to retain his intellectual investments in anthropology's symbolic structure as a human science by distancing his professional practice and part of his "self" from a biculturalism wherein the mapping of cultural diversity has become increasingly pervaded by the ambivalence of cultural difference. The same biculturalism that can, from a distance, be appreciated as a manifestation of cultural diversity, when negotiated locally manifests an irreconcilable tension and troubles the very conceptual basis of the unity and diversity of "man" that anthropology supposedly mirrors. Aotearoa/New Zealand is a particularly poignant location for an anthropology of anthropology. It is, to draw on the work of Simon During (1989:40), a place where the official policy of biculturalism rests on the colonized, indigenous people preserving a cultural identity supposedly "grounded in the era before the modern to which current needs and wants attach."

What appears in anthropology's mirror as the frontier between two cultures erupts in the heart and mind of this particular anthropologist into an open wound (to borrow a phrase from Gloria Anzaldua) that cannot easily be healed. What temporarily stops the bleeding is his reflexive recognition that anthropology's interface with Maori "culture" is made possible by the transfiguration of Maori things into "intellectual things." This transfiguration, he seems to indicate, is no mere face-lift but rather involves a radical disembodiment and reembodiment of constitutive meanings. To apply a certain anthropological language reflexively, this would suggest that the scientific/intellectual culture of anthropology involves its own form of fetishism. (Although strangely homologous to the mana an artifact holds within its native culture, the mana an artifact holds within the culture of anthropology is of a different order. The mana of archaeological artifacts for a scientist, for example, is quite different than the mana of one's grandmother's bones.) Despite his desire to retreat to the psychological haven afforded by unquestioned participation in the symbolic economy of anthropology, this anthropologist nevertheless manages to formulate the beginning of a meta-anthropological analysis of that symbolic economy and its relation to the "zone of occult instability" wherein he finds himself implicated.

Experiment Two

As in experiment one, the anthropologists in this experiment are working with cracked mirrors. The difference is that they arrived at the laboratory/field site for the specific purpose of theorizing the value of (some) such cracks. As actual or potential practitioners of "cyborg anthropology," they came to El Delirio in part to talk about how certain kinds of anthropological examinations of technoscientific institutions may also challenge the very invocation of "the human," which, through diverse iterations, continues to serve as the foundational gesture through which anthropological practice is linked to a coherent field of objects.

Curiously enough, it is in this second experiment, concerning fieldwork among anthropologists who are themselves engaged in ethnographic encounters requiring a high degree of sensitivity to the politics and pleasures of human science, that ethnographic protocol was most tricky. Just as participants stressed that feminist practices were central to the work of the cyborg anthropology seminar, Judith Stacey's (1988, 1994) influential (if embarrassingly so) feminist conclusion regarding ethnographic practice is coincident with my own. Greater intimacy does not necessarily lessen or resolve the dilemmas of the ethnographic encounter. Instead, greater intimacy between the subject and the object intensifies the potential for hurt and increases the risks of misunderstanding. Feminism multiplied by technology might, to modify a formula of J. G. Ballard, equal the future, but feminism times technology also definitely equals an acute sensitivity to vulnerability. This vulnerability was explicitly theorized by a number of participants in relation to their own fieldwork situations.

Seminar Participant (SP): Different senses of vulnerability always seem to iterate when you're talking about technology. Acknowledging my association with the powers of technoscience always raises questions about who I am as human. It raises my awareness of and tolerance for vulnerability, which I'm not sure I would feel working in another area.

SP: Do you see that coming out of feminist practice?

SP: Well yes, in a way. I mean this certainly doesn't happen in STS circles. I've been experiencing agonistic struggles for years. So I'm not speaking to some essential property of science and technology. Rather it's the angle we're taking here, acknowledging our participation inside [the institutions of science and technology]. I put that alongside the commitment to feminist practice.

Turning the seminar itself into a fieldwork situation meant that these vulnerabilities were doubled for all of us. Not only were we on the edge with ourselves and each other, but our work as "cyborg anthropologists" was forced to account for the overwhelming demands it inspired. The following excerpts suggest something of what was at stake when these demands were negotiated in terms of the ethnographic encounter that made them explicit.

SP: One of the things that I really like about there being ethnographies of anthropologists is that the people I study often will say, "So do you ever get studied?" When they get a little exasperated with me being around, or when I'm around when some heavy problem suddenly emerges and they're not entirely sure that they want me there, they say, "Do you know how we feel?" And I'm really happy to be able to say, "Yes. I know a couple people doing it, but it's really tricky." That's why I'm happy to see anthropology of anthropologists doing anthropology of science. We're specially vulnerable in really interesting ways.

SP: I want to say that it cuts both ways, because not only did you just make a wonderful object lesson for us on why the risks of doing the fieldwork were far superior to the risks of just writing the texts, but over the course of the five days—I feel very empowered to speak for everybody—I think we have so much better a sense of what you're doing, and what the role of an ethnographer might be in interrogating the emergence, to use the word of group practices, that is going on here. When fieldwork is working one likes to think that there are these moments when people are sort of co-constructing these communicative selves.

SP: Part of the resistance or trepidation on the part of some of us about being asked to agree to this project without prior discussion came from the fact that we didn't yet have the shared histories that create articulations between situated knowledges and form the foundation of trust. Part of the luxury of having this week together is that we have been able to develop something that's akin to the sort of intersubjective relationships that emerge on its best days from long-term fieldwork and its intimate engagements between mindful bodies.

However, the intimacies of long-term fieldwork and shared professional histories were not the basis of trust for all participants. And texts versus bodies had different metaphoric significance for each of us regarding the politics of cyborg subjectivity. Such differences were anticipated by the seminar's organizers. Not only did they themselves represent an unusual trio for SAR advanced seminar organizers, they explicitly chose participants who represented a cross-section of academic rank. My role as a resident participant observer and the way it developed were related both to the transgenerational issues of the seminar and the apparent feminist, ethnic, racial, and economic harmony, if not homogeneity, of the group. In the absence of any explicitly articulated or irreconcilable racial, sexual, ethnic, or class differences, the process of my acceptance as the group's anthropologist was predicated on the perception that I was someone the group could accept as a colleague, someone who, as a participant in peoples' professional lives, would support their work as well as its vulnerabilities. We discussed these vulnerabilities.

> SP: How many people felt at some point during the week that they don't belong here?
> SP: It's like asking whether you ever thought you were a fraud.
> SP: Have you ever thought anything else?
> SP: Self-loathing . . .
> SP: Belong?
> SP: I didn't really talk about it, but at Bandelier [National Monument] I felt very alone and isolated. But then X asked me if I felt the stress there. She said it was really intense. It felt good to say "Yes!" Then she told me that someone else was feeling out of place at the seminar. You sense that you're the only person who's experiencing something and it turns out you weren't.
> SP: Yes, and the way you work is wrong, because somebody else just described the way they work differently, and it's, oh my god, it's vastly better than the way I work.

The vulnerabilities participants felt were related both to the event of the seminar itself and to their struggles regarding their work in the larger world. The conversation turned to a discussion of symbolic domination in academic worlds. These battles, which were deplored by everyone, were referred to as "assassination attempts" within a "war murder game." Why, the reader might ask, does intellectual life get played out this way?

> SP: Because of truth. There's one truth, and everybody is competing for it, and for my truth to win, your truth has to lose.
> SP: Die.
> SP: Die!
> SP: Symbolic domination, as we keep showing ourselves, is at the heart of all the fields we cross, from our families, to educational process, to the research, to the places we work, and it's endless. And you can't erase it, it goes on here. It's a question of what you do with it, first being aware of it and conscious of it, and of what it's done to us, and what scars and wounds it's left. But then the resolution not to repeat as much of that as one can. I don't know where to go with it, I'm just seconding what's been said. The out of placements as opposed to who, who should be here?

SP: Well that's the point of feeling an outsider. In each case it's a different sense, yet it's those differences that are interesting, and acknowledging them is what holds us together.
SP: And they're painful.
SP: And they're painful.

We had talked about the symbolic economy of knowledge production, but I
wanted to ask about payment for fieldwork, an in-your-face kind of issue that
made me cautious.

SW: You can tell me if you don't want to talk about it, which I will totally accept. But I'm curious about the issue of payment. Whether blatant in terms of working in labs, or anthropologists working for
companies that say, "Write this report for us for X number of dollars,"
or "Come to this international conference, we've got lots of money,
come along on our plane with us," many of you are receiving payment
for your work. Will you talk with me on record about this? I know the
old model, and this is a stereotype, but the old model is that if you're being paid to do your work, it's not real work, it's not scholarly any more,
it's not valid, it's not . . .
SP: It's not fieldwork.
SP: It's not fieldwork! Maybe there's another language to bring to it from
feminism. Do we have to tell the same story, or can we begin to articulate a different way of understanding payment in terms of what an anthropologist can and can't do effectively in a fieldwork situation? Or is
it still too sensitive to have that be something talked about? To go on
record saying I'm paid X to work in this lab, will that immediately make
us nervous?
SP: I think that it's a very important question and that in order to answer it we need an ethnography of native categories of work, money,
spending, earning, value, volunteering. Because that's what we go into,
like it or not, and those are the terms in which our activities will be interpreted. So there's not one rule. It's just that you need to be able ethnographically to describe your situation.
SP: I think some of us are in situations that successfully avoid a patron-
age relationship. Because we're actually getting paid for real work done.
Anything else might be inappropriate. In a society where the cash nexus
clarifies certain kinds of things, unambiguous payment for work done
as a lab tech puts us in a clearer situation.
SP: That's how I felt about being a data producer for the scientists I
worked with. Truly for me, compared with other fieldwork experiences
I've had, lab work has felt like a more authentic kind of reciprocity than
the other forms we try to fabricate in the field. Using skills I've been paid
to learn has made cultural sense to the people I've worked with. And
even as I say this, I recognize that the issue is still more complex.
SP: In one of my field sites, I considered it my ethical responsibility as an
anthropologist to give people feedback since I sort of had the goods on
what was happening. I was happy to consult with them—as long as I

was on a grant to study them, they didn't have to pay me. But they all ignored that. Nobody took up my offer not to pay me. They all sent me some very small check, it's like automatic. My sense of ethics was, as long as company X was footing the bill and I was doing the research, my feedback mechanism was talking. But my community's native category is consultant fee.

Understood in terms of the same meta-anthropological language as was used to contextualize and theorize the comments of the anthropologist interviewed in the first experiment, this second experiment reveals the simultaneous operation of five levels of refetishization at work in the anthropological project:

1. Given the complex web of linkages developed over the course of centuries between the respective institutions and ideologies of modern technoscience capitalism and liberal humanism, it should come as no surprise that money continues to serve the "culture industry" of anthropology (to reapply the phrase of Max Horkheimer and Theodor Adorno) both as general equivalent and as unrecognized fetish object. "Before, the fetishes were subject to the law of equivalence. Now equivalence itself has become a fetish" (Horkheimer and Adorno 1972:17). As the discussion excerpted above begins to indicate, the troubling of the cultural/anthropological refetishization of money is easier said than done.

2. As is invoked in the discussion concerning "symbolic domination" and the zero-sum game of competition for unitary truths, one promise of cyborg anthropology is to pose a theoretically acute, practically negotiated, and politically engaged challenge to the kind of symbolic economy that enables power to circulate unnoticed when passed along under the imprint of truth, knowledge, data, and so forth. Hence, to use a language adopted by some cyborg anthropologists, competitions for truth are refigured as struggles over sites/cites for enactments of biotechnopower. The point I want to make here is that enactments of biotechnopower—in the world of technoscience *and* in the world of cyborg anthropology—are fetish objects. As Dumit (this volume) would say, they are instances of "objective self-fashioning." As Durkheim (1965:269) would say, "[A] collective sentiment can become conscious of itself only by being fixed upon some material object; but by virtue of this very fact, it participates in the nature of this object, and reciprocally, the object participates in its nature."

Anthropology, in order to represent such fetish objects, must reobjectify their sacredness. At the same time that anthropology defetishizes these objects in relation to anthropology's judgments concerning their place in their native culture, it also refetishizes them in terms of their placement in anthropology's own symbolic economy. Often this refetishization goes unnoticed. Like all fetish objects, those of anthropology require that we forget the labor and costs of their signification. Moreover, by focusing on the defetishization accomplished through the locating of these objects within a schema of cultural diversity, anthropology tends to blind itself to how this same fixation marks a refetishization at the point of intersection between schematics of cultural diversity and enunciations of cultural difference.

3. Concomitant with the fetishization of the object of desire—the achievement of truth, the gathering and consumption of data—is the fetishization of a reliable, reproducible, standardized pathway for reaching this goal. Such fetishization, whereby a formalized and instrumentalized mode of practice is institutionalized as a socially legitimated method, is remarkably prevalent in modern scientific cultures, including those of anthropology, despite the conceit that such cultures are deritualized. One reason for this prevalence of ritualized practice is no doubt the ease with which the person and the path, identity and practice, become mutually reinforcing. Hence the anxiety that obtains and is usually displaced back upon the self when you find, amongst others who "work differently," that "the way you work is wrong." Although no one in the discussion excerpted above mentions anything like "method-loathing," all seemed cognizant of a certain indefinite yet ominous "self-loathing."

4. And what better mechanism is there to cope with the anxieties generated by the fetishization of method than the fetishization of identity? As Durkheim would certainly remind us, what Dumit calls "objective self-fashioning" is nothing unless accompanied by an equally important process. Let's call it subjective object-fashioning. An ethnography of the SAR seminar ought, perhaps, to focus on just this, the work of subjective object-fashioning—the intersection of the formation of ego and text—by which participants negotiated their identities through the objects of their ethnographic desires. In general, the mode of fetishization through which enactments of biotechnopower are valued and circulated as truths, according to the conventions of scientific or ethnographic method, is concomitant with the mode of fetishization whereby identities are valued and circulated as names, according to the conventions of authorship.

5. But what then to do with that further anxiety, that which inevitably haunts any symbolic economy wherein fetishized identities are negotiated through investments of ego in objects of desire? As is well known, one standard mechanism available for coping with this anxiety over ego identity is the refetishization of group identity. The SAR seminar was explicitly structured in terms of a collectivist, feminist, group ethos. A motivation for the seminar was to develop, articulate, and grant membership to a new kind of anthropology: a politically astute and collegial anthropology of science and technology. Participants mediated this intersection of the political and the personal quite differently. If reflexivity remains merely theoretical, thus serving merely to refetishize ego investments in one's work and the world, then an anthropology of anthropology easily becomes a specter of and for anxiety, fear, and distrust. If reflexivity is politicized, thus providing a means for ego displacement, then meta-anthropology becomes a means for new formations of work, self, and community.

Participants' behavior during the seminar and after, during the process of participating or not participating in the writing and rewriting of their voices in this report, demonstrated the full range of response. But reflexivity was not the only issue. That the meta-anthropologist was someone whose own identity as a professional anthropologist was in question surfaced in oblique ways. Who could trust that I was a "real" anthropologist—one of them—if I had completed (or abandoned) my initiation into the field by objectifying not an Other tribe, but my own?

Discussion

> As students of anthropology we are taught to pursue our research with the dispassionate objectivity of natural scientists; yet our actions are constrained by ethical codes necessitated by the "unnatural" (human) nature of our subjects.
> —Richard Daly and Antonia Mills

The following statements, which are drawn from one participant's response to an earlier version of this paper, demonstrate that the desire to make conscious the pleasures and politics of the ethnographic encounter actually complexified the ambivalent structure of our encounter.

> We participants were searching for ways of diagnosing the present by building positive relations of exchange that nonetheless acknowledged and accepted internal power relations. Your [the meta-anthropologist's] very presence helped make this acknowledgment possible by getting us to assess the power differences we felt when we were researchers versus when we were objects of research. But then I felt us rehearsing something like the following position: "We have achieved some sort of transformation here and we don't want someone, i.e., Sarah, using some agonistic strategy, to dismiss or downplay that accomplishment or make us look foolish or some such thing. We have been assured because you contributed so actively to what we are hoping to achieve, and so all is well." The problem for me is that this very defensive position is drawn from the world of agonistic politics. I thought, "What if Sarah hadn't shown herself so much to be one of us, indeed a coauthor of the piece that provided the basis for the seminar application?" Would we not have experienced the same transformation in our interactions? And at the same time, what did it say about ethnography when everything seemed to be working out so well for Sarah's project only because she had contributed to the transformation we all experienced? Who was writing her text?
>
> In what ways do you, Sarah, feel bound not to reject the seminar, and wherein lies the room to move? By suggesting in a few places that the group, as people, determined the form of your ethnography, you present both the field worker and the author as occupying fairly powerless positions that do not acknowledge authorial power in controlling the content of your text. How might doing cyborg anthropology be different than allowing the people we study to write our texts?

The protocol for ethnography about human scientists and for their ethnographies about others, whether they are engaged in meta-anthropology, cyborg anthropology, or positivist "dirt archaeology," rests, in part, on the possibility and the impossibilities of including in science (and its texts) the desire of the subject. "The behaviorist experiments of the rat psychologists," Gregory Ulmer

(1985:200) has argued, "are interesting not in terms of the rats' behavior . . . but in terms of the scientists' behavior—their relation to the labyrinths they build." Indeed, who *has* written this text? Who were/are the rats, and who were/are the scientists? Whose desires have structured the cultural labyrinths of these experiments and this representation of it? Who is being protected, and who benefits, from this experimental (non)ethnography?

The fact that the participant-readers of this paper (including you) do not know how many times it has been rewritten or by whom means neither that this text is a fiction nor that it in any way approximates that much-debated holy grail of ethnography, the dialogic text. Rather, my reflexive spoofing of the established values of the scientific and ethnographic genres is intended to re-frame, not resolve, the dilemmas of writing culture.

This paper describes fieldwork/experimentation concerned precisely with how the culture of anthropology, a culture ethically committed to protecting if not benefiting those it studies, reproduces itself in the face of the open-ended uncertainties and instabilities of widespread cultural change, including those articulated through the representational politics of cultural texts. Contemporary anthropologists who began their careers carrying out a laudable mission of liberal humanism now stand accused of carrying out a neocolonial form of cultural imperialism. And the success of the concept of culture, anthropology's hallmark, has globalized and simultaneously privatized social life such that anthropologists, for many people, are no longer the authentic or legitimate experts regarding the politics of culture or identity.

This, the quite fascinating cultural state of anthropology, is perhaps what makes cultural studies of science and technology so compelling. Not only do anthropologists find a tribe who "deserve" to be studied, but when anthropologists engage in fieldwork with scientists the cultural privileging of science itself, of experts, and of technological progress becomes a knowledgeable cultural practice. If the success of science now makes it possible to see science itself as an indigenous knowledge tradition, it is no wonder that anthropologists are feeling extremely vulnerable about their profession as well as their sense of self and community. The concept of culture—linked as it is with Durkheim, Weber, Marx, and Freud—is a remarkable prosthetic device for essaying the constitutional distinctiveness of anthropology and the computational values of its practitioners.

So, what would have to happen before a scientific paper could do more than present the occult instabilities of anthropological life as raw data? How can we move from a symbolic economy that values the human desire to understand (grasp, commodify, represent) cultural diversity to one that encourages the incompleteness, the excessive and alienable desire—the ambivalence—of cultural difference?

In order for a cyborg anthropology to begin to answer these questions, it must do more than provide new modalities and new field sites through which to unreflexively reiterate the fetishizing desires and object-effective investments institutionalized in anthropology as a science of culture. Insofar as cyborg anthropology is a practical/theoretical activity oriented toward the examination of "relations among knowledge production, technological production, and subject production," it ought to be consciously involved in the reflexive enunciation of critical challenges to the humanistic presuppositions of anthropological discourse (Downey, Dumit, and Williams 1995). In other words, the practice of cyborg anthropology ought to be meta-anthropological as well as anthropological.

Coda

After the SAR event I was invited to attend a conference where indigenous knowledge experts and Western scientists were to interact on equal terms and discuss the cultural politics of science understood as an indigenous knowledge tradition. My application proposal was explicit ("I want both to participate in and observe this remarkable event"). And this proposal resulted in a research fellowship to attend the conference. But several weeks before the event I was told that "due to political struggles between indigenous and European populations [in the country of the conference], any talk of anyone studying anyone else would be likely to cause much misunderstanding." So I went to the conference and told my own ("fetishistic") stories about the cracked mirrors, irreconcilable tensions, and reflexive impossibilities inherent in the practice of anthropology in an age when the old categories of "native," "primitive," "the Other," and "the magical" are both dead and strangely resurrected.

Afterward, a prominent sovereignty activist told me he identified with the anthropologists whose stories I was telling because of the tensions he felt in trying to be both a native and a lawyer in this (post)modern world. I wondered then and now about how so many cracks on so many mirrors interconnect. How does the blood of so many open wounds flow together? Is the scientific report a mimetic device for essaying the emotional literacy noted but not observed in our fetishistic rubbing against biotechnopower?

Notes

Cecile Stein, SAR's academic programs coordinator, was of great assistance to many of us in making travel arrangements. Duane Anderson, SAR vice president, did a remarkable job of managing resources during our stay. One of the primary pleasures we all enjoyed during the week was the excellent food prepared and served for us by Sarah Wimett, Jennifer McLaughlin, and Sarah Sandoval.

1. Due to the birth of my son, it is a miracle that there is now any text at all. That there is, is due largely to the help of Fredrick Klemmer, partner in writing and in parenting. Many thanks to Claudia Brugman, Tom Maddox, Joseph Dumit, Gary Downey, and the reviewers of previous versions of this text for their invaluable contributions.

2. The point here isn't that there haven't been ethnographic studies of the ruling class, but that such studies have not reflexively examined anthropology's role in reproducing the proprieties and governmentalities of the ruling class.

3. The role of individual difference, however, relative to the determination of speaking position and the negotiation of complicities of knowledge and politics was immense at the SAR seminar. The ambiguous position I found myself in on arriving at SAR was due in part to one organizer, whose own commitment to the open negotiation of authority and vulnerability made it perhaps impossible for him to have even anticipated that any of his colleagues would find my study of them problematic.

4.   Although Aotearoa is widely used to acknowledge not only the Maori word for New Zealand but to support indigenous rights, Aotearoa properly refers only to the north island. This issue is indicative of the many tensions within Maoridom.

5.   My personal favorite example of this anthropological dilemma is Kurt Vonnegut's decision to leave a graduate program in anthropology in order to be able to write about his experience of culture (Ruby 1982).

6.   Consider, for example, the popularity of a text like *Reading National Geographic*. This book about the sexual and racial systems of meanings that infuse the humanity represented on the pages of *National Geographic* is an extraordinary example of anthropology's refetishization of racial and sexual difference. This ethnography tells us nothing about the racial and sexual circumstances of its production. In fact, it tells us nothing at all about the political negotiations that occurred in the ethnographic encounter between anthropologist and *National Geographic*.

7.   Henrietta Moore (1988) begins her book, *Feminism and Anthropology*, a book that does *not* incorporate poststructuralist or postcolonial visions of a feminism of difference, with these words of Bronislaw Malinowski: "Anthropology is the study of man embracing woman."

SCIENCE AS A PRACTICE / The Higher Indifference and Mediated Curiosity / **Paul Rabinow**

> The moment faith in the God of the ascetic ideal is denied, a *new problem arises*: that of the *value* of truth . . . the value of truth must for once be experimentally *called into question.*
> —Friedrich Nietzsche

Site 1: A Select Few

IN THE MID-1990s I ATTENDED an international conference on a cutting-edge topic at which leading figures from several disciplines and two continents met in a basement room at La Villette, a massive, high-modern complex comprising offices, a science museum, and a postmodern park in northeast Paris. La Villette was one of the anchors of President Mitterand's project to renew eastern Paris, in effect to build a monument to his fourteen-year reign as well as to establish a new axis of speculative development anchored by an opera house at the Bastille, the biggest library in Europe across the Seine, and La Villette in the north. Outside lay a rectangular reflecting pond and a giant, silver-coated globe that housed a planetarium. Inside, well below ground level, was our designated room, shaped like a modified surgical theater.

We began the conference, starting late, with the usual perfunctory greetings. Each presenter was allotted the standard thirty minutes; this guideline, it was immediately established, was to be honored in the breach. One way to pass the time was to ponder whether successive speakers, as they surpassed their allotted time, were (1) displaying run-of-the mill arrogance, (2) straightforwardly unprepared and disorganized, or (3) enacting a postmodern performance (it doesn't matter where I start this paper or end it or how long it goes on, I will keep reading until someone stops me, or I grow weary, or simply lose interest). Although the afternoon session began with a reminder that the morning's time schedule had not been respected, the afternoon was much the same. I was told at dinner that the French moderators would never intervene to cut someone off, and that among the Italians it was not uncommon for there to be much milling around and conversation among the audience during such events. Finally, one distinguished commentator, who had not attended the first two days of the conference as he had other more pressing business at a government commission and his country house, arrived late, took a seat, looking harried, and began writing out his comments. When the time came for his presentation, he strode to the platform and talked on and on and on—about his own work, not the paper at hand.

The core group at the conference constituted a "network," and in fact "networks" were one of the main areas of inquiry and discussion. At the breaks there was a good deal of talk about planning the next conference in Berlin. Clearly the actual event itself was of little or no importance except that it had taken place and could consequently form the material for a paragraph in a funding proposal submitted to finance the next one. The network had excellent connections to funding commissions; one could therefore assume that the chances of a positive response were high. The turf was being occupied. Indeed, the papers might well be published.

I was irritated and bored. First: bored. In his montage-parody of an autobiography, *Roland Barthes par Roland Barthes*, Barthes (1975:28) includes a photo of himself and three other men sitting at a table, each staring off into space while (presumably) someone else speaks. They are bored. The caption of the photo is "Ennui: La Table Ronde." Boredom, daydreaming, and restlessness are common at academic events. Their consistent presence, however, suggests that these moods are not merely accidental accompaniments to the occasional bad panel, or the speakers would exit in droves and not attend the encore performance with the same script and players (including themselves). As this rarely happens, I take these moods as constitutive elements of such events. They are a significant aspect of what Pierre Bourdieu (1988) has named the *habitus* of academic life in the late twentieth century, a tacit but important dimension of that life form's emotional tone, power relations, subjectivations, kinesthetics. As a transdisciplinary form they might be called the antisymposium, except that such a formulation is too negative; modern forms of power, Michel Foucault taught us, are productive. Modes of interaction, ways of talking, bodily praxis, are simultaneously inscribed through institutional custom and enforced through long-term civilizational practices of auto-policing and self-fashioning. To invoke the name of Norbert Elias is to indicate that manners are not marginal to cultural formations. Even Roland Barthes was disposed to play by the rules. It takes a lot of cultural work to produce a pervasive boredom and inner drift in an auditorium filled with researchers.

Then: irritated. The experience at La Villette led me to reflect on some of the constitutive elements of such events, their moods, and my reactions. In order to organize these reflections, I use the device of constructing—and thereby contrasting—two "types" (in the Nietzschean-Weberian sense) of science as a practice: (1) the vigilant virtuoso (mood = pathos or failed indifference), and (2) the attentive amateur (mood = attentiveness or reserved curiosity).

I take the work of Pierre Bourdieu as the most accomplished and successful example of the first type. Bourdieu has not only carried through an organized corpus of monographic investigations but has also achieved a powerful theoretical reflection on his own work. He believes the two to be inextricably bound, a belief that is contestable. I use the conference experience as a takeoff point for these thoughts and not as a metaphor for Bourdieu (although some of his disciples did play a role in the conference). I refract my boredom and irritation into an agonistic relation to Bourdieu's work. My aim is to learn from this exercise, to weigh the effects, the affects, the consequences of such a logically and teleo-logically consistent attitude. The aim of this engagement is not victory (we are not engaged in sports or war), revenge (for what?), or refutation (there is much to learn from him, although that is not foregrounded here). Rather, reflecting on my own ethnographic practice with the help of a set of distinctions developed by Michel Foucault, I assemble the elements of a second and alternate type, one embodying a different practice. My goal is the clarification and cultivation of that practice.

Practice

Although the theme of "practice" has been central to American cultural anthropology for almost a decade now, it is rarely defined with any rigor. Sherry Ortner (1984:149) makes this point: "What is a practice? In principle, the answer to this question is almost unlimited: anything people do. Given the centrality of domination in the model, however, the most significant forms of practice are those with intentional or unintentional political implications. Then again, almost anything people do has such implications. So the study of practice is after all the study of all forms of human action, but from a particular— political—angle." I take a different approach, one that is less general and that takes up practices from a different angle, the ethical rather than the political. How does such a shift reconfigure the practice of knowledge?

Alasdair MacIntyre (1981:175) defines a practice as

> any coherent and complex form of socially established cooperative human activity through which goods internal to that form of activity are realized in the course of trying to achieve those standards of excellence which are appropriate to, and partially definitive of that form of activity, with the result that human powers to achieve excellence, and human conceptions of the ends and goods involved, are systematically extended.

The language of the definition is full of terms (form, coherence, excellence, and so forth) one rarely if ever finds in contemporary social sciences. The reason for this unfamiliarity is that MacIntyre draws these terms from an older vocabulary and tradition, one that exists today as a minor current in moral philosophy, that of the virtues. A virtue, for MacIntyre (1981:178, 140), is

> an acquired human quality the possession and exercise of which tends to enable us to achieve those goods which are internal to practices and the lack of which effectively prevents us from achieving any such goods. The exercise of the virtues is not . . . a means to the end of the good for man. For what constitutes the good for man is a complete human life lived at its best, and the exercise of the virtues is a necessary and central part of such a life, not a mere preparatory exercise to secure such a life. The immediate outcome of the exercise of a virtue is a choice which issues in right action. . . . Virtues are dispositions not only to act

195

in particular ways, but also to feel in particular ways. To act virtuously is not, as Kant was later to think, to act against inclination; it is to act from inclination formed by the cultivation of the virtues. Moral education is an "education sentimentale."

My wager is that returning to this vocabulary will prove to be especially fruitful for understanding science as a practice. Although practices and virtues continue to exist, the discourse about them was marginalized by the rise of modern moral philosophy (Kantian, utilitarian, etc.). My advocacy of practice and virtue is an ethnographic and anthropological call to be attentive to existing minor practices that escape the dominant discursive trends of theorists of modernity and postmodernity alike. Therefore, somewhat unexpectedly to me at least, it can be taken as yet another critique of modernity. However, my intent is to contribute to what Hans Blumenberg (1983) has called *The Legitimacy of the Modern Age* by making some already existing practices and virtues more visible, more available, thereby contributing to their reinvention.

Type I. The Virtuoso of the Rational Will

Pierre Bourdieu and Loic Wacquant (1992) chose the following quotation to open their primer on Bourdieu's work, *An Invitation to Reflexive Sociology*, tailored to charm and conquer—i.e., to civilize—the American social scientific audience/market: "If I had to 'summarize' Wittgenstein, I would say: He made changing the self the prerequisite of all changes." Although it is not evident why Wittgenstein needs to be mentioned at all except as a mark of distinction, the quote draws our attention to the question: What kind of self is it that Bourdieu seeks to change? What subject does he want to produce? It also draws our attention to another question: What does he produce?

Illusio

Despite the complexity and analytic power of Bourdieu's sociological oeuvre, the answer to this question is relatively straightforward. Fundamentally there are only two types of subjects for Bourdieu: those who act in the social world and those who don't. Those who do, do so on condition that they are essentially blind to what they are doing; they live in a state of *illusio*, to introduce a fundamental concept in Bourdieu's system. The other possible subject position is the sociologist who studies those who act, those beings who take their lives seriously, those who have "interests." The scientist, through the application of a rigorous method preceded and made possible by the techniques of asceticism applied to the self, frees himself from the embodied practices and organized spaces that produce the illusio and sees without illusion what everyone else is doing (maximizing their symbolic capital while mistakenly believing they are leading meaningful lives).[1]

Bourdieu (1982) puts it this way:

> What I have called *participant objectivation* (and which is not to be mistaken for participant observation) is no doubt the most difficult exercise of all because it requires a break with the deepest and most unconscious

adherences and adhesions, those that quite often give the object its very "interest" for those who study it.

Or:

The sociologist unveils the *self-deception* [English in the text], the lies one gives to oneself, collectively kept up and encouraged, which, in every society is at the base of the most sacred values, and for that reason, at the basis of social existence in its entirety.

There is one sacrifice required by the sociologist in order to achieve the clarity occasioned by his radical change of consciousness and of ontology. That sacrifice is to refuse all social action, all "interest," in the meaning and/or stakes of social life. Bourdieu, again:

One must in a sense renounce the use of science to intervene in the object which is no doubt at the root of her "interest" in the object. One must . . . carry out an objectivation which is not merely the partial and reductionist view that one can acquire from within the game . . . but rather the all-encompassing view that one acquires of a game that can be grasped as such because one has retired from it. (Bourdieu and Wacquant 1992:253, 259)

For Bourdieu, all social actors are (always) self-interested insofar as they act. However, Bourdieu goes beyond rational actor theory, economism, and sociobiology because he takes great pains to show that self-interest is not a presociological given (he shares the critique of individualism with Louis Dumont and the classic tradition of French sociology). Self-interest is defined by the complex structure of overlapping sociological fields to which the actors must be blind in order to act. Bourdieu is absolutely unequivocal that social actors while acting in terms of their sociologically structured self-interest can never know what that self-interest is precisely because they must believe in the illusion that they are pursuing something genuinely meaningful in order to act. Only the sociologist is capable of understanding what is really and truly going on, and the reason for his success is the sacrifice he makes of his own social interests on the altar of truth. The sociologist's capacity to perform this miraculous act is based on method, rooted in his claim to be able to occupy a position of exteriority to the social field and the interests at play within it. That miracle is made into a mundane practice through a set of ascetic techniques.

For Bourdieu and Wacquant (1992:185) this practice of asceticism is shared by all scientists:

Indeed, I hold that, all the scholastic discussions about the distinctiveness of the human sciences not withstanding, the human sciences are subject to the same rules that apply to all sciences. . . . I am struck, when I speak with my friends who are chemists, physicians, or neurobiologists, by the similarities between their practice and that of a sociologist. The typical day of a sociologist, with its experimental groping, statistical analysis, reading of scholarly papers, and discussion with colleagues, looks very much like that of an ordinary scientist to me.

There are a number of debatable points made in the quote (Are the experimental gropings really the same in sociology and biochemistry? Is all science holistic? Why hasn't Bourdieu science become paradigmatic?, etc.), but none more so than Bourdieu's central point that the sciences are unified. Or more accurately, that ultimately there are no sciences in the plural, there is only science in the singular. If there is only one true scientific practice, then it follows that there is only one scientific vocation and one scientific ethos.

Indifference

Where does this manly bracketing of social life, this methodically relentless self-purifying, this self-imposed obsessive tracking down of any and all residual illusio leave Bourdieu? He is, as always, unflinching in drawing conclusions: It leaves him in a state of utter indifference.

> To understand the notion of interest, it is necessary to see that it is opposed not only to that of disinterestedness or gratuitousness but also to that of *indifference*. To be indifferent is to be unmoved by the game: like Buridan's donkey, this game makes no difference to me. Indifference is an axiological state, an ethical state of nonpreference as well as a state of knowledge in which I am not capable of differentiating the stakes proposed. (Bourdieu and Wacquant 1991:115–17)

Bourdieu invokes the Stoic goal of ataraxy (*ataraxia:* not being troubled). Illusio is the very opposite of ataraxy: it is to be invested, taken in and by the game. To be interested is to accord a given social game that what happens in it matters, that its stakes are important and worth pursuing. One can and should—qua social scientist—achieve ataraxia, indifference, about anything and everything. Not to be taken in, not to care: these are the necessary conditions for scientific practice.

What are the consequences, the rewards, the meaning—the value—of Bourdieu's hard-won indifference? He closes his inaugural speech at the Collège de France with a peroration that brings his distinctive style to its apogee:

> The paradoxical enterprise which consists in using a position of authority in order to say with authority what it means to speak with authority, so as to give a lesson, but a lesson in liberty with respect to all other lessons, would be simply inconsequential, almost self-destructive, if the very ambition to construct a science of belief didn't pre-suppose a belief in science. Nothing is less cynical, less machiavellian, than those paradoxical proclamations who announce or denounce the principle of power which they exercise. There is no sociologist who would take the risk of destroying the thin veil of faith or of bad faith which makes the charm of all institutional pieties, if he didn't have the faith in the possibility and the necessity of universalizing the freedom in regard to institutions which sociology has procured for itself; if he did not believe in the liberating virtues of that which is the least illegitimate of symbolic powers, that of science, especially when that science takes the form of a science of symbolic powers capable of restoring to social subjects the

mastery of false transcendentals which *la méconnaissance* never stops creating and recreating. (Bourdieu 1982:56)

Bourdieu professes his belief in science and defends that belief by arguing for its potential universality. Sociology triumphant will restore a "mastery of false transcendentals" which society constantly reinstates, as it is its function to do. Were sociology ever to triumph, were we ever all to become sociologists—at that moment, what would happen to society?

For Bourdieu, illusio and faith are the preconditions of science. Society could not exist without an ontologically rooted epistemological blindness. Given his premises, he is thoroughly consistent in arriving at an understanding in which humans are universally blind to the deep meaning of their own acts, a stance Alan Pred (personal communication) calls epistemologically rooted ontological blindness. From his position of achieved indifference, Bourdieu does not despise humans but expresses instead a kind of unmotivated solidarity with their plight. His position is unmotivated in the sense that there is no scientific reason to sympathize; one could just as consistently adopt any one of a series of attitudes, for example Claude Lévi-Strauss's *View From Afar* (1985), which expresses a kind of resigned contempt, a slightly tempered retake of the utter scorn he had asserted forty years earlier for anyone who would be interested in a "shop-girl's web of subjectivity."

We see this ethical, emotional, and eventual political impasse, perhaps a sign of a surviving shred of Bourdieu's own illusio, in the two-page caution "to the reader" with which Bourdieu (1993) opens *La Misère du Monde*. The first-person accounts of suffering, hurt, and the degradations of ordinary life in France contained in the volume, Bourdieu writes, have "been organized in order to obtain from the reader his accord to give them a gaze (*regard*) as comprehensive as that which the exigencies of the scientific method imposes on them." It is not sufficient, the reader is reminded, to quote Spinoza's precept "Don't deplore, don't laugh, don't detest, but understand" if one does not ensure the means to respect it. A rigorous sociological science provides those means.

Nonetheless, how could one not feel, Bourdieu observes, a certain "uneasiness" about publishing these private observations (*propos*) that had been given in confidence, even though the purpose of collecting them was never hidden from those who gave them? Not surprisingly, this uneasiness can only be addressed, if not entirely quelled, through taking every precaution that the observations not be misunderstood, that there be no "*détournements de sens.*" Surprisingly, Bourdieu (1993:8) apologizes for the possibility that he risks sounding clinical—though he never explains precisely what is wrong with sounding clinical, since "whether one likes it or not" these people are, and can only be, "objects" for the sociologist. He is, in point of fact, being perfectly clinical, and given his diagnosis, appropriately so.

Bourdieu's pages *are* unquestionably tinged with sympathy and solidarity for the people he analyzes. But Bourdieu's personal sentiments, sincere and noble though they may be, have no scientific role to play in his system. They remain private, personal, accidental—ultimately unreflective and unreflected. His sentimental education has been to deny the legitimacy of all sentiments;

rather, he has learned the lesson of illusio and indifference, of a certain ascetic scientific practice, one he teaches and practices brilliantly. But the brilliance of his analysis illuminates his own practice (as well as those of the social actors he sees as shrouded in illusio). For that reason his analyses are also informed by the (self-imposed ascetic) distance of what Hans Blumenberg (1983) has called the "missionary and didactic pathos" characteristic of a certain tradition of French thought. Against the grain of his own system, Bourdieu does sympathize, does find the pervasive reproduction of social inequalities that he documents so scrupulously both fascinating and intolerable. He *does* respect his subjects—that is clearly why he focuses book after book on these themes. However, he "knows" better and therefore must engage in the constant battle to overcome these sentiments, so as to become, like Buridan's ass, indifferent. Hence, his (unrecognized) pathos. An ethos of logos without (an accounted-for) pathos yields virtuosity, not virtue.

Site II. *Philia* or Friendship

I had been bored and irritated at La Villette. Why? The causes are multiple: poor socialization (I don't go to enough of these events), temperament (restless and exigent), the knowledge that there are other ways of doing things. Commenting on a draft of this paper, a friend pointed out that in the same passage that I mentioned, Barthes says that the boredom that used to dominate his existence had gradually decreased thanks "to work and to friends."

I propose that a primary site of thinking is friendship (*philia*). Such a formulation sounds strange, as friendship is a relationship and not a physical place. Today we have lecture rooms and conference rooms and meeting rooms and classrooms and offices and studies; the Romans had rooms for friends. In the ancient world, friendship figured in different ways as an essential component of the good life (*eudaimonia*), a prime site of human flourishing. Conceptions and practices of philia took many different forms, from the honor-based culture described in the *Iliad* through the classic Greek tragedies and philosophy. The *locus classicus* of this tradition is Aristotle's *Nicomachean Ethics*. In Book VIII Aristotle says,

> Friendship is a virtue . . . most necessary with a view to living. For without friends no one would choose to live, though he had all other goods; even rich men and those in possession of office and of dominating power are thought to need friends most of all. . . . For with friends men are most able both to think and to act.

Aristotle distinguished between three kinds of friendships: those based on utility, pleasure, and the good. The first is no doubt the most common, both then and now. Utility/philia relationships are those in which self-advantage figures prominently for both parties. Consequently, Aristotle observed, when the motives or occasions of mutual profit are loosened, such friendships are readily dissolved. For Aristotle there are a great number of different types of such relationships, and, good observer that he was, he details them at length.

Today the daily life of the sciences is saturated with personal ties that serve diverse functions. It is sometimes forgotten that mutual advantage needs to be identified and negotiated as much as anything else. Hence professional self-interest, the accumulation of symbolic capital and the like, is more than a question of predefined roles or "objective interests" or even dispositions, although those dimensions constitute the preconditions for utility/philia. The utility/philia dimension remains largely unexplored and is thematized more in soap operas and talk shows than in scholarly books. Observation suggests it is a contemporary site of ethical reflection where goods are balanced.

Friendships based on pleasure (of commensality or bodily acts, from erotic to athletic) are probably less cultivated among scientists, although certain pleasures are included in professional circles. Scholarly and scientific conventions play a significant role in the American economy; the face-to-face encounters they foster, while hard to justify in quantifiable economic terms, clearly continue to be valued.

For Aristotle, the highest form of friendship is the philosopher's philia. "Perfect friendship is the friendship of men who are good and alike in virtue, for those wish well alike to each other qua good, and they are good in themselves." The best friendships require time and a long familiarity to develop and solidify. "A wish for friendship," Aristotle writes, "may arise quickly, but friendship does not" (Aristotle 1941). Friendship is mutual, social, and quasi-public. It is ecstatic in that its practice draws one out and toward a friend. Philia primes the bond, the among and the between. In contrast, as Marilyn Strathern (1992) has pointed out, in modern Western societies it is rare to conceive of anything other than individuals and contractual relationships.

How to situate the philia-site?

> Imagine for a moment a marginalized intellectual, who has been engaged in philosophy and the intellectual life since his teens, and is almost entirely lacking in political power or influence, sitting down to determine what the best kind of life and the best kind of polis are. Imagine that and you have to a large degree imagined Aristotle in his ethics and politics class in the Lyceum. For Aristotle entered Plato's Academy at the age of 17, staying there for twenty years. . . . Moreover, he was a metic—a resident alien—in Athens and was barred from playing an active role in Athenian political life. (Reeve 1992:195)

Aristotle did, however, find the space to do a good deal of participant observation before leaving Athens rather urgently in order, or so the legend goes, to prevent the Athenians from committing a second crime against philosophy.

First the Stoics, and then the Christians, displaced friendship from the center of the good life. For the Stoics, friendship was one of the local and contingent relationships from whose ties they sought release in order to establish a connection to the cosmos, the universal. The place of friendship in Christianity is a complex one; although considered a minor good, it was theorized as a danger to the primary attachment of the faithful (often monks) to God. Friendships became inherently triangular. Though there are occasional references to friendship in more contemporary philosophic writings (e.g., Montaigne), it basically became a marginal practice in the sense of losing its philosophic centrality. Today friendship is largely seen as therapeutic. Against that trend, I view philia as an ethical and epistemological *practice*.

Type II. The Attentive Amateur—Cosmopolitan Curiosity

In his last works Michel Foucault produced a provocative framework for analyzing ethics. Ethics, for Foucault (1983:238), is "the kind of relationship you
ought to have with yourself, *rapport à soi*, . . . which determines how the individual is supposed to constitute himself as a moral subject of his own actions."
(Foucault identifies four distinct aspects in ethical self-constitution: ethical substance, modes of subjectivation, ethical work (askesis, in the sense of training or
exercise), or telos. This scheme was developed by Foucault to further refine his
interest in sexuality in relation to the "care of the self"—how the subject ordered
his sexuality in a general economy of the philosophic life. My interest here is scientific practice and the care of the self, the elements at play in the ethical elaboration of science as a vocation. I use Foucault's distinctions, with this major
reorientation of content, as an experimental means to explore what is in play
and what is at stake.

(1) Ethical Substance
What is the aspect of myself or my behavior which is concerned with moral
conduct?
—Michel Foucault

For Bourdieu, the ethical substance is the will. The will is that aspect that must
be mastered for science to be possible, and it is only through science that such
mastery is possible. In contrast, I take the ethical substance, the prime material
of moral conduct, to be reflective curiosity. Here I follow a long tradition stem-
ming from Aristotle—whose *Metaphysics* opens with the famous proposition
"All men by nature desire to know"—on through Hans Blumenberg (1983:347),
who provides an analysis of what he calls "The Trial of Theoretical Curiosity"
(including an erudite commentary on the history of interpretations and criticisms of Aristotle's opening line).[2] Blumenberg, like Aristotle, sees curiosity as a
natural disposition that is shared with animals but that takes a distinctive
reflective or mediated form in humans. That historically shaped form is the ethical substance.

It is precisely its open-endedness that makes reflected curiosity an ethical
substance. As Aristotle (1941:Book II, 339) argued, virtues (and the dispositions
that make them possible) are neither passions nor faculties, "for we are neither
called good nor bad, nor praised nor blamed, for the simple capacity of feeling
the passions; again we have the faculties by nature, but we are not made good
or bad by nature." Virtues then are a state of character which "makes a man
good and which makes him do his own work well." Ethical substance, understood in this way, is not a potential that could be actualized—that would be too
naturalistic—but rather a disposition to be shaped.[3] This intellectual disposition, if formed by the right practices and cultivated in the right institutions, can
become a virtue. Equally, given other contexts and other dispositions, it has the
possibility to take a degraded form—either a deficient state in which it withers
away or an excessive manner in which it seeks complete autonomy. It is precisely that malleability that makes the dispositions the material of intellectual
and ethical virtues. Seen in this light, it is possible to imagine a history of science and philosophy as a long series of experiments to determine the extent of

the dispositions' malleability, their elaboration, and their enduring possibility of corruption.[4]

One advantage of choosing reflected curiosity as the ethical substance is that it is at least partially shared by both the person studying science and the scientists themselves. Hence the inquiry begins on the basis of a shared disposition. This commonality does not mean that what molecular biologists and anthropologists, for example, do with this disposition, the kind of work they perform on it, the forms they give it, the norms and institutions within which they practice, are identical; clearly they aren't. Nonetheless a tacit sharing of curiosity makes possible, even encourages, an exchange, a situation that fosters mutual reflection on each other's practices. It follows that the choice of reflective curiosity as the ethical substance is itself an ethical one.

(2) Mode of Subjectivation

The way in which the individual establishes his relation to the rule and ac knowledges oneself to be a member of the group that accepts it, declares adherence to it out loud, and silently preserves it as a custom. But one can practice it too as an heir to a spiritual tradition that one has the responsibility of maintaining or reviving.
—Michel Foucault

Bourdieu's scientist freely takes up a kind of "this-wordly *mysticism*," in Max Weber's phrase. The "inner worldly mystic," Weber (1946:326) writes, "is in the world and accommodates to its orders, but only to gain a certainty of grace in opposition to the world by resisting the temptation to take the ways of the world seriously." The social scientific "this-worldly mystic" is a man of abiding conviction. He believes society has *a* meaning; to seize it he has to stay close to human actors (as the habitus is always specific) yet release from society's hold through scientific askesis to seize that meaning.

One could link experience and experiment, ethics and epistemology, in a more exploratory manner. John Dewey (1916:73–74) provided an excellent characterization of such a stance:

> An intellectual integrity, an impartiality and detachment, which is maintained only in seclusion is unpleasantly reminiscent of other identifications of virtue with the innocence of ignorance. To place knowledge where it arises and operates in experience is to know that, as it arose because of the troubles of man, it is confirmed in reconstructing the conditions which occasioned those troubles. Genuine intellectual integrity is found in experimental knowing. Until this lesson is fully learned, it is not safe to dissociate knowledge from experiment nor experiment from experience.

And, of course, this lesson has not been learned.

In my recent ethnographic work with molecular biologists in France and the US, I explicitly addressed the question of who I was, my affiliations, and my commitments with the scientists I was engaging. At Roche Molecular Systems, a California biotechnology company, I explained that as a citizen I was concerned with and interested in ethical and political implications of the Human Genome Initiative; as an anthropologist I was attempting to evaluate claims about human behavior coming from genetically oriented physical anthropologists; as a professor I thought I ought to know more about how the lines between the academy and industry had changed the practice of science. Although presenting myself as an ethnographer, I never pretended to be entirely an observer. Consequently I constantly engaged these scientists in discussions and debates, which they welcomed and from which I learned what to ask. In France, during my work at the Centre d'Etude du Polymorphisme Humain (CEPH), I was invited in as a "philosophic observer" by Daniel Cohen, the institution's driving force. Cohen wanted to experiment with ways to introduce a "social" interlocutor into his lab without introducing ethics committees, which he saw as too preemptive.

This slot of floating inquirer, which involved watching and commenting on an institution engaged in thinking through and acting on such unresolved but pressing issues as the place of national science within a new international arena, which paradigms for multifactorial genetic diseases should be pursued, and the ethical and strategic role of patenting, suited me. The extraordinary outpouring of views, reflections, demonstrations, and debates to which I became privy arose to a large extent because I explicitly presented myself in a tradition based on both philosophic questioning and empirical inquiry. The "social field" of the scientists at both Roche Molecular Systems and CEPH was in a state of change, as was my own. Making that state of uncertainty and flux explicit acknowledged and encouraged reflection on the different practices we were both engaged in.

My mode of subjectivation aligned me with those who start with the native's point of view. It separated me from that tradition, however, insofar as the natives did not have a stable point of view but were themselves engaged in questioning their allegiances and their dispositions. Their culture was in the making. Furthermore, it was partially my culture. Their self-questioning over how to shape their scientific practice, and the limits of their ability to do so, partially overlapped with my own scientific practice. I attempted to foreground both the overlap and the difference—and, over time, to make this emerging situation both a topic of curiosity and a mode of inquiry.

(3) Ethical Work
The work that one performs on oneself, not only in order to bring one's conduct into compliance with a given rule, but to attempt to transform oneself into the ethical subject of one's behavior.
—Michel Foucault

For Bourdieu, in order to become the ethical subject of one's behavior, one must overcome illusio. The work that one must perform to do so is participant objectivation. The participation is at the level of the self, a constant self-vigilance, a methodical, relentless hunting down of (social) life-enabling mystifications. The constant danger is to be naive, to be taken in. The self achieves its ethical status through two steps: the objectivation of society as a totality and the cognitive relationship it affects to that (constructed) objectivation. By so doing, one is able to "bring one's conduct into compliance with a given rule" and "transform oneself into the ethical subject of one's behavior."

The contrastive challenge is to overcome ressentiment. The work that one performs is participant observation. But not all participant observation will do. First, the participation is at the level of a relationship between subjects that are not identical but not radically different either. Second, the observation, at least in part, takes as its object the process of construction of that intersubjective relationship. At no point, therefore, is the curious participant observer either totally external to or totally identified with the field of study. Ressentiment requires, demands, fabricates, and defends clearly drawn boundaries between subjects and objects in order to operate. Hence a liminal placement provides a preliminary defense. The danger is in losing the balance, tipping too far toward the subjective or objective side. The ethical task is finding the mean. As Aristotle argued, there are no rules for achieving this mean; it is a question of experience and practical activity, the work of keeping track of the just proportion of things.

Today the critique of ressentiment is conspicuously present in some branches of feminism. Some feminist philosophers identify two dangers they find prevalent, often most acutely among other feminists. First, adopting a politics of ressentiment fixes subject positions and boundaries between subjects. Wendy Brown (1995:27) writes, "Ressentiment fixes the identities of the injured and the injuring as social positions and codifies as well the meanings of their actions against all possibilities of indeterminacy, ambiguity, and struggle for resignification or repositioning." This fixing is detrimental to forming alliances and impoverishes the political imagination, blocking other modes of subjectivation. Second, rigid boundary maintenance and the cursed couple of demonization/cupidization lead to the inability to affirm. Marion Tapper (1993) develops this line of argument, showing how a politics of "victimage" yields a politics of moral superiority. This politics produces, among other things, the inability to admire and respect, to the extent it is their due, those one is seeking to understand or to change. Following Nietzsche, Tapper argues that this demonization and its intimately associated contempt apply—reflexively and doggedly—to the knower as well as the known.

Although the identification of these critical points of ressentiment's entry into politics is pertinent, they require a twist to fit the ethical-epistemological practice. It is in the carefully chosen fieldwork site that one acquires and tests a sense of the ethical and intellectual limits of what is actual and what is possible, of virtue and corruption, of domination and growth of capacities. Every situation is historically and culturally overdetermined. Part of the work of fieldwork is to identify the particularities and generalities of the situation, of the contingent and less contingent, and to be concerned with both sides of these pairs. The ethical work is concerned less with being vigilant and more with an attentiveness, a reserved and reflected curiosity about what form of life is being made. It is through fieldwork, through experiential experimentation, that one establishes "partial connections," reflects on them, gives them an appropriate form.

Such normatively oriented fieldwork discovered arenas in which the scientists themselves expressed irritation and frustration about aspects of scientific work they considered violations of their practice; some instances embodied tensions, others corruptions, others betrayals. The scientists certainly assumed they were engaged in "normed" activity and were quite appreciative of being systematically questioned about it. In some instances they lacked a vocabulary to identify what troubled them; in others they were quite eloquent. In both the US and France it became clear to me that I was not expected to play an expert role of analyzing scientists sociologically, a therapeutic role of helping them work through problems, or a denunciatory role of identifying malevolent forces and actors. Rather, mine was a problematizing role in which observer status allowed me a certain overview of the situation, including its fluidity. Such a mode of subjectivation requires the ethical work of being attentive to one's own ressentiments, of claiming neither mastery nor ignorance, of publicly balancing identification and distance.

(4) Telos

That activity in which one finds the self. An action is not only moral in itself, in its singularity; it is also moral in its circumstantial integration and by virtue of the place it occupies in a pattern of conduct.
—Michel Foucault

Bourdieu's sociological subject attains the universal by achieving separation from all particularity and action, a self freed from illusio. The aim is mastery. The goal of the "dominated members of the dominated class" (Bourdieu's term for intellectuals) is discursive mastery. The failed form of discursive mastery is, as Nietzsche saw, "spiritual revenge." The successful form is the state of higher indifference toward the world, the overcoming of suffering understood as passion, as a way of life.

As long as one accepts the equation of human science with natural science, as Bourdieu does, then the search for technical mastery of society and self appears plausible. It appears plausible, and not in need of a defense, as long as one ignores the work of the social studies of science and technology and feminism in the last two decades on the status of "nature," "society," and "self." Even then, those who accept the equation must confront the Weberian line of argument most recently put forward by the philosopher Gyorgy Markus (1987), who states that there is no hermeneutics of the natural science because there is no need for one. The natural sciences have succeeded in evacuating meaning from their productions. Markus's conclusion is directly in line with Weber's claim that the logic of rationalization had in principle disenchanted the world but had provided no answers to questions of general cultural significance.

Growing technical mastery and specialization in the natural sciences has yielded both control and a progressive narrowing of meaning. In a strict sense, there is no self-questioning within molecular biology. From time to time there are debates about the ends to which results could be put, political projects that might be dangerous or beneficial; there are occasional discussions about the composition (gender, race, class) of the social body of scientists. But the normative parameters of the textual and nondiscursive practices of sciences like molecular biology are not in question, however fluid they may be in daily life. The relation of molecular biologists toward their objects of study may well be open to criticism, but not on the grounds that they are characterized by ressentiment. Most, perhaps all, of the practicing molecular biologists I have worked with accepted these limitations, some quite consciously—that was one of the reasons they were curious about my work. When natural scientists stop doing science and start producing "world views," ressentiment and illusio run rampant.

As historians such as Steve Shapin have shown, there is a history of the creation, stabilization, and maintenance of the figure of the natural scientist, but that history shows a socially successful displacement of self-examination into "experiments" and "nature." Many dream of replicating this historical feat in the human sciences, but they have never been successful in achieving and stabilizing the institutional and cultural conditions required to authorize and enforce such a consensus. It appears that it would only be through political means that such a consolidation could take place. Then mastery would be complete.

In my formulation, the telos of a reflectively curious practice of human science is a form of *Bildung*, a kind of individual and collective self-formation or *Lebensführung* (life-regulation), a type of care and cultivation. Such a Bildung turns on being attentive to the plurality and dignity of beings (humans and others) as well as to the limits of that pluralism and dignity. It is not a classical Bildung insofar as that tradition took the textual and cultural canon to be already fully identified and the arts of interpretation already codified, or sought that stability as an end. It is worth noting that how stable or hegemonic the identification and codification of the canon is—or was—is an open question, a topic of debate, a site of interpretation or struggle. Nor does the proposed Bildung take the form of a revolutionary high-modernism; it rejects the practice that proceeds as if it were either possible or desirable to start de novo as well as the one that seeks to totally remake the self and/or society.

The contrastive type might be called a cosmopolitan amateur. Although "amateur" is a somewhat clumsy term, it points to a practice that does not take mastery as its goal. The amateur stands back from the virtuosity of the expert; its excellence lies in the mean. Finally, the telos of the second type is not only amateur but cosmopolitan. A cosmopolitan ethos entails a perspective on knowledge, ethics, and politics that is simultaneously local and global, native and foreign. At one level this claim is quite commonsensical and follows directly from the voluminous literature on globalization (market, media, bureaucracy, arms trade, development, etc.). However, the "state of the world" can always be taken up, problematized in diverse ways.

Molecular biology, for example, has taken up the current conjuncture through an increased use of electronic means of communication, data storage, and internationally coordinated projects like the human (and other organisms) genome mapping projects. The circulation and coordination of knowledge has never been more rapid or more international. Articulating and sustaining these goals is extremely expensive. Heads of major laboratories may well spend the majority of their time raising money, making contacts, and forging alliances. The appearance in the last two decades of "start-up" biotechnology companies funded by venture capital and stock offerings, first in the United States and increasingly in Asia, India, and Europe, has reshaped both the financing of research and (probably) its directions. Capital is international. While the principle of the international status of science has been in place for a long time, the form that it is taking in the biomedical sciences today is quite distinct. What kind of scientific life is it that is constantly traveling, negotiating over resources, engaged in competitive claims of priority, and competing in multiple arenas? Is this the end of an era of a scientific type? Or is it an acceleration, a hyper-modernity, which is pushing the type to its limits, opening new possibilities of beneficial and appropriate control of things?

The questions one might well pose to the molecular biologist today have a refracted resonance for those of us involved in a different practice. Will it be possible to cultivate yet more complicated participation in larger and more elaborate projects, enabling the development of capacities we barely knew about previously, while minimizing our imprecations with new systems of domination? Are we thrusting toward some threshold of scale where hybrid networks of things and people, micro- and macro-knowledges and powers, make the very idea of practice archaic? Although I doubt the answer to the last question is "yes," the only way to find out is to experiment with the circumstantial integration of our current practices into new patterns of conduct. It is a question of finding the means.

Notes

Special thanks go to Joao Biehl and James Faubion. Helpful readings came from John Zu, Nadine Tonio, Mike Panisitti, Natasha Schull, Andrew Lakoff, Adryana Petryna, and Rebecca Lemov. Teaching, when one has such students, is a pleasure, a utility, a good, and a privilege. Hubert Dreyfus, Robert Bellah, and Roger Friedland provided helpful commentary as well.

1. On this point, see Dreyfus and Rabinow (1993). A parallel (and thoroughly elaborated) overall line of argument about Bourdieu's work is made by Luc Boltanski (1991), a former Bourdieu student.

2. For the thirteenth-century nominalists, "The first sentence of Aristotle's *Metaphysics* could have been modified so as to say that man has to strive for knowledge, not indeed 'by nature,' but as the being who is exposed to this uncertain world, whose ground plan is hidden from him" (Blumenberg 1983:347).

3. On the distinction of potentiality and disposition, see Kosman (1980:111).

4. The relation of the generality of the virtues and the particular ways in which they are given specific forms is treated at length in Nussbaum (1993). On the malleability and its limits, see Foucault (1984).

MICE INTO WORMHOLES / A Comment on the Nature of
No Nature / **Donna J. Haraway**

Mathematical Beginnings—A Proportion
Transuranic Elements:Transgenic Organisms ::
The Cold War:The New World Order

Kinship

TRAINED IN MOLECULAR AND DEVELOPMENTAL BIOLOGY, I identify professionally as a historian of science. I have applied for a visa for an extended stay in the permeable territories of anthropology. But my real home is the ferociously material and imaginary zones of technoscience, into which I and many millions of people on this planet have been interpellated, whether we like it or not.

The *Oxford English Dictionary* says that "to interpellate" means to break in on, to interrupt a person in speaking or acting. The term also means to appeal or petition, to hail, or to intercept, cut off, or prevent. Obsolete in English before 1700, "interpellation" was reimported back into anglophone practice from the French in the twentieth century in the context of a special kind of interrupting or hailing: calling on a minister in a legislative chamber to explain the policies of the ruling government.

Interpellation, then, has several tones that resonate among French and English speakers. These tones sound here in my warping of the French philosopher Louis Althusser's theory of how ideology constitutes its subjects out of concrete individuals by "hailing" them. According to Althusser (1971:171, 174), interpellation occurs when a subject, constituted in the very act, recognizes or misrecognizes itself in the address of a discourse. Althusser used the example of the policeman calling out, "Hey, you!" If I turn my head, I am a *subject in* that discourse of law and order; and so I am *subject to* a powerful formation. *How* I misrecognize myself speaks volumes about both the unequal positioning of subjects in discourse and about different worlds that might have a chance to exist.

With a double meaning typical of most interesting words, interpellation is also an interruption in the body politic that insists that those in power justify their practices, if they can. It is also best not to forget that "they" might be "we." Whoever and wherever we are in the domains of technoscience, our practices should not be deaf to troubling interruptions. Interpellation is double-edged in its potent capacity to hail subjects into existence. Subjects in a discourse can and do refigure its terms, contents, and reach. In the end, it is those who misrecognize themselves in discourse who thereby acquire the power, and responsibility, to shape it. Finally, technoscience is more, less, and other than what Althusser meant by ideology; technoscience is a form of life, a practice, a culture, a generative matrix. Shaping technoscience is a high-stakes game.

The nonhyphenated energy of technoscience makes me adopt the term.[1] This condensed signifier mimes the implosion of science and technology into each other in the last couple of hundred years around the world. I want to use technoscience to designate dense nodes of human and nonhuman actors that are brought into alliance by the material, social, and semiotic technologies through which what will count as nature and as matters of fact gets constituted for—and by—many millions of people. All the actors in technoscience are not scientists and engineers, and scientists and engineers are an unruly lot. They are not pawns in a morality play about modern damnation or apocalyptic salvation put on for the benefit of scientifically illiterate critical theorists or euphoric, jacked-in apologists for technohype. Perhaps most important, technoscience should not be narrated or engaged only from the points of view of those called scientists and engineers. Technoscience is heterogeneous cultural practice that enlists its members in all of the ordinary and astonishing ways that anthropologists are now accustomed to describing in other domains of collective life.[2]

Technoscience also designates a condensation in space and time, a speeding up and concentrating of effects in the webs of knowledge and power. In what gets politely called modernity and its afterlife (or half-life), accelerated production of natural knowledge pervasively structures commerce, industry, healing, community, war, sex, literacy, entertainment, and worship. The world-building alliances of humans and nonhumans in technoscience shape subjects and objects, subjectivity and objectivity, action and passion, inside and outside in ways that enfeeble other modes of speaking about science and technology. In short, technoscience is about worldly, materialized, signifying, and significant power. That power is more, less, and other than reduction, commodification, resourcing, determinism, or any of the other scolding words that much critical theory would force on the practitioners of science studies.

I belong to the "culture" whose members answer to the "Hey, you!" issuing from technoscience's authoritative practices and discourses. My people answer that "Hey, you!" in many ways: we squirm, organize, revel, decry, preach, teach, deny, equivocate, analyze, resist, collaborate, contribute, denounce, expand, placate, withhold. The only thing my people cannot do in response to the meanings and practices that claim us body and soul is remain neutral. We must cast our lot with some ways of life on this planet and not others. We cannot pretend we live on some other planet where the cyborg was never spat out of the womb-brain of its war-besotted parents.

The cyborg is a cybernetic organism, a fusion of the organic and the technical forged in particular, historical, cultural practices. Cyborgs are not about

The Machine and The Human, as if such things and subjects universally existed. Instead, cyborgs are about specific historical machines and people in inter-action that often turns out to be painfully counterintuitive for the analyst of technoscience. Manfred Clynes and Nathan Kline (1960) coined the term "cyborg" for the enhanced man who could survive in extraterrestrial envi-ronments. A designer of physiological instrumentation and electronic data-processing systems, Clynes was the chief research scientist in the Dynamic Simulation Laboratory at Rockland State Hospital in New York. Kline, director of research at Rockland State, was a clinical psychiatrist. They imagined the cyborgian man-machine hybrid would be needed in the next great technohu-manist challenge—space flight—and presented their paper at the Psychophys-iological Aspects of Space Flight Symposium sponsored by the US Air Force School of Aviation Medicine in San Antonio, Texas.

Enraptured with cybernetics, Clynes and Kline (1960:27) thought of cyborgs as "self-regulating man-machine systems." One of their first cyborgs was a stan-dard white laboratory rat implanted with an osmotic pump designed to inject chemicals continuously.[3] Exchanging knowing glances with their primate kin, rodents will reappear in this essay at every turn. Beginning with the rats who stowed away on the masted ships of Europe's age of exploration, rodents have gone first into the unexplored regions in the great travel narratives of Western technoscience.

Consequently my people are kin to field mice who have entered the anom-aly in an evolutionary space, or wormhole, called the laboratory. Like the science-fictional wormhole in the *Star Trek* "Deep Space Nine" episodes, the labo-ratory continues to suck us into uncharted regions of technical, cultural, and po-litical space. Passing through the wormhole of technoscience, the field mice emerge as the finely tailored laboratory rodents—model systems, animate tools, research material, self-acting organic-technical hybrids—through whose eyes I write this essay. Those mutated murine eyes give me my ethnographic point of view.

Cyborg anthropology attempts to refigure provocatively the border relations among specific humans, other organisms, and machines. The interface between specifically located people, other organisms, and machines turns out to be an excellent field site for ethnographic inquiry into what counts as self-acting and as collective empowerment. I call that field site the cultural practice and practi-cal culture of technoscience. The optical tube of technoscience transports my startled gaze from its familiar, knowing, human orbs into the less-certain eye sockets of an artifactual rodent, a primal cyborg figure for the dramas of techno-science. I want to use the beady little eyes of a laboratory mouse to stare back at my fellow mammals, my hominid kin, as they incubate themselves and their human and nonhuman offspring in a technoscientific culture medium.

The relocated gaze forces me to pay attention to kinship. Who are my kin in this odd world of promising monsters, vampires, surrogates, living tools, and aliens? How are natural kinds identified in the realms of late-twentieth-century technoscience? What kinds of crosses and offspring count as legitimate and il-legitimate? Who are my familiars, my siblings, and what kind of livable world are we trying to build?

Sibling Logic

Let me pursue technoscience's blasted family pedigrees by means of the epigraph for "Mice into Wormholes," namely, my mathematical joke about transgression in the form of a statement of proportion:

> Transuranic Elements:Transgenic Organisms ::
> The Cold War:The New World Order

I translate this proportion to read:

> The transuranic elements (such as plutonium produced by nuclear reactors) are to transgenic organisms (such as the genetically engineered mice and tomatoes produced in biotechnological laboratories) as the Cold War (fueled by its core generator of nuclear culture) is to the New World Order (driven by its dynamic generator of transnational enterprise culture).

In *Secrets of Life, Secrets of Death*, Evelyn Keller (1992) explored the scientific and psychoanalytic connections between the midcentury search for the "secret" of the atom that resulted in nuclear physics and weapons and the search for the "secret" of life that issued in molecular genetics and genetic engineering. Plumbing all those "secrets" is a major narrative of erotic transgression in technoscience. Walking through the museum of the Los Alamos National Laboratories in New Mexico on a field trip during the SAR seminar that initiated the present volume of essays, I was arrested by the exhibit about the first atomic bombs built at Los Alamos during the Manhattan Project. The display was mouse-nibbled and timeworn; it looked like old news. The more glitzy projects in recent years in and around Los Alamos have been informatics development for GenBank© as part of the Human Genome Project at the National Labs and the artificial life research associated with the nearby Santa Fe Institute. In the national science policy of the New World Order, nuclear weapons research—albeit still quite a going concern—is almost, but not quite, an embarrassment even at the birth place of the atomic bomb.[4] National security discourse in the 1990s turns on creating a chain reaction between technoscience and enterprise. The National Laboratories are supposed to become breeder reactors for competitiveness, whose decay products are at least as world threatening as those of plutonium^{239}.[5]

What interests me about the proportion that links plutonium with genetically engineered organisms and situates them in their historical chronotopes, World War II and the Cold War of the 1940s through the 1980s and the New World Order of the early 1980s to the present, is the question of taxonomy, category, and the natural status of artifactual entities—of kinship, in short. Kinship is a technology for producing the material and semiotic effect of natural relationship, of shared kind.

In 1869 the Russian chemist Dimitri Ivanovich Mendeleyev published his work on the periodic law and the periodic table of the elements that ordered the sixty-three elements then known by properties that seemed to repeat as a function of atomic weights. Later, chemists argued that the table is ordered by atomic number, or the number of protons in the nucleus, and not by atomic weights (neutrons plus protons). Then Niels Bohr's early-twentieth-century atomic model interpreted the recurring properties of the elements as a function of quantum numbers—the number of electrons in the "outer shell" of an atom.

The important issue here is that in all of its interpretations the periodic table predicted several unknown elements that were subsequently discovered, or made, to occur and whose properties fit prognostications nicely. Setting up relationships diagonally, vertically, horizontally, and transitionally, the table stood for traditional family values in the culture of chemistry. The periodic table of the elements still hangs in every chemistry lecture hall I have ever seen. More than merely an authoritative historical artifact that graphically displays the power of science to order fundamental properties of matter, the periodic table continues to generate knowledge in the experimental way of life. It is a potent taxonomic device for what my people understand as nature. The kinship relations of the elements are a natural-technical object of knowledge that semiotically and instrumentally puts terrans, or earthlings, in their proper place.

Uranium is the naturally occurring earthly element with the highest atomic number, 92. Uranium is where the evolution of the elements that make up the solar system stopped. In that sense, uranium represents a kind of "natural limit" to the family of terran elements as well. But every child who has bitten into an apple in the Atomic Cafe knows that elements with higher atomic numbers than uranium have existed on earth since 1940, when Glenn Seaborg and his associates made the first transuranium elements, including plutonium, whose atomic number is 94. In order to make explosive Pu^{239}, the first breeder reactor was built by Enrico Fermi and others on a squash court at the University of Chicago in 1942 in the context of the Manhattan Project. Pu^{239} fueled the device that was tested at Alamogordo, New Mexico, on 16 July 1945 and the bomb called "Fat Man" that exploded over Nagasaki on 9 August 1945.[6] U^{235} fueled the bomb dropped on Hiroshima. When I wrote the first draft of this essay in 1993, bomb-grade Pu^{239} was refueling threats of war on the Korean peninsula, as North Korea refused inspection of its nuclear power reactor refueling process. International regulatory mechanisms are not containing the rogue element's production and use in the post–Cold War era.

Two things stand out simultaneously in the presence of the transuranic elements. First, they are ordinary, natural offspring of the experimental way of life, whose place on the periodic table was ready for them. They fit right in. Second, they are earthshaking, artificial productions of technoscience, whose status as aliens on earth, and indeed in the entire solar system, has changed who we are fundamentally and permanently. Nothing changed, and too much changed, when plutonium joined the terran family. The transuranic elements—embedded in the semiotic, technical, political, economic, and social apparatus that produces and sustains them on earth—are among the chief instruments that have remade the third planet from the sun into a global system. The transuranic elements have forced humans to recognize their problematic kinship with each other as fragile earthlings at a scale of shared vulnerability and mortality barely suspected on that squash court in Chicago but explicitly ritualized at Alamogordo when J. Robert Oppenheimer quoted from the Bhagavad Gita, "I am become Death, the shatterer of worlds" (Kevles 1977:333). Now a nuclear fuel disseminated worldwide and one of the deadliest toxic substances ever encountered, plutonium has done more to construct species being for hominids than all the humanist philosophers and evolutionary physical anthropologists put together. And as the dog-eared exhibit at Los Alamos brought home to me, this is old news.

The shiny news in the 1990s, as every Business Monday section of the important newspapers attests, is transgenic organisms produced in another kind of breeder reactor, the biotechnological laboratory, in transnational enterprise culture. In the mid-eighteenth century the Swedish naturalist Linnaeus constructed a hierarchy of taxonomic categories above the level of the species (genus, family, class, order, kingdom) and introduced the binary system of nomenclature that gives all living terrans a genus and a species name. Species, whether regarded as conforming to an archetype or descending from a common stock, were taken to be natural taxonomic entities whose purity was protected by a natural envelope. In 1859, in *The Origin of Species*, Charles Darwin provided both an evolutionary narrative and a plausible mechanism that unified diverse bases for classification and accounted both for the transformation and relative constancy of species. In the mid-twentieth century the neo-Darwinian synthesis powerfully imported population genetics into evolutionary thinking. In that potent account, genetic change is evolutionary change. Evolutionary theory and genetics unified life on earth, as the periodic table placed earth's elements into stable families. Humans are interpellated into both of these species-defining kin networks.

The techniques of genetic engineering developed since the early 1970s are like the reactors and particle accelerators of nuclear physics: their products are "trans." They themselves cross a culturally salient line between nature and artifice, and they greatly increase the density of all kinds of other traffic on the bridge between what counts as nature and culture for my people. Transported, terran chemical and biological kinship gets realigned to include the extraterrestrial and the alien. Like the transuranic elements, transgenic creatures, which carry genes from "unrelated" organisms, simultaneously fit into well-established taxonomic and evolutionary discourses and blast widely understood senses of natural limit. What was distant and unrelated becomes intimate. By the 1990s, Genes R Us, and we seem to include some curious new family members at every layer of the onion of biological, personal, national, and transnational life. What could be more natural by now than worldwide commercial, familial, biotechnical, and cinematic genetic traffic?

Transgenic organisms are at once completely ordinary and the stuff of science fiction. I use them metonymically to mark world-shaping changes in biology since the 1970s. Thus transgenic organisms are indicator species, or perhaps canaries in the gold mines of the New World Order. In 1993 the first issue of a new journal, *Transgene*, noted that more than 2,500 titles in the current MEDLINE database used the word "transgenic" in the title, up from ten to twenty papers per year in the early 1980s (Cruse and Lewis 1993). More than 60 percent of all of the biological and biomedical research funded federally in the United States by the mid-1990s used the techniques of molecular biology/molecular genetics. Two conclusions from that statistic are obvious: (1) molecular biology has major creative importance in practically every area of biology and medi-

Table 11.1
Percent Share of World Technoscientific Literature

Nation or Region	All Scientific and Technical Fields	Biomedical Articles	Biological Articles
United States	35.1	38.9	37.6
United Kingdom	7.5	7.6	6.9
Germany	6.8	6.3	5.4
France	4.8	5.1	3.3
Italy	2.8	2.3	1.4
Rest of Western Europe	10.7	13.3	11.0
Japan	8.5	7.9	7.5
Near East and Africa	1.6	0.9	3.1
Israel	0.9	0.8	1.1
India	2.0	1.4	2.1
Central and South America	1.4	1.5	2.3
Australia and New Zealand	2.5	2.2	6.1
Former Soviet Union	6.7	6.9	2.2
Other Eastern and Central Europe	2.1	2.0	1.2
East Asian Newly Industrialized Countries	1.1	0.6	0.7

Source: Adapted from NSB 1993:423–25.

cine, and (2) fundable questions in the life sciences have conformed drastically to those compatible with the practice of biology as molecular biotechnics. The organism has been materially as well as semiotically retooled in the New World Order.

The implications of US, western European, and Japanese hegemony in this process are global. Based on articles published in the worldwide scientific literature in 1991, table 11.1 gives a minimal comparative picture of scientific power. Without invoking any notions of conspiracy, I think the conclusion that the technoscientific agenda for everybody is set by the economically dominant powers, especially the United States, is inescapable. It is also inescapable that sizable resources go into technoscience in every area of the planet and that, dominant or not, many actors are on the stage. The story is not closed.

"Developing" nations and the major world financial and political powers alike perceive that the stakes in biotechnology in general and genetic research in particular are high (Juma 1989; Shiva 1993). Egypt, for example, is building the Mubarak City for Scientific Research, modeled after the European Molecular Biology organization (ScienceScope 1994a). Significantly, the first of eight planned institutes to be constructed is the Institute of Genetic Engineering and Biotechnology. The government has budgeted 100 million Egyptian pounds ($36 million), compared to the less than $1 million per year it spends on academic scientific research. (That $1 million does not include foreign grants, the main source of research money in Egypt, another index of who sets the worldwide scientific agenda.) The scramble for the control of genes—the sources and engines of biological diversity in the regime of technobiopower—drives venture capitalists, crafters of international treaties, makers of national science policies, bench scientists, and political activists alike. The control of genes means both access to naturally occurring diversity and access to the material, social, and semiotic technology to recraft its riches to produce beings new to earth. Which new beings, for whom, and out of whom, are pressing questions at the heart of democracy, social justice, the economy, agriculture, medicine, labor, and the environment.

For example, focusing on US agricultural science and biodiversity politics, Glenn Bugos (1992, 1994) explores in exquisite detail and analytical rigor the historical periodization and the dynamic division of labor that characterize the interplay among changing industrial structure, intellectual property conventions, and the methods and results of technoscience research in the movement from natural genetic diversity to finished commodity in the food and pharmaceutical domains of capital accumulation. Narrating how germ plasm becomes database, where the question of who owns biodiversity gets worked out in material detail, Bugos's story puts biotechnology, especially genetic engineering, into rich historical perspective.

As the apparatus for the production and sustenance of high-atomic-weight fissionable materials interpellated diverse peoples into a kind of global species on the "whole earth" or "spaceship earth," so also the semiotic, technical, and social systems for conceiving and propagating transgenic organisms interpellate diverse peoples into a transnational enterprise culture that I call the New World Order, Inc. In this timescape, species being is technically and literally brought into being by transnational, multibillion-dollar, interdisciplinary, long-term projects to provide genetic catalogs as maps to industrial, therapeutic, conservationist, military, ethical, and even cosmetic action.

Furthermore, the "trans"-action is not limited to splicing among the genomes of organisms. Marked with the stigmata of a dream, a symptom, and an ordinary research project, in a kind of ultimate genetic transspecific cross, scientific efforts to splice carbon-based life forms to silicon-based computer systems take many shapes. A college biology textbook, for example, opens its chapter on the nervous system with a photomicrograph of a nerve cell growing on the surface of a Motorola 68000 microprocessor chip (Campbell 1993:982). That particular "trans" join, producing a classical cyborg in the dimensions of microns, is unadulterated pedagogical ideology. More technically functional in its approach, merging silicon-patterning techniques borrowed from microelectronics with combinatorial biochemistry, Affymetrix, a biotech start-up company in Palo Alto, California, is developing a chip that anchors arrays of nucleotide sequences. The chips will be tools for detecting aberrant genetic sequences in large-scale automated diagnostic tests, a major investment area for current biotechnology (Alpers 1994).

Two related considerations emerge for me from this idiosyncratic meditation on a mathematical proportion. One concerns the problem of purity of type and the thematics of the mixed and the alien in US culture; the other touches on the issue of how to represent technoscience.

216

Pure Life

A transgenic organism contains genes transplanted from one strain, species, or even taxonomic kingdom to another—for example, from fish to tomatoes, fire flies to tobacco, or bacteria to humans, or vice versa. Transgenic border-crossing signifies serious challenges to the "sanctity of life" for many members of Western cultures, which have been historically obsessed with racial purity, categories authorized by nature, and the well-defined self. The distinction between nature and culture in the West has been a sacred one; it lies at the heart of the great narratives of salvation history and their genetic transmutation into sagas of secular progress.

In hoary terms, what seems to be at stake is the story of the human place in nature, that is, genesis and its endless repetitions. It is a mistake in this context to forget that anxiety over the pollution of lineages lies at the heart of racist discourse in European cultures as well as of linked gender and sexual anxiety. The discourses of transgression get all mixed up in the body of nature. Transgressive border-crossing pollutes lineages. In the case of a transgenic organism, it pollutes the lineage of nature itself, transforming it into its binary opposite, culture. The line between the acts, agents, and products of divine creation and human engineering has given way in the sacred-secular border zones of molecular genetics and biotechnology. The revolutionary continuities between natural kinds instaurated by the theory of biological evolution seem flaccid compared to the rigorous couplings across kingdoms (and nations) produced daily in the genetic laboratory.

Opponents of the production, and especially the patenting, of transgenic organisms appeal to notions such as the natural telos, or self-defining purpose, of all life forms.[7] From this perspective, to mix and match genes as if organisms were legitimate raw material for redesign is to violate natural integrity at its vital core. Transferring genes between species violates natural barriers, compromising species integrity. Other objections to biotechnological practice in the New World Order, Inc., include increased capital concentration and the monopolization of the means of life, reproduction, and labor; appropriation of the commons of biological inheritance as the private preserve of corporations; the global deepening of inequality by region, nation, race, gender, and class; inadequately assessed and potentially dire environmental and health consequences; misplaced priorities for technoscientific investment funds; intensified cruelty to and domination over animals; depletion of biodiversity; and undermining established practices of human and nonhuman life, culture, and production without engaging those most affected in democratic decision making. I take all of those objections very seriously, although I do not think simply naming the concerns either decides the direction of effects or describes the cross-cultural polyphony through which scientific practice is constituted worldwide. Effects and practices are multilayered and context-specific, and it is too easy for all parties to fall into dogma when fundamental cultural and material values are both not shared and at stake.

For the moment, however, I want to focus only on the theme of purity of type, natural purposes, and transgression of sacred boundaries. The history of racial and immigration discourses in Europe and the United States ought to set off acute anxiety in the presence of these supposedly high ethical and ontological themes. I cannot help but hear in the biotechnology debates the unintended tones of fear of the alien and suspicion of the mixed. In the appeal to intrinsic natures, I detect a mystification of kind and purity akin to the doctrines of white racial hegemony and US national integrity and purpose that so permeate American culture and history. The appeal to other organisms' inviolable, intrinsic natures aims to limit turning all the world into a resource for human appropriation. But it is a problematic argument resting on unconvincing biology. History is erased, for other organisms as well as for humans, in the doctrine of types and intrinsic purposes, and a kind of timeless stasis in nature is piously narrated. The ancient, cobbled-together, mixed-up history of living beings, whose long tradition of genetic exchange will be the envy of industry for a long time to come, gets short shrift.

More fundamentally, in a nation where race is everywhere reproduced and enforced, everywhere unspeakable and euphemized, and everywhere deferred and treated obliquely—as in talk of drug wars, urban underclasses, diversity, illegal aliens, wilderness preservation, terrorist viruses, immune defenses against invaders, and crack babies—I cannot hear discussion of disharmonious crosses among organic beings and of implanted alien genes without hearing a racially inflected, xenophobic symphony. Located in the belly of the monster, I find the discourses of natural harmony, the nonalien, and purity unsalvageable for understanding our genealogy in the New World Order, Inc. Like it or not, I was born natural kin to Pu[239] and to transgenic, transspecific, and transported creatures of all kinds; that is the family for which and to whom my people are accountable.

Perhaps it is perverse for me to hear the dangers of racism in *opposition* to genetic engineering and especially transgenics at just the moment when national and international coalitions of indigenous, consumer, feminist, environmental, and nongovernmental development organizations have formed to oppose "patenting, commercialization and expropriation of human, animal and plant genetic materials."[8] Although the moral, scientific, and economic issues are far from simple, I oppose patenting of animals, human genes, and much plant genetic material. Genes for profit are not equal to science itself, or to economic health. Genetic sciences and politics are at the heart of critical struggles for equality, democracy, and sustainable life. The global commodification of genetic resources is a political and scientific emergency, and indigenous peoples are among the key actors in biopolitics, just as they have had to be in nuclear culture. But the tendency of the "left"—my area of the political spectrum—to collapse molecular genetics, biotechnology, profit, and exploitation into one undifferentiated mass is at least as much of a mistake as the mirror-image reduction by the "right" of biological—or informational—complexity to the gene and its avatars, including the dollar.

Tunneling into my collective racial anxieties in the midst of thinking about Calgene's Flavr Savr tomatoes with a long shelf life and fissionable heavy elements with distressing half-lives points to a wormhole into the poorly charted and contested semiotic practices for representing technoscience. Resisting the separation of science and technology, the word itself makes clear that category fusions are in play. There is one other category separation, in particular, that seems ill-fitted for useful work in representing technoscience: that between science and politics, science and society, or science and culture. At the least, one such category cannot be used to explain the other, and neither can be reduced to the status of context for the other. But the taxonomic trouble goes deeper than that. The bifurcated categories themselves are reifications of multifaceted, heterogeneous, interdigitating practices and their relatively stable sedimentations, all of which get assigned to separate domains for mainly ideological reasons. Fortified with this belief, I want to insist on four matters in my own efforts, which are perhaps less committed to *represent* technoscience (as if such a copying practice were possible) than to *articulate* clusters of processes, subjects, objects, meanings, and commitments.

Representing Technoscience

First, I call attention to the figures and stories that run riot throughout the domains of technoscience. Not only is no language, including mathematics, ever free of troping, not only is facticity always saturated with metaphoricity, but also any sustained account of the world is dense with storytelling. "Reality" is not compromised by the pervasiveness of narrative; one gives up nothing, except the illusion of epistemological transcendence, by attending closely to stories. I am consumed with interest in the stories that inhabit us and that we inhabit. Such inhabiting is finally what constitutes this "we."

Second, I am convinced that technoscience engages promiscuously in materialized refiguration; that is, it traffics heavily in the passages that link stories, desires, and material worlds. Materialized refiguration is a solid process, not some textual dalliance. An Operon Technologies, Inc., advertisement in *Science* magazine from 9 April 1993 makes the point. The ad's text announces, "At $2.80 per base, Operon's DNA makes anything possible." The manifest content is that this company, "the world's leading supplier of synthetic DNA," will cheaply manufacture specific nucleic acid sequences custom-tailored for your lab. The latent content is that this product promises marvelous transformations. The point of technical virtuosity and infinite possibility is orthographically emphasized by the use of three different font styles—as well as the bold, underline, caps, italics, and shadow features—to highlight elements in a mere nine-word sentence. Like a genie from *Arabian Nights*, Operon will grant you your wishes; anything is possible.

Synthetic DNA bears those kinds of promises. If DNA signifies "life itself"[9] in the semiotic orders of biotechnology, synthetic DNA is especially open to realizing the future and to realizing profit from your investment in that future. The company promises "speed, purity, and savings," all technical matters of great moment for the bench scientist. The center of the full-page, color ad is filled by three genetically engineered mutants, each of which is at once ordinary and fantastic. The "applorange" is a spliced apple and orange; the "zucchana" is a spliced zucchini and banana; and best of all, and most "real" of all, the $2.80 is spliced to the DNA sequence provided by Operon Technologies, Inc.

Here the double helix, sign of life itself, is spliced perfectly to the words "one dollar" under George Washington's portrait—a seamless join between the textual systems of nucleotide base pairing and English denominations. The manifest content of the splicing of the dollar and the DNA helix is to highlight the specific savings from using a particular supplier of a commodity needed for your research. The latent content is the graphic literalism that biology—life itself—is a capital accumulation strategy in the simultaneously marvelous and ordinary domains of the New World Order, Inc. In the processes of materialized refiguration of the kinship between different orders of life, the generative splicing of synthetic DNA and money produces promising transgenic fruit. Specifically, natural kind becomes brand or trademark, a sign protecting intellectual property claims in business transactions; we will meet this corporeal refiguration again in the score for the technoscience fugue.

Third, with many others in technoscience studies, I believe that science is practical culture and cultural practice.[10] The laboratory is a special place, not for epistemological reasons that still comfort a few nostalgic philosophers and dyspeptic mathematicians, but because it is an arrangement and concentration of human and nonhuman actors, action, and results that change entities, meanings, and lives on a global scale. And the laboratory is not the only site for shaping technoscience.[11] Far from depleting scientific materiality, worldliness, and authority in establishing knowledge, the "cultural" claim is about technoscience's presence, reality, dynamism, contingency, and thickness.

The British biotechnology firm Quadrant, at least, seems unworried by a picture of science as practice and culture. Its *Science* advertisement from 1993, "Molecular Biology made simpler," is a cartoon depiction of multiracial laboratory workers, both male and female, old and young, who are cutting, sawing, gluing, sweeping up after themselves, measuring, weighing, inspecting, and otherwise manipulating macromolecules. One laggard scientist lying in a crook of his molecule appears to be smoking a joint, and a business-suited salesman (or is it the lab chief?) with a briefcase is scurrying out the door marked "Genetic Research." The lab is patently a place for the collective craft work of knowledge making, where Quadrant's restriction enzymes for cutting up nucleic acids might be welcome tools. It is an ordinary picture of specifically located practice and culture, except for one detail: the molecules are so macro that they are giant indeed. The scientists have stepped through Alice's looking glass and have become so small that they are dwarves in a gigantic world of helical objects.

The tiny people and the giant molecules inhabit this consummately ordinary scene of daily work. Again we see the simultaneously mundane and fantastic truth of technoscience, where a change of scale refigures fundamental relationships (Latour 1983). A final touch of magic completes the picture: nowhere to be seen among the pulleys, saws, and magnifying glasses are the chief tools that are the functional equivalent of the air pump in every molecular biology laboratory at the end of the twentieth century, namely, the gaggle of computerized instruments without which all the workers in this lab might as well take their DNA to the beach.[12]

Yet I think it is not the thickness, fantasy, or ordinariness but the contestability of science as practice and culture that galls the guardians of orthodoxy. I suspect that some scientists and philosophers are dismayed by the insistence that science is practice and culture because that account makes ample room for a motley crew of interlopers to take part in shaping and unshaping what will count as scientific knowledge, for whom, and at what cost.[13] In the "culture and practice" account, maintaining boundaries can no longer be rendered invisible but is hardly proscribed. Far from it. Boundary maintenance, as well as splicing and breaking, requires work, including but not limited to the semiotic and rhetorical work of convincing people who are both like and different from oneself; such labor is cultural practice in action.

The lines between the inside and the outside of science, or between the goodness or badness of specific technoscientific accounts, remain important. The lines simply no longer appear to be prethought in the minds of the gods or drawn once and for all by heroes in mythic times like the scientific revolution. The gods might still think in numbers and draw in geometries, but if so they are in for the same kind of rude culture and practice analysis as that meted out to dabblers in slimy biological brews or professional watchers of furry mammals. As Xerox Palo Alto Research Center computer scientist and philosopher Brian Smith (1994, in discussion following his paper) put it in relation to the ongoing work it takes to establish and maintain the identity of a microprocessor such as Intel's Pentium chip, "You have to stop being what you were when you start paying attention to the work it takes to maintain your clear distinctions." Establishing identities is kinship work in action. And, as Quadrant knows too, playfulness and pleasure are part of the practice and culture of technoscientific boundary making, erasing, and testing. The labor and play link humans and nonhumans—technological, chemical, and organic—in a vastly underdetermined drama.

So in the practice and culture account, the worlds of science and technology have many more movers and shakers, and what counts as too many or the wrong kind of participants and interlocutors has to be established through multifaceted engagement where the sites of action, power, interpretation, and authority are at stake. The fantastic and the ordinary commingle promiscuously. Boundary lines and rosters of actors—human and nonhuman—remain permanently contingent, full of history, open to change. The relations of democracy and knowledge are up for materialized refiguring at every level of technoscience, not just after all the serious epistemological action is over. I believe that last statement is a fact; I know it is my hope and commitment.

Fourth and last in my score for orchestrating the action in technoscience is the dubiously mixed physical and biological metaphor of the force of implosion and the tangle of sticky threads in transuranic and transgenic worlds. The point is simple: The technical, textual, organic, historical, formal, mythic, economic, and political dimensions of entities, actions, and worlds implode in the gravity well of technoscience—or perhaps of any world massive enough to bend our attention, warp our certainties, and sustain our lives. Potent categories collapse into each other. Analytically and provisionally we may want to move what counts as the political to the background and to foreground elements called technical, formal, or quantitative, or to highlight the textual and semiotic, while muting the economic or mythic. But foreground and background are relational and rhetorical matters, not binary dualisms or ontological categories. The messy political does not go away because we think we are cleanly in the zone of the technical, or vice versa. Stories and facts do not naturally keep a respectable distance; indeed, they cohabit the same, very material places.

Determining what constitutes each dimension takes boundary-making and maintenance work. In addition, many empirical studies of technoscience have disabled the notion that the word "technical" designates a clean and orderly practical or epistemological space. Nothing so productive could be so simple.

Any interesting being in technoscience, like a textbook, molecule, equation, mouse, pipette, bomb, fungus, technician, agitator, or scientist can—and often should—be teased open to show the sticky economic, technical, political, organic, historical, mythic, and textual threads that make up its tissues. "Implosion" does not imply that technoscience is "socially constructed," as if the "social" were ontologically real and separate; "implosion" *is* a claim for heterogeneous and continual construction through historically located practice, where the actors are not all human. While some of the turns of the sticky threads in these tissues are helical, others twist less predictably. Which thread is which remains mutable, a question of analytical choice and foregrounding operations. The threads are alive; they transform into each other; they move away from our categorical gaze. The relations among the technical, mythic, economic, political, formal, textual, historical, and organic are not causal. But the articulations are consequential; they matter. Implosion of dimensions implies loss of clear and distinct identities but not loss of mass and energy. Maybe to describe what gets sucked into the gravity well of a massive unknown universe we have to risk getting close enough to be warped by the lines of force. Or maybe we already live inside the well, where lines of force have become the sticky threads of our bodies.

I think that is where I live, beyond warping and committed to mucking about in the biological, and so I want to introduce my sibling figure who has been covertly informing the fugue of this essay from the start: OncoMouse™.[14] Our exchange of glances structures my point of view. We have been commercially, biologically, textually, and politically interpellated into the same public and private family networks. By August 1973 DNA from *Xenopus laevis*, the South African clawed frog that had inhabited embryology laboratories for many decades, was being transcribed into messenger RNA in a bacterium, *Escherichia coli*. Promising that soon genes from one creature could be made to function in the bodies of vastly different organisms, these experiments were the direct ancestor to those that gave terran existence to OncoMouse™, whose public debut as rodent intellectual property and transgenic breast cancer model came in 1988 (Morrow et al. 1974).

Sisterly Conversations

"Available to researchers only from Du Pont, where better things for better living come to life."

OncoMouse™ is my sibling, and more properly, male or female, s/he is my sister. Her essence is to be a mammal, by definition a bearer of mammary glands, and a site for the operation of a transplanted, human, tumor-producing gene— an oncogene—that reliably produces breast cancer. Although her promise is decidedly secular, s/he is a figure in the sense developed within Christian realism: s/he is our scapegoat; s/he bears our suffering; s/he signifies and enacts our

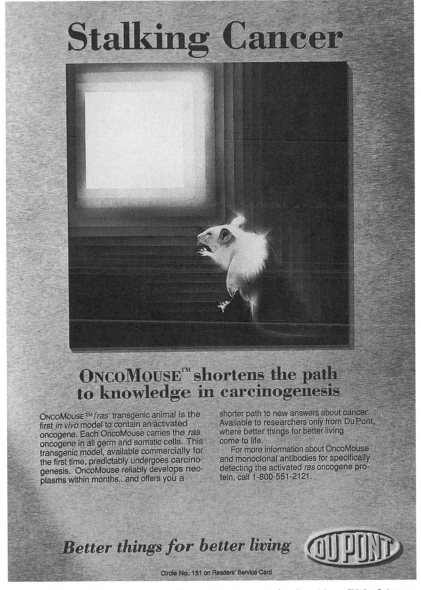

Figure 11.1. Stalking Cancer. Du Pont advertisement for OncoMouse™ in Science, 27 April 1990. Reprinted with the permission of Du Pont NEN Products. On 19 May 1995 Du Pont announced its intent to divest its Medical Products businesses. The former Du Pont NEN Products business will become NEN Life Science Products.

223

mortality in a powerful, historically specific way that promises a culturally privileged kind of salvation—a "cure for cancer." Whether or not I agree to her existence and use, s/he suffers, physically, repeatedly, and profoundly, that I and my sisters might live. In the experimental way of life, s/he is the experiment. S/he also suffers that we, that is, those interpellated into this ubiquitous story, might inhabit the multibillion-dollar quest narrative of the search for the "cure for cancer."

If not in my own body, surely in those of my friends, I will someday owe to OncoMouse™ or her subsequently designed rodent kin a large debt. So who is s/he? Gestated in the imploded matrices of the New World Order, OncoMouse™ is many things simultaneously. S/he is an animal model system for a disease, breast cancer, experienced by more than 10 percent of women in the United States sometime in their lives.[15] Self-moving in Aristotle's defining sense, s/he is a living animal and so fit for the transnational discourses of rights emerging from green social movements, in which the consequences of the significant traffic between the materialized, ethnospecific categories of nature and culture are as evident as they are in patent offices and laboratories. OncoMouse™ is an ordinary commodity in the exchange circuits of transnational capital. A kind of machine tool for manufacturing other knowledge-building instruments in technoscience, the useful little rodent with the talent for mammary cancer is a scientific instrument for sale like many other laboratory devices.

Above all, OncoMouse™ is the first patented animal in the world.[16] By definition, then, in the practices of materialized refiguration, s/he *is* an invention. Her natural habitat, her scene of bodily/genetic evolution, is the technoscientific laboratory and the regulatory institutions of a powerful nation-state. Crafted through the ordinary practices that make metaphor into material fact, her status as an invention who/that remains a living animal is what makes her a vampire, subsisting in the realms of the undead.

Vampires are narrative figures with specific category-crossing work to do. The essence of vampires, who, like Victor Frankenstein's monster, normally do their definitive labor on wedding nights, is the pollution of natural kinds. The existence of vampires tropes the purity of lineage, certainty of kind, boundary of community, order of sex, closure of race, inertness of objects, liveliness of subjects, and clarity of gender. Desire and fear are the appropriate reactions to vampires. Figures of violation as well as of possibility and of escape from the organic-sacred walls of European Christian community, vampires make categories travel. From the points of view crafted in their Christian narrative sources since at least the end of the eighteenth century, vampires are ambiguous—like capital, genes, viruses, transsexuals, Jews, gypsies, prostitutes, or anybody else who can figure corporate mixing in a rapidly changing culture that remains obsessed with purity (Gelder 1994; Geller 1992). No wonder queer theorists and novelists alike find vampires to be familiar kin (Case 1991; Gomez 1991). So do Du Pont's advertising copy writers. Whether s/he proves to be otherwise productive or not, OncoMouse™ has already done major semiotic work.

Buying and selling, breeding and selecting, experimenting on, and contesting the treatment of lab animals are not new activities, but the controversies surrounding the patenting and marketing of "the Harvard mouse" were densely covered in the popular and scientific press in Europe and the United States. The heightened sense of controversy around OncoMouse™ is the fruit of the New World Order's floridly regenerated narratives of original transgression in the Garden of the Genome, even if the universal singular (*the* genome) polluted here belongs to a genetically compromised mouse, or rather to the licensee of the patent-holder. Inventions do not have property in the self; alive and self-moving or not, they cannot be legal persons, as corporations are. On 12 April 1988 the United States Patents and Trademarks Office (PTO) issued a patent to two genetics researchers, Philip Leder of Harvard Medical School and Timothy Stewart of San Francisco, who assigned it to the President and Trustees of Harvard College. In an arrangement that became a trademark of the symbiosis between industry and academia in biotechnology since the late 1970s, Harvard licensed the patent for commercial development to E. I. Du Pont de Nemours & Co. With an unrestricted grant to Philip Leder for the study of genetics and cancer, the Delaware-based Du Pont had been a major sponsor of the research in the first place.

Du Pont was interested in transgenic mice, or, more broadly, in lines of animals genetically predisposed to cancer, in three main ways: as research projects in their own right, as test systems for toxicology, and as vehicles for crafting cancer therapies. Du Pont issued research licenses to academic and other nonprofit investigators without fee to use its patented process to produce transgenic animals in exchange for keeping Du Pont informed of scientific developments.

Du Pont also made arrangements with Charles River Laboratories in Wilmington, Massachusetts, to market OncoMouse™. In its 1994 *Price List*, Charles Rivers listed five versions of these mice carrying different oncogenes, three resulting in mammary cancers. Oncomice can get many kinds of cancer, but breast cancer has been semiotically most potent in news stories and in the original patent. Costs ranged from $50 to $75 per animal, an amount that could not recoup the original investment even if sales were brisk, which they have not been for many reasons. In its view, Du Pont's pricing was conservative because its long-range goals were for effective cancer therapies toward which the corporation hoped transgenics would be a step, but only if researchers could afford to use them.[17]

Altered in their germ line, the offspring of transgenic mice bear the transplanted genes in all their cells. Continued testing is necessary to make sure the new genes are not lost or mutated. Testing transgenic creatures to ensure their identity as a technoscience product is similar in principle to the testing that a microprocessor like Intel's Pentium or Motorola's 68000 must undergo. Charles River provides a host of services critical to sustaining the identity and utility of its mice: colony maintenance and development, genetic analysis by polymerase chain reaction, sample collection, cryopreservation and storage, rederivation, and customized projects.

The mice at Charles River, and in laboratories everywhere, are also sentient beings who have all the biological equipment, from neuronal organization to hormones, that suggests rodent feelings and mousy cognition, which, in scientific narratives, are kin to our own hominid versions. I do not think that fact makes using the mice as research organisms morally impossible, but I believe we must take responsibility for using living beings in these ways and not talk, write, and act as if OncoMouse™, or other kinds of laboratory animals, were simply test systems, tools, means to brainier mammals' ends, and commodities. Like other family members in Western biocultural taxonomic systems, these sister mammals are both us and not-us; that is why we employ them. Exceeding the economic traffic, there is an extensive semiotic-corporeal commerce between us. Because patent status reconfigures an organism as a human invention, produced by mixing labor and nature as those categories are understood in Western law and philosophy, patenting an organism is a large semiotic and practical step toward blocking nonproprietary and nontechnical meaning from social sites like labs, courts, and popular venues. Technoscience as cultural practice and practical culture, however, requires attention to *all* the meanings, identities, materialities, and accountabilities of the subjects and objects in play. That is what kinship is all about in my "ethnographic" fugue.

In its 27 April 1990 advertisement for OncoMouse™ in *Science* magazine (see fig. 11.1) , Du Pont featured its artifactual rodent under the title for a series of ads called "Stalking Cancer." The series played on the fundamental, if numbingly conventional, biopolitical metaphor of war and the hunt. Diseases are targeted in an ever-escalating arms race with infectious alien invaders and treasonous selves. OncoMouse™ is a weapon in a specific long-term campaign— the war on cancer declared by Richard Nixon in 1972.[18] Propelled by federal money through the National Institutes of Health and later by substantial corporate investment, this material-semiotic conflict has lavishly underwritten the last quarter century's exploits in molecular biotechnology. In that sense transgenics are as much a war baby as plutonium. From conception to fruition, both these millennial offspring required massive public spending, insulation from market forces, and innovations in the practices of major corporations.

OncoMouse™ is a technological product whose natural habitat and evolutionary future are fully contained in that world-building space called the laboratory. Denizen of the wonderful realms of the undead, this little murine smart bomb is also a cultural actor. A tool-weapon for "stalking cancer," the bioengineered mouse is simultaneously a metaphor, a technology, and a beast living its life as best it can. This is the normal state of the entities in technoscience cultures, including ourselves. In science, as Nancy Stepan (1986) pointed out for nineteenth-century studies of sex and race, a metaphor may become a research program. I would only add that a research program is virtually always also a very mobile metaphor.

In the advertising image, as if s/he were in a diorama in a natural history museum, a radiant white laboratory mouse seems to be glancing back to lock her gaze with that of the reader of the ad. Intent on the goal ahead, the mouse is climbing steps leading to a square of blinding light above her. S/he might be inside a camera, moving toward the open shutter. S/he is our surrogate on a quest journey, but s/he is also in the dark passages of a birth canal before s/he emerges into the light of pure forms. An enlightenment figure who belongs in the genre of scientific revolution narratives, OncoMouse™ could also be a character in "Hystera," Luce Irigaray's (1985) feminist psychoanalytic and philosophical commentary on Plato's allegory of the cave. Irigaray rereads Plato's myth to figure the womb passage for the treasured Western masculine fantasy of the second birth, of children of the mind rather than children of the body, or, here, of legitimate corporate issue rather than unauthorized natural offspring.

Marx had a great deal to say about such rebirths into the realm of pure capital, as well.

The ad multiplies the stigmata of the kinds of property that this significant white mouse grounds, naturalizes, and normalizes in her origin story. The ad itself is copyrighted by the corporate person and, therefore, author, Du Pont. Indeed, Du Pont is credited with inventing the form of the modern corporation, and, no stranger to the laws of literal kinship, the giant company was run by Du Ponts for well over a hundred years.[19] The mouse itself is patented and licensed. And OncoMouse™, the name under which the animal is marketed, is trademarked under the Federal Trademark Act of 1946, as amended in 1988: "A trademark is a distinctive mark, motto, device, or emblem that a manufacturer stamps, prints, or otherwise affixes to goods so that they may be vouched for" (OTA 1989:44). Such marks brand one form of intellectual property important in technoscience generally and biotechnology specifically.

Du Pont's mutated famous slogan—OncoMouse™ is "available to researchers only from Du Pont, where better things for better living come to life"— signals precisely a recent metamorphosis of the industrial chemical giant. In a complex pattern of diversifications, acquisitions, and investments, Du Pont, like other large chemical and oil companies, began about 1980 to commit sizable resources to biotechnological research in both pharmaceuticals and agriculture, including an $85-million, in-house, agricultural research lab that was one of the largest in the country (Wright 1986:352). Following its first entry into pharmaceuticals in 1964, in the last quarter of the twentieth century Du Pont began dealing seriously in the promisingly undead entities proper to the regime of biotechnopower, in a New World Order that depends on strategies of flexible accumulation. David Harvey (1989:147) elaborated the theory of flexible accumulation to describe the emergence of "new sectors of production, new ways of providing financial services, new markets, and above all, greatly intensified rates of commercial, technological, and organizational innovation." Biotechnology and genetic engineering make most sense in this framework.[20]

In 1991 Du Pont was the largest chemical producer in the United States and, with $40 billion in total sales, the country's seventh largest exporter. Pharmaceuticals and medical products represented one of six principal business segments of the huge corporation. Du Pont's total 1990 research budget for all categories was an impressive $1.4 billion, up from $475 million in 1980. In 1981 Du Pont acquired New England Nuclear (NEN), which brought the chemical company into medical radioisotopes and other biotechnology research products. Valued at about $1 billion in 1995 (about 2% of the total value of Du Pont), the medical products division was the unit that housed OncoMouse™. In 1991 Du Pont and Merck entered a joint venture to establish an independent drug company involved in, among other things, *in vivo* diagnostic agents. New Jersey-based Merck is the world's largest pharmaceutical company, with eighteen drugs that generated over $100 million each in sales in 1991. Besides a huge domestic market in the US, pharmaceuticals have continued to show a trade surplus of exports over imports since the 1980s, when the US became a net importer of high-technology products (NSB 1993:xxix).

"Drugs" are important to national policy in more ways than one. In 1990 Merck spent 11 percent of sales on research and development—$854 million, an amount equal to 5 percent of money spent on all global pharmaceutical research. Technoscience is not cheap. Besides its joint venture with the very established Du Pont, Merck is also paired up with one of the new breed of biotechnological firms, Repligen, to develop an AIDS vaccine (see Hoover, Campbell, and Spain 1991:221, 378 and Moskowitz, Katz, and Levering 1980: 229–32). OncoMouse™ has had powerful godparents in the extended company family.

Signifying Synthetics

I narrate the exploits of Du Pont and its mousy acquisition because they can *signify* and *incarnate*, perhaps more than explain, the world into which I have been interpellated. OncoMouse™ and its academic-corporate family are like civic sacraments: signs and referents all rolled into one fleshy mystery in a secularized salvation history of civilian and military wars, scientific knowledge, progress, democracy, and economic power.

In that frame of reference I risk telling an allegorical story of Du Pont as a history of the semiotic-material production of the key synthetic objects and processes that characterize the last century of the second Christian millennium: nylon, plutonium, and transgenics,[21] made possible, respectively, by synthetic organic chemistry, transuranic nuclear generation, and genetic engineering. A constantly self-reinventing Du Pont figures centrally in all three theaters of action.

Du Pont's roots were nourished in 1811 with the sale of blasting powder to Thomas Jefferson to clear the forest from Monticello and of the same substance to the US government in the War of 1812. Throughout the nineteenth century the company made the explosive nitrogenous powder that blasted the railroad tunnels and gold mines that undergirded the conquest of the continent by the United States. In the context of competitive crises and the invention of the corporate forms of monopoly capital, Du Pont reorganized in 1902–03; by 1906 it controlled 70 percent of the American explosives market. But with the founding of the Eastern Laboratory in New Jersey in 1902 and of the Experimental Station outside Wilmington soon after, the enterprise was already mutating from an explosives manufacturer to a diversified chemical company. Subject to various antitrust litigation as well as internal investment decisions, Du Pont energetically diversified and divested parts of itself throughout the twentieth century and became one of the first (after AT&T and General Electric), and one of the most powerful, innovators of industrial technoscientific research and development practice in the country.

Du Pont entered polymer technology before 1900 with its production of cellulose nitrate as smokeless gun powder. In the first decades of the twentieth century, the company made several important cellulose-based products, including celluloid and cellophane. Du Pont's research strategy changed fundamentally in 1926–27 when it invested $300,000 in a new research pattern that included $20,000 for "pure," rather than only "applied," chemical research in materials science. In the new laboratory called Purity Hall, condensation polymerization yielded a fiber that played an important role in World War II and changed the texture of the everyday world after the war—nylon, first commercialized in 1938. With the Manhattan Project and the subsequent reorganization of national science, the dominance of industrial funding of US science decisively ended, only to begin to be reasserted in the last years of the twentieth century. Throughout these transitions the elemental nitrogen in explosives, textile fibers, and DNA fibers has circulated many times over, turning a profit with each cycle.

Du Pont had its part to play in the Manhattan Project too, but a part in which plutonium, not nitrogen, was the key explosive element. The company did not want to get mired in the short-term profits and headaches of war production at the long-term cost of its highly advantageous new research products, and planned for its postwar reconstruction even before the United States had joined the conflict. Nonetheless, as requested, Du Pont took on an alternate track for the production of bomb-grade plutonium from the works at Oak Ridge, Tennessee. It built the Hanford Engineering Works in Washington, employed 40,000 people, carried off a major engineering and production feat, and had an unparalleled understanding of atomic power in all its scientific and managerial complexities by the end of the war. But Du Pont got out of nuclear production as soon as it could, wanting no part of the postwar atomic power industry with its inevitable national-security limitations and its permanent dependence on the government.

Ultimately becoming one of the most polluted places on the global nuclear map, the Hanford facility continued to produce plutonium for decades after the war. But having gleefully ceded the plutonium-making business at Hanford and atomic power generation generally to General Electric, that story was no longer Du Pont's problem. Du Pont would go nowhere where patents would not smooth the way; the company did not want markets dominated by the government, especially in an uncertain new industry. The science-based products emerging from organic chemistry provided Du Pont's steadier star.

At the end of the 1980s OncoMouse™, the third key synthetic being midwifed by Du Pont's changing research and investment policies, joined its nylon and plutonium older siblings. Like transuranics, however, transgenics had no permanent place in Du Pont's corporate family. On 19 May 1995 Du Pont announced its intention to divest its medical products businesses, which contained the transgenic mammals and their authorizing patent. The corporation reinvents itself again, but my narrative must return to the patent story and its context for more insight about the anatomy of citizenship in technoscience. Dissecting OncoMouse™ uncovers important aspects of the history of patenting practice in biology and sharpens the focus on the difficulty of achieving or preserving a multicultural, democratic, biotechnological commons.

Patent Acts

The Committee Reports accompanying the US Patent Act of 1952 made clear that Congress "intended patentable subject matter to include 'anything under the sun that is made by man'" (OTA 1989:5). The 1952 Act changed the original 1790 patent law language, substituting "process" for "art" in the broad intellectual property protection provided by the 1790 Act for "any new and useful art, machine, manufacture, or composition of matter, or any new and useful improvement [thereof]" (OTA 1989:4). The legal power to enclose nature, if only it were mixed with human labor, was broad indeed in the founding documents of the United States. In European-derived worlds, nature and labor (culture) have a hoary pedigree as salient categories, held together in relations of transformation and foundation. Even so the Patent and Trademarks Office did not always consider living organisms—which could be owned and manipulated in a myriad of legally recognized ways, not least in the system of human slavery—to be patentable under the law. Improvers of agriculture and husbandry were not authors and inventors until very recently.

In 1930 the Plant Patent Act changed that status for producers of nonsexually generating plants. The point was not the transcendental power of sex to guard its practitioners from being considered patentable material. Rather, adequate control of the patentable process was precluded by the inability of such seedy plants to reproduce true to type. When that technical difficulty was overcome, intellectual property protection, embodied in the Plant Variety Protection Act of 1970, was not far behind. Advances in biological specificity and control over reproduction have shaped the evolution of intellectual-property protection for plants in the US. In the absence of specificity and control over the germ plasm of plants, "private breeders were content to let their public counterparts bear the principal costs of plant innovation and to exploit the public product for market purposes. The greater the degree of specificity and control, the stronger the incentive for private breeders to invest in innovation, because they could define it and thus seek to protect and enforce their rights in it"(Bugos and Kevles 1992:103).

Control of sexual reproduction was hardly the stopping point in deciding just when to enclose the commons in germ plasm in this particular way. Food crops are perhaps the most lively area of transgenic research worldwide in the 1990s. In late 1991 federal agencies had applications for field-testing about twenty transgenic food crops (*The Gene Exchange* 1991:6). Techniques are being widely adopted for fine-tuning agriculture to the productive processes of transnational agribusiness and food processing. Herbicide-resistant crops are probably the largest area of plant genetic engineering. I am especially drawn by such engaging new beings as the tomato with a gene from a cold-sea-bottom-living flounder, which codes for a protein that slows freezing, and the potato with a gene from the giant silk moth, which increases disease resistance. DNA Plant Technology, Oakland, California, started testing the tomato-fish antifreeze combination in 1991, as described in a untitled news item in *Science* (253:33, 5 July 1991).

Mostly involving questions about safety and consumers' rights to know (e.g., through product labeling at the point of marketing), controversies surrounding these beings may be followed in *The Gene Exchange*, put out by the National Wildlife Federation. Safety (at least for consumers if not for workers, if the trouble the United Farm Workers have had in making anyone care about farm laborers' safety in pesticide use in the California grape fields is any evidence) and rights-to-know are established liberal discourses in the US. Of course such issues are strongly shaped by class and race formations. Whose safety and whose right to know, and to know what and when, have everything to do with whether it is easy or hard for regulators to hear various social actors. Going another giant step into the sacred spaces of the laboratory and the technoscience curriculum, putting the questions at the point of research design as well as at the point of recruitment and training of knowledge producers, rather than at the point of product testing and marketing, provokes the most amazing defensive reactions among the elites of technoscience.

The struggle is over who gets to count as a rational actor, as well as an author of knowledge, in the dramas and courts of technoscience. It is very hard to ask directly if new technologies and ways of doing science are instruments for increasing social equality and democratically distributed well-being. Those questions are made to seem ideological, while issues of safety and labeling can be cast as technical and so open to rational (objective, negotiated, adjudicated, liberal) resolution. The power to define what counts as technical or as political is at the heart of technoscience. To produce belief that the boundary between the technical and the political, and so between nature and society, is a *real* one, grounded in matters of fact, is a central function of narratives of the scientific revolution and progress. My goal is to help put the boundary between the technical and the political back into permanent question as part of the obligation of building situated knowledges inside the materialized narrative fields of technoscience.

In a more Puritan vein, my scopophilic curiosity about, and frank pleasure in, the recent doings of flounders and tomatoes must not distract attention from what is entailed by such new kinship relations in the conjoined realms of nature and culture. Large commercial stakes, with attendant national and international intellectual property issues, are involved. Hunger, well-being, and many kinds of self-determination—implicated in contending agricultural ways of life with very different gender, class, racial, and regional implications—are at stake (Hobbelink 1991). Like all technoscientific facts, laws, and objects, seeds only travel with their apparatus of production and sustenance. The apparatus includes genetic manipulations, biological theories, seed-genome testing practices, credit systems, cultivation requirements, labor practices, marketing characteristics, legal networks of ownership, and much else. These apparatuses can be contested and changed, but not easily. Seeds are brought into being by, and carry along with themselves wherever they go, specific ways of life as well as particular sorts of dispossession and death. Such points should be second nature to any citizen of the republic of technoscience, but they bear repeating. Genes R Us in ways that have nothing to do with the narrow meaning of genetic determinism and everything to do with entire worlds of practice. It's all in the family.

Here my story must leave the critical struggle for the germ plasm of seeds and turn back to the trajectory that made a white mouse into an invention. Up to 1980, although biotechnical *processes* such as alcohol or acetic acid fermentation and vaccine production were patented, the Patent and Trademarks Office ruled that microorganisms themselves, even if modified by the gene-splicing techniques developed in the 1970s, were still "products of nature" and so not patentable. But in 1980 the Supreme Court overruled the PTO in the case of *Diamond v. Chakrabarty* (see Krimsky 1991; OTA 1989; Wright 1986). The result was a patent for a genetically modified bacterium that breaks down petroleum. A living organism became a patentable "composition of matter." The court saw Chakrabarty's bacterium as a product of human ingenuity, of labor mixed with nature in that magical, constitutional way that legally turns the human being into nature's author or inventor, and not simply its inhabitant, owner, or steward.

Several other significant events in the US around 1980 marked the status of biotechnology in the transition from the economies and biologies of the Cold War era to the New World Order's secular theology of enhanced competitiveness and ineluctable market forces. Intensifying changes begun in the Carter administration, which in 1979 emphasized an economic-incentive-oriented approach to environmental regulation, the Reagan administration immediately began to dismantle statutory controls, including those affecting recombinant-DNA technology. While the National Institutes of Health dismantled mildly restrictive, safety-oriented controls on recombinant-DNA research, which never applied to industry in any case, the National Science Foundation initiated several grants programs for fostering university-industry cooperation in research and development.

In 1980 Congress passed the Patent and Trademarks Amendments Act, which granted title to nonprofit and small businesses whose research was federally funded, opening the way for universities to benefit commercially from tax-supported research performed on campus. Also in 1980 Stanford University and the University of California at San Francisco were awarded the Stanley Cohen–Herbert Boyer patent (applied for in 1974) on the basic technique of gene splicing, which has undergirded all genetic engineering. In 1980 Genentech—the California biotechnology firm founded in 1976 by Herbert Boyer, an academic geneticist, and Robert Swanson, a venture capitalist—made its initial public stock offering, an event that substantially raised general awareness of the commercial significance of genetic engineering.[22] In 1981 the Economic Recovery Tax Act gave economic incentives to cooperative arrangements between academia and industry, and in 1982 the Department of Commerce "began to promote the use of tax shelters for joint research and development ventures for investors and industry" (Wright 1986:338). In addition, new export markets for high-technology goods developed in the 1980s, and chemicals and pharmaceuticals were areas in which the US had a growing surplus in a generally dismal balance-of-trade picture.

Susan Wright's (1986) densely documented and incisively argued paper ties together the technical, economic, political, and social dimensions of the major transformation that has taken place in molecular biology since the 1970s. Wright named the period from 1979 to 1982 "the cloning gold rush," when large investments poured into genetic engineering directly from multinationals based in Europe and the United States as well as through the rapidly appearing small biotechnological enterprises. Although the biotech firms have received a great deal of the credit and blame for the rapid commercialization of molecular biology, Wright (1986:304) argues that they have been "highly dependent on universities for expertise and on multinational oil, chemical and pharmaceutical corporations for capital." The story of Du Pont, Harvard, and OncoMouse™ is a little piece of this story.

As rates of increase of federal support for basic science declined, direct industrial support of university biological research strengthened. In 1980 the federal government funded 68 percent of academic research and development in science as a whole; by 1993 the figure was 56 percent. In constant dollars, all academic research and development directly funded by industry between 1980 and 1993 grew 265 percent (NSB 1993:xviii). Although industry performs 68 percent of all US technoscientific R&D, universities still do 62 percent of what gets classified as basic research, much of it in biology. About 54 percent of all university R&D dollars go to the life sciences, which have been leaders in reorganizing the institutional form of scientific practice in the last fifteen years.

Industrial support of biology has taken many forms, including major commercially funded research institutes connected to the scientifically powerful campuses. In the early twentieth century, American biological research in universities was funded by capital accumulated by giant corporations but mediated through philanthropies like the Rockefeller and Carnegie Foundations. After World War II the huge increase in size of American basic science was funded overwhelmingly by federal tax dollars. In 1981 the Massachusetts Institute of Technology accepted $125 million from a private businessman to host the Whitehead Institute for molecular biological research.[23] For many academic biologists then, the Whitehead Institute had troubling implications in relation to autonomy, intellectual integrity, and conflicts of interest.

By the 1990s, arrangements like the Whitehead Institute were avidly sought, and hardly a serious molecular geneticist existed without some commercial connections. In 1994, for example, the University of Maryland announced plans to build a $53-million Medical Biotechnology Center to house both academic and industrial researchers under one roof. The idea was "to give scientists at startups cheap access to equipment and advice . . . In exchange, Maryland will collect rent and receive stock in participating firms" (ScienceScope 1994b:1071). The university researchers would be free of academic duties. Harvard planned a similar facility to open in 1996.

From the mid-1970s on, the social norms in biological research and communication changed from expert-communal and public ideals (if hardly always practice) to approved private ownership of patentable results, widespread direct business ties of university biological faculty and graduate students to corporations, marked convergence of "basic" and "applied" contents of research questions, and greater secrecy in research practice. Between 1987 and 1992 the number of university-industry licensing agreements more than doubled, and one-quarter of all patents awarded to universities since 1969 were awarded in 1990–91. The hundred largest universities got 85 percent of the patents (NSB 1993:xxvii, 152–53).

Formal cooperative research and development agreements between federal labs and private industry increased from 108 in 1987 to 975 in 1991 (NSB 1993: 119). In 1993, showing a huge increase across the 1980s, there were more than one thousand university-industry research centers in all scientific areas, spending about $3 billion a year on R&D, 41 percent of that for chemical or pharmaceutical research. Federal or state tax dollars contributed to building 72 percent of those centers (NSB 1993:xxii, 121). In 1994 the new director of the National Institutes of Health, Nobel prizewinner Harold Varmus, as he looked for new ways to link NIH, academia, and industry, was quoted as saying, "We're not interested in giving grants to Merck. We're interested in giving grants to small businesses" (Schrage 1994:3D). I think that was supposed to reassure worried radical science activists who think economic competitiveness might be getting out of hand as a goal of national health research policy. Meanwhile health-related research and development commanded 13 percent—about $28 billion—of the total US R&D budget in 1993 (NSB 1993:105).

Capital also squirts directly into industrial biotechnology. Every year between 1990 and 1994 "more money has been invested in new biotechnology and health-care companies in [California's Silicon Valley] . . . than in any of the industries that currently dominate the economy" (Wolf 1994:1D). Indeed, in this region famous for its computer and information technoscience, twice as much venture capital flowed into biotechnology and the life sciences in 1993 as into all computers, peripherals, semiconductors, and communications combined (Wolf 1994:1D). The original biotechnology companies, like Genentech, have spun off several other start-ups and joint ventures. There were 29 companies in the area in 1980 developing drugs and diagnostic products; there were 129 such firms in 1993. Nationally, in the third quarter of 1993, for the first time more venture capital sloshed into the trough to feed the life sciences than the information sciences (Wolf 1994:9D).

Although biotechnology has not yet produced many successful products, and the economic dream nourishing the huge investments is more luminous than the results so far, molecular biology, including the Human Genome Project, has germinated its share of millionaire scientists since Genentech's Herbert Boyer in 1976. In 1992, for example, J. Craig Ventor left NIH, where he researched technology for DNA sequencing, to help found Human Genome Sciences, Inc., of Bethesda, Maryland, to commercialize the technology. Ventor's shares were valued at $9.2 million in November 1993, when the company began to offer shares on the public stock exchange, and $13.4 million by January 1994. Other Human Genome Project scientists have also founded companies based significantly on tax-supported research results. The names of the companies fuse the magical and the mundane, just as the Alice-in-Wonderland scene of laboratory work in Quadrant's ad image did: Millennium Pharmaceuticals, Darwin Molecular Technologies, Mercator Genetics, Inc.[24]

By the mid-1990s the biotechnology industry faced major restructuring that involved buy-outs of the most promising companies and failures of many others. "The billions in investment have created an industry choked with copycat, capital-hungry companies lacking the critical mass of technology to survive" (Hamilton 1994:84). About $8 billion of public and private capital poured into US biotechnology in 1991–92. But a biotech company needs seven to ten years and $100 to $150 million to bring a new drug to market, and the majority of the companies raised only enough capital during the "biobull run" to last about three years. Venture capitalists tended to score disproportionately high returns on their biotech investments, while public stock offerings have yielded far less for investors. Missing the sober restructuring of an industry, the venture capitalists go in early, then cash out when companies go public.[25]

The corporatization of biology is not a conspiracy, and it is a mistake to assume all its effects are necessarily dire. For example, I believe technology transfer from academic research to other areas of social practice is important. I also insist that research priorities *and systems of research* must be shaped *from the start* by people and priorities from many areas of social practice, including, but not dominated by, profit-making industry. Each issue merits careful analysis and interrogation of one's own assumptions as well as those of others. Nonetheless I agree with Sheldon Krimsky (1991:79), who argued on the basis of his Tufts University Biotechnology Study from 1985 to 1988 that "the greatest loss to society is the disappearance of a critical mass of elite, independent, and commercially unaffiliated scientists . . . The stage is set for what University of Washington Professor Philip Bereano aptly described as 'the loss of capacity for social criticism.'"

Public Actors

Federal policy is clear about using science and technology to achieve national competitiveness goals. The government established a $12.5-billion budget for cross-cutting interagency initiatives in 1993, with $4.3 billion of that for bio technology (NSB 1993:xix). That compares with $1 billion for interagency computing and communications. The capacity for independent criticism and vision *that fundamentally shapes the way science is done* hardly seems to be on the political agenda, much less in the R&D budget of universities, in-house government labs, or industries.

Not surprisingly, the section on public attitudes and knowledge about science and technology in the 1993 edition of *Science and Engineering Indicators* (NSB 1993) did not even try to conceptualize or measure democratic participation in technoscience. Studies asked how many citizens follow the science and technology news (maybe 15%), whether folks had a high regard for US scientific leadership (seems so), and whether or not people understood the ozone layer and DNA (sort of). The "public" was conceptualized as a passive entity with "attitudes" or "understandings," but not as a bumptious technoscientific actor. No measurements or analyses were reported for such things as serving on science policy bodies, participating in workplace or community design projects, engaging in debates in education about science and technology, contributing to, formulating, and following up on impact statements, organizing technoscience-oriented action groups, writing novels or composing music that refigure beliefs and practices in technoscience, articulating technoscientific issues in race and gender justice goals, and so on.

Indeed, the spectrum of science policy discourse in the United States in the 1990s makes even mentioning such things appear to be evidence of hopeless naiveté and nostalgia for a moment of critical, public, democratic science that never existed. Whether it existed in the past or not, such a technoscience—committed to projects of human equality, modest, universal material abundance, and multispecies flourishing—must exist in the future. And lots more is going on in the present than the National Science Board knows how to count. I believe wealth is created by collective practice, figured by Marx as labor, but needing a messier metaphoric descriptive repertoire. Even a narrow view, however, which looks only to tax dollars feeding technoscience instead of to all collectively produced wealth that is eaten, digested, expanded, and excreted by technoscience, must insist on radically reconstituted public participation and critical discourse. If technoscience is to develop situated knowledges and strong standards of objectivity that take account of its webs of human and nonhuman actors and consequences, then at a minimum questions about content and availability of jobs, richness of what counts as scientific knowledge, cultural breadth among scientists and engineers and their constituents, distribution of wealth, standards of health, environmental justice, and biodiversity ought to vie with "competitiveness" for sexy luminosity in the eyes of molecular biologists and other politicians.[26]

Cooperating Mice and Molecules

The corporatization of biology could not have happened if mice and molecules did not cooperate too, and so they and their kin were actively solicited to enter new configurations of biological knowledge. The technical and intellectual success of the new biology is stunning by any measure. Much has been written about how the reconstitution of biological explanations and objects of knowledge in terms of code, program, and information since the 1950s has fundamentally recast the organism as a historically specific kind of technological system (see especially Haraway 1991a; Kay 1993; Keller 1992; Martin 1994; Spanier 1991; Wright 1986; Yoxen 1981). Nineteenth-century scientists materially constituted the organism as a laboring system structured by a hierarchical division of labor and an energetic system fueled by sugars and obeying the laws of thermodynamics. For us, the living world has become a command, control, communication, intelligence system (C^3I in military terms) in an environment that demands strategies of flexible accumulation.[27] Artificial life programs as well as carbon-based life programs work that way.

These issues are about metaphor and representation, but they are about much more than that. Not only does metaphor become a research program, but also, more fundamentally, the organism for us *is* an information system and an economic system of a particular kind. For us, i.e., those interpellated into this materialized story, the biological world *is* an accumulation strategy in the fruitful collapse of metaphor and materiality that animates technoscience. We act and are inside *this* world, not some other. We are subject to, subjects in, and accountable for *this* world. The collapse of metaphor and materiality is not a question of ideology but of modes of practice among humans and nonhumans that configure the world—materially and semiotically—in terms of some objects and boundaries and not others. The world might be different, but it is not. The heterogeneous practices of technoscience are not deformed by some ontologically different "social" bias or ideology from the "outside." Rather, biology is built from the "inside"—both the kind of inside pictured by Quadrant in its magical ad and the kind of inside I have tried to signal with the term "implosion"—into materialized figurations that are life as it is really lived.

OncoMouse™ makes technical and semiotic sense in the world of corporate biology, where the author of life is a writer of patentable (or copyrightable) code. Such authors and innovators might be naturally evolving organisms or the scientists who interact with critters to nudge their codes in directions more useful to (some) people. Because they provide a manipulable, mammalian model for human biology and disease, mice have been especially valuable as genetic research organisms for a long time.[28] That fact is evident in the *Encyclopedia of the Mouse Genome I*, a special 1991 issue of the journal *Mammalian Genome*. Playing on the belief that everything that really matters to an organism is in its "program," the *Science* magazine advertisement for the *Encyclopedia* offered "The Complete Mouse (some assembly required)."

Patents are only one form of intellectual property protection for transgenic animals, and not the most common form. No US patents were granted after OncoMouse™'s debut in 1988 prompted protest from animal-rights groups and environmentalists. The European Patent Office initially rejected the application for a patent for the Harvard mouse but granted it on the second round, in 1992. On 29 December 1992, ending a self-imposed moratorium on patenting transgenic animals, the US Patent and Trademarks Office granted patents to three organizations for novel transgenic mice. By January 1993 over 180 applications for transgenic animal patents were pending.

Custom-tailoring transgenic mice for specific projects is both routine, for procedures already established, and a leading-edge research area capable of providing tools to address some of the most interesting questions in biology. Intricately engineered "knockout mice," with particular genes eliminated and various control mechanisms installed, have become indispensable tools in genetics, immunology, and developmental biology (Barinaga 1994). Researchers who make a useful mouse have been inundated by their colleagues with requests for the beasts. "Since the researchers were reluctant to get into the mouse breeding business, their universities awarded companies, including GenPharm International, a biotech firm in Mountain View, California, licenses to market the animals" (Anderson 1993:23). David Winter, the president of GenPharm, considers the technique of custom-making a rodent so routine that he calls it "dial-a-mouse" (Cone 1993:A16).

Around 1990, laboratories began cranking out custom-made research mice in significant numbers, and firms like GenPharm began buying the rights. "Marketing gimmicks, complete with catchy names, have emerged. Scientists can call (800) LAB-RATS to take their pick of regular rodents or seven strains of transgenic ones" (Cone 1993:A17). Business writer Michael Schrage (1993:3D) quotes GenPharm corporate development director Howard B. Rosen: "'We do "custom-tailor" mice. We view them as the canvas upon which we do these genetic transplantations.'" Using mice as model systems for genetic engineering in biomedicine, instead of bacterial or yeast systems, matters. "This transition will have as big an impact on the future of biology as the shift from printing presses to video technology has had on pop culture. A mouse-based world looks and feels different from one viewed through microorganisms" (Schrage 1993: 3D). The analogy to inscription technologies and conventions of literacy could not be more apt.

As genetically engineered mice diversify to fit research protocols and biomedical production, the ubiquitous technoscientific object called a database accompanies the fleshy rodents in a higher-order mimesis of their biochemical genomes. In postmodernity's practices of flexible accumulation, the database is to the filing systems of monopoly capital as the computer is to the typewriter and cyberspace is to mundane space. Oak Ridge National Laboratories is creating a "computer database for mutated mice" so that researchers can find the animals they need (Cone 1993:A17).

More fundamentally, the entire mouse genome is a central research object in the context of the Human Genome Project. Recursively miming each other at every level, mice and humans are siblings in these projects. A biochemical genome is already a kind of second-order object, a structure of a structure, a conceptual structure of a chemical entity. The electronic genome databases represent still another order of structure, another structuring of information. The genome is a collective construct; it is built. Developing computer databases for handling data from the various genome sequencing projects, with their Niagara Falls of sequence information and physical and genetic maps at finer and finer degrees of resolution, requires advanced informatics research and complex interdisciplinary negotiations.[29] Materially, like the human genome, the mouse genome is part of the technical-semiotic zone called cyberspace.

Cyberspace is the spatio-temporal symbol of postmodernity and its regimes of flexible accumulation. Like the genome, the other higher-order structures of cyberspace, which are displaced in counterintuitive ways from the perceptual assumptions of bodies in mundane space, are simultaneously fiercely material realities and imaginary zones. These are the zones that script the future, just as the new instruments of debt scheduling and financial mobility script the future of communities around the globe.[30] The genome is a figure of the "already written" future in which bodies are displaced into proliferating databases for repackaging and marketing in the New World Order, Inc. The promise of the genome is its capacity to occupy the future. Contesting for the shape and content of such promises is the job of displaced, uncanny, undead figures in technoscience.

Constitutions

One important category of such contestation is teaching contemporary biology. Drawing from scholarship in science studies and feminist theory, Flower (1994, n.d.) argues for a promising constitutional premise for the republic of technoscience; he calls it "technoscientific liberty." Flower thinks of liberty as "relational power . . . seeking to reconfigure the possibilities of action" in the practical world of science. "Articulate" and "communal," liberty is achieved in solidarity. For Flower, liberty is at stake in science-in-the-making, not in the realm of the already settled. Technoscience is about "world-binding narratives that connect humans and nonhumans" into consequential patterns. Liberty is not the principle of stripped-down choice animating the "free market" of the New World Order, Inc., but "the struggle within and about the 'politics of technoscientific truth' of our world." Technoscientific liberty takes shape in strong, contestatory democratic practice, "and in the creation of technoscientific ends achieved by citizen activity. This means that creation of *politicoscientific community* is one of the chief tasks of participatory public action and a goal toward which a liberty-tuned science pedagogy would be directed." Technoscience is civics, in the strong sense, at the heart of what can count as knowledge. "If constitutive technoscience is a source of fresh politics, it always operates . . . [by changing the] human/nonhuman polity" (Flower n.d.:5).

My uncanny sibling figures would do well in Michael Flower's science classes at Portland State. They would engage in the natural acts laid out in his and his colleague William Becker's Science in the Liberal Arts Curriculum, a project funded by the National Science Foundation. The NSF officer for the grant called its approach "deep reform" (personal communication). Doing science is the focus, and that means doing the work of boundary maintenance and boundary crossing that does not ask permission from the border police guarding the line between the technical and the political as well as the human and the nonhuman. Students are hailed, interpellated, into technoscience, where they are subject to and subjects in a world-making discourse, but within an apparatus committed to culturally rich and historically specific liberty. The power-knowledge nexus is called to account at the heart of doing science, not in the leisure time reserved for "social relevance." From the first year, at all levels of difficulty, and for those who specialize in science and those who do not, the curriculum emphasizes "investigative, 'hands-on' and data-rich labs; collaborative inquiry; alternatives to lecture; facil[itation of] students' coming to know how scientists know; themes common to several sciences; and situ[ating] the questions and aims of science in social, political, historical, and ethical contexts" (Becker and Flower 1993).

Students bind worlds of humans and nonhumans together with disregard for what is supposed to be politics and what science but with a high regard for the hard and sustaining work of problem development, inquiry that depends on colleagues, struggles for meanings and goals, and building multidisciplinary and practical knowledge. There is no public with attitudes to measure here but an emerging pedagogical wormhole for transporting the citizens of technoscience into unexplored regions of a truly new and democratic world order.

Thus, "choice" is less the metaphor I seek for how to behave in technoscience than "engagement," or even, at the risk of piety in the permanently contingent games of mimesis that I want to play, "commitment." Commitment cannot take place in the empty spaces of Nature™ and Culture™, and the all-too-full spaces of foundational, unmarked nature and culture have been permanently sucked out of the world. Such foundations are unlamented by those they marked as nonstandard or as resource for the action of the hero. So commitment after the implosions of technoscience requires immersion in the work of materializing new ways of being in an always contingent practice. Refigured as an undead rodent looking back at us as it climbs toward the promising and blinding light of technoscience, the new actors in scientific narratives have got to do better than disappear into the culture of no culture.

Following Star's (1991) lead, I want to ask my sibling species, a breast-endowed cyborg like me, a simple question: *Cui bono?* For whom does OncoMouse™ live and die? If s/he is a figure in the strong sense, then s/he collects up the whole people. S/he is significant. That makes such a question as *Cui bono?* unavoidable. Who lives and dies—human, nonhuman, and cyborg—and how, because OncoMouse™ exists? What does OncoMouse™ signify when, between 1980 and 1991 in the US, death rates for African American women from breast cancer increased 21 percent, while death rates for white women remained the same? Both groups showed a slight increase in incidence of the disease.[31] Who fits the standard that OncoMouse™ and her successors embody? Does s/he contribute to deeper equality, keener appreciation of heterogeneous multiplicity, and stronger accountability for livable worlds? Is s/he a promising figure, this utterly artifactual, self-moving organism? These questions cannot have simple, single, or final answers. However, a serious commitment to refusing both the culture of no culture and the nature of no nature means these questions have to be asked, *as a constitutive part of technoscientific practice*. That is what science in action must mean.

Notes

1. Bruno Latour (1987: chap. 4) is responsible for the common adoption of the word "technoscience" in science studies. Latour argued that the "inside" of the world-changing site called the laboratory constitutes itself by extending its reach "outside" through the mobilization and reconfiguration of resources of all kinds. Stressing that academic scientists are a small part of the "armies of people who do science" (1987: 173), Latour concentrates on enrolling, enlisting, mobilizing, and aligning from the point of view of the powerful center. In Althusser's terms, he concentrates on hailing.

2. In the context of feminist method and science studies, Star (1991, 1994) addresses the question of the membership of objects in communities of practice that web together historically situated humans, nonhumans (natural and artifactual), and actions. See also Callon (1986), Downey, Dumit, and Williams (1995), Haraway (1985, 1992), Latour (1987, 1993). In her social-network approach, Oudshoorn (1994) argues against an overly exuberant sense of the agency of things. David Hess (discussion notes, SAR seminar on cyborg anthropology, October 1993) cautions that "granting membership" to things can be a fancy phrase for the fetishism of commodities. Things have always been luminous sources of fascination in capitalism. Hess points out that corporations have the legal status of persons, and such "membership" is crucial to the reproduction of capitalist relations, which extract liveliness from people and embed it in things and abstractions. Yet appealing to the subject is the worst way to deal with the disturbing half-lives of undead objects. Located in society and outside nature, the bounded individual with property in the self is the chief fetishized object (thing mistaken for a living being, while the actual living beings and processes that produce and sustain life are effaced) in Western political and economic writing after about 1700. One can hardly invoke that individual and his stripped-down, body-phobic societies to object to the liveliness of mice, microbes, narrative figures, lab machines, and chimerical collectives of humans and nonhumans. How to "figure" actions and entities nonanthropomorphically and nonreductively is a fundamental theoretical, moral, and political problem. Figuration and narration are more than literary decoration. Kinds of membership and kinds of liveliness—kinship, in short—are the issues for all of us.

3. Human mental patients were part of psychiatric research on neural-chemical implants and telemetric monitoring at Rockland in the 1960s, a fact I learned while researching nonhuman primates as model systems for human ills (Haraway 1989: 109). Kline was associated with the Psychiatric Research Foundation in New York, which promoted controversial psychopharmacology investigations. Marge Piercy (1976) used research at Rockland State Hospital as background for the brain-implant experiments practiced on psychiatric patients in her feminist science fiction novel.

Chris Gray (1991) interrogates real-life military cyborgs. Gray first called my attention to the Clynes and Kline paper.

4. In 1993, 59 percent of the federal research and development budget went directly for defense, including nuclear weapons, down from 67 percent in the peak year of 1987 (NSB 1993:xviii).

5. The superscript 239 designates the atomic weight of fissionable, that is, explosive, plutonium, Pu^{239}. Fissionable uranium has an atomic weight of 235; 99 percent of naturally occurring elemental uranium has an atomic weight of 238. A breeder reactor uses small amounts of U^{235} to produce Pu^{239} from the abundant U^{238} in the reactor mix. The original Fermi reactor at the University of Chicago (1942) used the naturally occurring U^{235} in six tons of purified uranium oxide to get a chain reaction. The purpose of Fermi's breeder reactor was to get fissionable plutonium for the Manhattan Project. The U^{235}-enriched isotope mix for the Manhattan Project was produced by gas diffusion plants.

6. Due to the decay of radioactive uranium in ore, tiny amounts of plutonium and neptunium form spontaneously, a process described after the laboratory production of plutonium.

7. Jeremy Rifkin (1984a, 1984b), and his Foundation for Economic Trends, and Michael Fox (1983, 1992) have been outspoken about purity of type and natural integrity. See also Krimsky (1991:50–57) and Office of Technology Assessment (OTA 1989:98–102, 127–38). Under the banner of the Pure Food Campaign, Rifkin leads the opposition to Calgene's Flavr-Savr tomato, approved for the US market by the Food and Drug Administration in May 1994, and Monsanto's genetically engineered bovine growth hormone. Pure food is a curious concept to invoke for the tomato, a member of the deadly nightshade family. Well before genetically engineered fruits joined the fray, the tomato was at the center of struggles over immigration, science, food, and labor in California's agribusiness fields, research institutions, grocery stores, and kitchens (Hightower 1973). On biotechnology and world agriculture see Hobbelink (1991), Shiva (1993), and Juma (1989).

8. Bereano (1995). The press release covered meetings in the Adirondack Mountains to plan oppositional strategies. The group issued a position statement called the "Blue Mountain Declaration." Working with indigenous organizations to eliminate funding for the Human Genome Diversity Project emerged at the meeting as a major priority. The coalition's statement did not evoke arguments about purity of natural kinds, but the sanctity of life and opposition to manipulation of the natural world remained important ideological resources. I recognize, and often share, the power and importance of those commitments and languages; but I wish my fellow travelers seemed more nervous and less self-certain in their presence. The historical pedigree, for both "indigenous" and "Western" speakers, of those languages, ideologies, and associated actions hardly gives cause for unruffled calm. I think progressive politics have to be rooted in more fraught, unsettled, dirty, hybrid languages and expressions of belief, hope, and action.

9. I owe "life itself" to Sarah Franklin's (1993) distillation of Foucault's idea of biopower.

10. Mutating Pickering's title (1992) highlights the nonseparability of culture and practice.

11. On the many sites of action in technoscience, see Clarke and Montini (1993), Escobar (1994), Hess and Layne (1992), Martin (1994), and Rouse (1993).

12. John Law (personal communication) pointed out the absence of computers in this advertisement.

13. The most dyspeptic recent complaint is Gross and Levitt (1994). For a cogent critical review, see Berger (1994).

14. OncoMouse™ should be joined by another sibling figure, FemaleMan©, mutated from Joanna Russ's *The Female Man* (1975). For that argument, which explores the commercial circuits of transnational feminist theory and the history of copyright in tandem with biotechnology patent law, see Haraway (1997).

15. Oncomice get many kinds of cancer, but breast cancer was semiotically most potent in news stories and in the original patent.

16. For the initial part of the story of OncoMouse™ and evolving patent rights in relation to genetic technologies, see Krimsky (1991:43–57). For a fundamental early analysis, see Yoxen (1984); for further references, see Woodman, Shelly, and Reichel (1989). For oncogene research as "do-able science," see Fujimura (1992:168–211; 1996). One is at least as likely to find the latest news on transgenic animals on the business pages of the newspaper as in the science and medicine section. Bioengineered, transgenic farm animals captured much of the early attention, but the present stress is on biomedical products that are likely to be crucial for biotechnology companies to raise capital in the 1990s (Andrews 1993:1A).

17. I am indebted to officers of Du Pont, who preferred not to be named, for generous and time-consuming discussions of these and related matters in 1994 and 1995. Du Pont people saved me from many errors of fact; I remain responsible both for remaining errors and for interpretations.

18. Teitelman (1994:50, 184) points out that the splicing of biology and medicine—and academic research and the drug industry—at both a verbal level (*biomedicine*) and an organizational level began in the 1970s, the same decade that saw *E. coli* genes working in frog cells. "The factors driving this process were quite involved, reflecting the social complexity of the modern scientific enterprise: from government (the war on cancer), academia (the development of genetic engineering and the rise of the immunotherapies), and the economy (the inflation of the 1970s, the deregulation of Wall Street, various tax reforms)" (184).

19. See Moskowitz, Katz, and Levering (1980:606–10) for a history of Du Pont the company and du Pont the family before the acquisition of Conoco in 1981. That acquisition complexified Du Pont culture significantly, and by the 1990s the Du Ponts, whose power in the company was already diluted over three generations, do not hold even a significant minority interest.

20. Martin (1992, 1994) discusses flexible accumulation and the biological, especially immunological, body.

21. My sources for the following allegory are Hounshell and Smith (1988), Noble (1977), Teitelman (1994), and Du Pont's own 1995 brochure, *The World of Du Pont: Better Things for Better Living*.

22. OTA (1989:30). Venture capital was greatly encouraged from the mid-1970s on by cuts in capital gains taxes from 48 percent to 28 percent (Wright 1986:332).

23. Yoxen (1984:182). Landscape architect Martha Schwartz designed a rooftop "splice garden" for the Whitehead Institute. The synthetic garden splices together design elements of Japanese and French gardens (Johnson 1988). Thanks to landscape architect Anne Spirn for the tip. In the 1990s the federally funded Whitehead Institute/MIT Center for Genomic Research is the largest US genome research center. Its director, Eric Lander, and the director of France's genome effort, Daniel Cohen, were founding scientific advisors to the biotech firm Millennium Pharmaceuticals (Fisher 1994:9A).

24. Fisher (1994:9A). Deborah Heath pointed out that Microsoft cofounder Paul Allen is on the board of directors at both Affymetrix and Darwin Molecular Technologies. Informatics and biologics are ubiquitous siblings in the New World Order, Inc.

25. Hamilton (1994:85, 88). Beardsley (1994) discusses the reconfiguration since 1973 of the culture of biology as a result of its commercial transformation.

26. See Haraway (1988) and Harding (1992) on situated knowledges and strong objectivity in feminist science theory.

27. Dawkins (1982). For contrast, see Margulis and Sagan (1995) for life as autopoiesis. Gilbert (1994) consistently gives a nonreductionist treatment of molecular biology and development.

28. For mice in the material culture of science, see Reder (n.d.). Mice are *The Right Tools for the Job* (Clarke and Fujimura 1992). Grosveld and Kollias (1992) survey how transgenic animal technology has been applied to a wide range of biological problems. Michael and Birke (1994) discuss how biologists involved in animal experiments defend their practices and see those who do not share their commitments.

29. See Cuticchia et al. (1993) and Hilgartner (1994) on informatics development in the genome project.

30. See Christie (1993) on cyberspace, flexible accumulation strategies, and temporality.

31. Centers for Disease Control statistics, reported on National Public Radio, 22 April 1994.

CORRIDOR TALK / Gary Lee Downey, Joseph Dumit,
and Sharon Traweek

cor•ri•dor talk *n* **1:** the practice of passing on tips, insights, and strategies about the means of production of academic work (as at professional
conferences, where, it is frequently remarked, the most important business
takes place "out in the corridor" rather than inside the meetings rooms)
2: nonascribable (off-the-record) but necessary information; practical gossip
3: common-sense, informal (not publicly taught) mentoring; the unsaid, but
frequently said anyway (though not to everyone)

ORRIDOR TALK, for which we propose the above definition, is a collo-
quial phrase for the kinds of knowledge one should have in order to
live in the academy but which are often discussed as if they were of
little formal value. In this chapter we offer some corridor talk for anthropological projects that use cultural perspectives and ethnographic fieldwork
to study and intervene in emerging sciences, technologies, and medicines.

Corridor talk is a key dimension of conference friendships, those professional and personal relationships among scholars that are renewed annually
through interactions ranging from the chance encounter to the intense discussion over drinks or a meal. Electronic mail has extended the corridor talk of the
conference back to the office or home while expanding its boundaries outward
in both public (conversations on Listservs) and private (intimate, one-on-one
exchanges) directions. Corridor talk carries no formal authority; we tend to receive it as an offering of resources or advice. We listen and then take away whatever seems valuable, leaving the rest behind. Participants in the SAR seminar
agreed that, despite its informality, corridor talk is a key means of intellectual
(re)production, and its strategies for helping people move into and through
scholarly arenas deserve more visible acknowledgment, consideration, and explicit discussion.

Corridor talk was constantly present during the seminar week. Some of
our more sensitive conversations were particularly valuable, broaching issues
that we all face regularly but that are not normally part of formal, professional
interactions. We welcomed the opportunity to share stories and worries about
our work with colleagues and fantasized about a time when such sharing
might be taken for granted. We also drew up lists of journals, presses, academic

programs, and funding sources and contacts, and discussed issues we faced in doing fieldwork, getting grants, and finding jobs. Many of the issues clustered around the question, "How do you do that?" We began calling it all "corridor talk," referring generally to information that gets passed on from colleague to colleague and mentor to mentored through interactions that are usually private, rarely show up in published form, and can make the difference between the competent person and the canny, networked, competent person.

Publishing corridor talk is risky—it goes out of date fast, is often more local than general, and is always provisional—but we decided to go ahead anyway and offer some discussion of our career practices. We had two incentives for doing so. The first is that seminar participants came to study areas of science, technology, and/or medicine from widely differing circumstances and career positions and without the mentorship traditionally available to disciplinary workers; each of us would have benefited from greater guidance. Second, mentoring appears to be increasingly sporadic as programs become more interdisciplinary and as the demographics of academic researchers change. For example, as David Schneider revealed to Richard Handler (Handler 1995), Schneider came to realize that many of his early career "choices" were in fact determined behind the scenes by his mentors, for reasons he knew little or nothing about at the time. Today such direct or indirect intervention by mentors is more the exception than the rule. We feel that mentoring is an important part of the academic life, but that this traditionally private and privileged mode of transmitting the social knowledge, habitus, or savvy in negotiating academic careers should be shifted into more open and public channels.

In this spirit, we offer the following information and advice, which we would have liked to have access to earlier in our moves to conduct research on science, technology, and medicine. We hope it will inspire further development of corridor talk archives.

Career Narratives

During the seminar we were amazed by the different pathways participants had traveled to study science, technology, and/or medicine through cultural perspectives and ethnographic fieldwork. Some had made this choice in advanced stages of their careers after they had already benefited from the securities and comforts of tenure, others at earlier stages when career opportunities in interdisciplinary areas seemed worth the risks, and some, more recently, at the graduate level in a time when the job market is frighteningly uncertain. Motivating passions ranged from long-standing advocacy for social movements to bewilderment about past or present experiences, curiosity about centers of power, and experiences of pleasure in previous interventions. In making such moves, all of us had tried to map the interdisciplinary worlds in which we might travel. Each of us had made explicit efforts to sort out mazes of disciplinary boundaries, academic programs, professional organizations, and societal interests. Then, working in idiosyncratic ways, we had written speculative histories of the future—"career narratives"—that trace how choosing different projects might locate our travels on those maps and shape the career opportunities that might come our way.

Career narratives are a variation on what anthropologist Edward Bruner (1990) termed the "tenure narrative": an integrated story of an individual's career trajectory, extending through the present into the future. After serving for many years as a participant observer in promotion and tenure cases for university faculties, Bruner concluded that successful cases generally offered coherent and persuasive narratives that located the candidates in valuable places and as valuable members of the academy. These narratives provided tenure committees with easy handles for remembering who the applicants were, what they had accomplished, and where they were going.

Researchers undertaking cultural or ethnographic studies of science, technology, and medicine must find their place amidst a collection of overlapping interdisciplinary and intradisciplinary boundaries. At this writing, no academic program has made such work its main focus. While interdisciplinary work and collaboration are often important for these studies, and individual researchers often find themselves moving across disciplinary boundaries in their reading and writing practices, educational institutions still tend to rely on disciplinary categories in their hiring practices. Crossing academic boundaries may offer great opportunities, but it can also be risky.

Subdisciplinary categories are particularly important in career narratives since they often appear in job descriptions. Within anthropology, for example, subdisciplinary divisions include cultural anthropology, medical anthropology, anthropology of work, anthropology of computing, and applied anthropology. Increasingly, cultural anthropology is turning to organized science as a site of significant cultural emergence and meaning production and, hence, locating it as worthy of sustained study (Fischer 1995; Franklin, Lury, and Stacey n.d.; Hess 1995; Marcus 1995; see also chap. 1 and Hess, this volume). Medical anthropology is branching out beyond an earlier emphasis on patient perspectives to examine the cultural positioning of medical expertise and institutions (B. Good 1994; M. Good n.d.; Heath and Rabinow 1993; Romanucci-Ross, Moerman, and Tancredi 1991). Anthropology of work is extending beyond a traditional focus on blue-collar perspectives to explore the experiences of technical work wherever these are located (Dubinskas 1988; Forsythe 1993b; Hess and Layne 1992; Julian Orr 1990; Suchman 1987, 1992, 1994). The anthropology of computing, a relatively new subdisciplinary activity, contributes to growing interdisciplinary interests in user perspectives and the positioning of computing expertise in society (Downey 1992; Forsythe 1992, 1993a; Hakken 1993; Hakken with Andrews 1993b; Nyce 1996; Nyce and Loewgren 1995; Pfaffenberger 1988, 1990). Applied anthropology has gained increased status as a fertile and active area for thinking through and enacting the practices of participation and intervention (see the journal *Human Organization*). How you locate yourself in relation to subdisciplinary boundaries can affect whether or not you make the first cut in a job search or are judged appropriate for inclusion in a conference or other professional activity.

Locating yourself in relation to interdisciplinary categories can be tricky. Where you are situated in the midst of ongoing intellectual and political debates can have significant implications for how you are read, received, and located as a person. For example, what counts as science and technology studies has been contested heavily over the years (see Hess, this volume), and an individual's account of what has been taking place is an indicator of his or her location in relation to it. Consider, for example, what is opened up, closed down, or otherwise positioned by the following science and technology studies narrative:

The earliest label, "science, technology, and society," fell out of favor because it suggested work that begins by separating science and technology from society and, hence, limited itself self-consciously to the study of impacts. The label "science studies" shifted focus to knowledge contents but, in the process, abandoned the goal of intervening in society to solve problems involving science and technology in favor of building a new interdisciplinary discipline. At the same time it emphasized scholarship on science alone, to the exclusion of technology and medicine. Meanwhile the label "technology studies," which once suggested a policy interest in technological development and innovation, came during the 1980s to mean academic work on the social construction of technology. Those who held onto the label "STS" were generally indicating a continuing interest in science, technology, and medicine as social problems and advocating work that participated and intervened in such problems, often with explicit policy considerations. Recently "science studies" has reemerged as a shorthand for any work that explores, and intervenes in, the positioning of sciences, technologies, and medicines in society.

One key to locating work on science, technology, and medicine is to position it in terms of the future of the academy as a whole. Some seminar participants speculate, for example, that the research university system in the United States may be undergoing fundamental change. This narrative goes as follows: Researchers and teachers almost everywhere seem to be experiencing new demands for accountability, demands to show how their knowledge makes a difference. While postmodernist critiques of science may have added to the calls for accountability by bringing attention to the power dimensions of official knowledges, a significant source of pressure appears to be the rise of nationalist concern about economic competitiveness since the end of the Cold War. The quasimilitary logic of competitiveness seems to reach deep into everyday lives and selves, turning every act into an economic defense of the nation. State legislatures are pressuring universities to focus on undergraduate teaching, with employment for students as the primary objective. The National Science Foundation is shifting its emphasis to strategic research and spending great sums on educational reform to beef up the work force. "Competitiveness" could become a key measure of value in teaching and research.

From one perspective, a rise of interest in how knowledges make a difference could help legitimize activities that focus on such questions and seek to improve both how they are formulated and how they might be resolved. Interdisciplinary work focusing on science, technology, and medicine could thrive. A second possibility is that an expanding logic of competitiveness could produce a dogma that undercuts the legitimacy not only of traditional research that pursues truth and purports to be value free but also of work that aims to contribute through critique and critical participation.

As one maps boundaries and builds stories that cross them, it is crucial to understand that different disciplines and subdisciplines often have radically different customs, protocols, and criteria of competence when it comes to citation practices, conference submissions and presentations, publications, letter recommendations for jobs and grants, grant reviews, and so on. Many researchers undertaking anthropological projects on science, technology, and medicine are collecting successful grant proposals, syllabi, bibliographies, and other documents, as well as experiences in job hunting and negotiations. These are the kinds of materials that both junior and senior scholars need if they are to move between disciplines or enter new disciplines for the first time. By sharing this kind of information, we can help one another map important differences and build successful career narratives.

Managing the Field

Although many fieldwork issues and strategies in studies of science, technology, and medicine are common to other ethnographic work, some appear to be particularly salient. It can be very difficult, for example, even to pretend to be a faceless observer, a person who might have meaning only as participant observer or a scientific voice from nowhere. At the SAR seminar many of us told stories of informants who were well-read in the social sciences, had invested considerable time and energy in reflecting on and analyzing the matters at hand, and engaged our work critically. Some had read our published work. It may be worthwhile to consider how such heightened levels of visibility might sharpen issues of access and confidentiality.

The strategies seminar participants used to establish initial contacts varied from highly personal connections to formal letters of introduction, sometimes including an outline of the project and list of publications. How we positioned ourselves was crucial to these interactions. Occasionally the very idea of an anthropologist in a high-tech field was bizarre enough to permit admission, though the ethnographers involved found they had to negotiate very carefully the purpose and practice of fieldwork with virtually everyone involved. Also, our status as a junior (presumably naive) student or as a senior (presumably wise and experienced) scholar inflected whether and how the people we encountered were willing to incorporate us into their lives. Some participants explicitly inhabited the position of journalist, especially with regard to scientific conferences where the press is often accorded special treatment. Acquiring a press pass by volunteering to write for a professional newsletter or other academic news publication, for example, can give access to news releases, background documents, and interview rooms, not to mention free conference admission.

Other participants took on the role of historian in addition to that of anthropologist. Since most researchers in science, technology, and medicine do not have projects important enough to merit Nobel-like recognition and histories, many are happy for the opportunity to go down in history or otherwise set the record straight. In this case the fieldworker must negotiate with the subjects of study: Does one commit, for example, to transcribing tapes and preparing the material for deposit in an official archive or simply to mentioning someone by name in subsequent publications?

The seminar touched only briefly on the question of confidentiality, probably because of its enormity. Most of the contributions to this volume involve multiple field sites and, therefore, negotiation and positioning with regard to more than one group. Such groups may indeed be in competition or have narrow channels of communication with one another. As an ethnographer of science, technology, and medicine, when and how do you preserve the confidentiality of your informants? Just as extended involvement in a field site offers you the opportunity to make visible experiences that might otherwise be ignored, it also tends to build your sense of commitment to those whose experiences you are studying. There are occasions when you might want to name names, such as when the study recounts historical developments or sheds light on a scientific controversy. On other occasions you might be committed to preserving someone's anonymity, such as when you are helping a subordinate perspective gain legitimacy.

During the seminar we learned that participants had used several different strategies for grappling with problems of confidentiality and had negotiated ethical issues on an ongoing basis. In other words, ethics became a practical problem with flexible boundaries. How much control should informants have over how they appear in our texts? Some participants permitted interviewees to edit texts of interviews or to approve written materials before publication. Some attempted to tape everything, others taped only when something interesting happened, and still others avoided tape recordings completely. Some dealt with the issue entirely on an informal basis; others sought approval for their projects from university or institutional review boards and made written agreements with every informant. Some relied on written materials as much as possible to maximize their critical distance; others used them mainly to support ethnographic observations. Most important, no one was entirely consistent; each of us varied strategies a bit from project to project, trying to stay aware of the ethical commitment to acknowledge and respect the rights of informants. This issue is, of course, a subject of ongoing negotiation among ethnographers both inside and outside of anthropology.

Getting Published

Seminar participants also shared stories and strategies concerning negotiations in the world of publishing. Most of us agreed that articles in peer-reviewed, disciplinary journals remain essential to professional advancement. What seemed patently unclear, and therefore contested, was the value of articles in non–peer-reviewed disciplinary journals, articles in interdisciplinary journals even if peer reviewed, and chapters in edited collections. We were surprised that our small group of ten participants had published in over forty different journals and with thirty different presses. This diversity appeared to indicate both a relative lack of traditional publishing outlets for work on science, technology, and medicine and a desire to reach audiences who did not read our disciplinary journals.

Refereed journals tend to be idiosyncratic, and successful submission generally requires becoming familiar with each one's publishing practices, including common reference points, intended audience, and the writers they publish. The period from submission to response and from acceptance to publication varies dramatically, from six months (rapid) to three years (unfortunately not uncommon). A few journals, such as *American Ethnologist*, publish this information, but most do not, and often it can be gleaned only first-hand from the editorial board or previously published authors. Furthermore, many key disciplinary journals place the highest value on the work of senior scholars. Some do publish excellent graduate work, but the standards may be higher for these junior scholars, and requests for revisions more common. Some disciplinary journals emphasize case studies; others publish theoretical advances. Some interdisciplinary journals welcome papers that serve as placeholders for whole disciplines or fields, in which theoretical approaches and objectives are described in language that is as clear as possible, with minimal use of jargon. Other interdisciplinary journals focus on particular research topics, actively encouraging crossover and collaboration among humanities, social science, and science contributors.

Mapping exercises often prove helpful in selecting journal outlets and preparing written submissions. For example, one can investigate the publishing decisions of specific scholars via curriculum vitae, citation indexes, *Current Contents*, and interviews. Mapping subdisciplinary developments this way also proves useful as a writing strategy for career narratives. What roles have different kinds of journals, special issues, edited books, and single-authored books played in the development of specific theoretical areas and collections of researchers? How and where do you want to fit?

Book publishing raises a different set of issues, many of which are related to the objectives of publishers and a given book's potential sales. Book publishers tend to classify themselves as either "academic" or "commercial," but both groups sell books to make money. A key difference is commercial presses need to generate profits for their investors, whereas academic presses are probably trying just to break even. Many academic presses are affiliated with universities, and to varying degrees they contribute to, complement, and draw on the reputations of their namesakes. Often divided internally by discipline, academic presses emphasize particular areas, or lists, as shown in their catalogs. Commercial presses, by contrast, tend to organize internally according to the subject areas, or niches, they stake out in the marketplace. Although such divisions may have disciplinary content, commercial presses may be more likely to feature interdisciplinary, problem-oriented fields of research.

Editors of publishing houses can be found at most conferences, where they may seek out particular panels or presenters. Many editors encourage the submission of written book proposals that outline the objectives and structure of the book and its probable audience. Particular emphasis is placed on a book's potential classroom use; this is often what makes manuscripts most attractive. It now appears standard for authors to contact several presses to get a sense of how interested particular editors might be, as well as how they might market a book. Some presses, however, won't consider a project that has been presented simultaneously to multiple publishers. Try to interview other authors about their experiences with various book publishers and editors. It is important to remember that an editor is not likely to be the one making final acquisition or publishing decisions (this role may fall to an editorial board or the publisher), but instead serves as the book's advocate within the publishing organization.

The particulars of contracts are rarely discussed outside of corridors, even though academic authors appear to be the only ones without literary agents, due to the relatively small proceeds they and their publishing houses make on their books. A signed contract commits the author to the publisher, but the publisher's commitment is generally more flexible: the publisher can reject the completed manuscript if it is found unsatisfactory. Nevertheless a contract does establish a working relationship between author and editor and commits the editor to working with the manuscript. Many items in a typical book contract are negotiable; it's always worthwhile to investigate your options.

You might want to ask the following questions: How large will the first print run be? (For an academic monograph, a small first run of 800 to 1,000 copies is common; multiple thousands are rare, especially for a first-time author.) How long will the review process take? (Three months is very good; over a year is too long but not uncommon.) Once the manuscript is accepted, how long will it take before it is released? (Many press contracts promise to take not more than twenty-four months to get a book into print; the actual time your book takes could be less, or more.) How much time and expense will be budgeted for copy-editing? What efforts will the publisher make to market the book, including attending annual meetings, sending copies to journals, and advertising by mail or in the media? Do you want to request extra copies to send to journals yourself? Authors unhappy with their publishers' efforts in getting books out have been known to buy extra copies themselves to send to journals. How much will the book cost? Will the publisher issue your book in hardbound, paperbound, or both? If both, will the two editions be released simultaneously? How are author royalties structured? Is it possible to get an advance on royalties? (Advances are common with commercial and crossover presses, less likely—and almost invariably small—from academic presses.) Finally, who owns electronic rights, translation rights, and other subsidiary rights? Who decides how much is charged for subrights? And what is the author's cut? There are some helpful guides to publishing with academic presses; Paul Parsons's *Getting Published: The Acquisition Process at University Presses* (University of Tennessee Press 1989) and Robin Derricourt's *An Author's Guide to Scholarly Publishing* (Princeton University 1996) are good places to start.

Attending Conferences and Being Seen

Interdisciplinary scholars are faced with a multitude of professional conferences that may be relevant to their work. "Being seen" can be a key part of "gaining membership." At what point should you begin to attend conferences, and how deeply involved do you want to get? Should you present a paper, organize a session, serve as a discussant, or seek election to an office? The answers will vary according to career stage and research interests, and with subdiscipline, conference organizers, department conventions, and so on. They also always depend on the availability of travel money. An inexpensive way to establish a high level of involvement and make difficult choices is to read the relevant association newsletters.

Those new to the conference scene should recognize that session organizers are subject to a number of constraints. The organization might distinguish between "submitted" and "invited" sessions; if so, how do you gain the greater visibility that goes with invited status? In the American Anthropological Association you'll need to locate, contact, and lobby the program representatives for the constituent societies. Representatives are appointed every year at the annual meeting in late November or early December; though often new to the tasks involved, they must make their final decisions on the following year's program by the end of February. Attending business meetings is one way to collect the names of representatives, and early contact can help get a proposed project accepted.

If you are planning a session, how do you strike a balance between relatively prominent names and people just breaking into the area? Gaining commitments from a few prominent scholars, perhaps as discussants, can help attract others. How do you secure a good room and a good time on the program? Consider writing a cover letter with your proposal that outlines likely interest and attendance.

Negotiating a Job Offer

Today, in a time of widespread restructuring and downsizing, securing any job anywhere can be considered a success. The offer of a position carries with it the task of negotiating its details. The short period between receiving an offer and accepting or declining it is likely to be your only opportunity to negotiate the terms of employment, and small changes made then could prove highly significant later. If you handle negotiations in a reasonably positive manner, it is likely that any worries about offending someone or appearing greedy will dissolve after you start working with your new colleagues on a daily basis. Asking questions and paying attention to the details is entirely legitimate. The following bits and pieces, which focus almost entirely on academic positions, were collected from seminar participants and a dozen or so other people. We offer these observations with the assumption that readers will make their own judgments about whether or how to use them.

In negotiating for a position, try regarding the person with whom you are dealing (the department chair, for instance) as being on your side. The people interviewing you are genuinely seeking a new colleague. In addition, the very existence of the new position (and any additional resources that might come with it, like new computers) is a direct benefit to the hiring unit. If the department is unable to fill the position, it might lose it.

You might begin with this stance: "I have an intellectual reason to move, but don't want to lose what I have." Another option: "My situation, my needs, and my goals have evolved since the last job/last move/last promotion." One approach to negotiating is to explain your situation and goals and then ask the other person for help in solving your problems together. Clarify up front that you don't want to pit one institution against the other, and that you don't intend to add issues later. Be both persistent and polite, never talk or act as if an issue were closed, and present your goals as information requests rather than as demands, asking, for example, how it might be possible to do or get something. Knowing your priorities in advance, especially what is essential and what might be conceded, is important. You might also consider saying explicitly, "This is the understanding I want . . ." We've found that men and women often have very different negotiating styles; consult with your more experienced colleagues of both genders for advice in dealing with common miscues and miscommunications; also see Heim (1993).

Although many of the seminar participants agreed that money should be the last item discussed, there was much debate about whether or not you should ask for more than you expect to receive. Some suggested always asking for at least $10,000 more than you are getting now, or securing a promise for the following year. Others recommended seeking to match rather than increase current resources. It could be helpful to provide evidence of your current salary and likely increases. You might request information regarding merit raises and cost-of-living increases. Since freezes on hiring or salary increases could emerge or continue, it is important to start out at a good level.

During academic budget crises, consider trading money for time, such as time off or a reduced teaching load. Depending on your priorities, the goal might be either to match or reduce a current teaching load. You could ask for reduced teaching for the next year or two—but be sure to demonstrate a strong and permanent commitment to teaching. What are the local values regarding teaching? What has been common practice, and has it changed in recent years? How are courses and course times assigned? Are course equivalents available, such as independent studies, graduate student advising, committee activities, or center leadership? Find out if getting outside funding guarantees a research leave. Clarify any sabbatical calculations for the future; you might request, for example, that time spent at another university apply toward sabbatical leave.

What about seeking a joint appointment? This will give you access to two sets of resources, but at the cost of two sets of responsibilities, such as faculty meetings and committee assignments. At the senior level, a joint appointment might be a vehicle for greater financial support. At the junior level, fulfilling two different sets of expectations could make tenure more difficult to achieve.

With nine-month positions, you might want to inquire about routine practices for getting summer money. Are any housing allowances available, such as support for a down payment or mortgage assistance? What is included in the retirement and benefits packages; for instance, what percentage of your salary does the school contribute to a retirement fund? Can you negotiate your arrival date?

Think about other resource issues that might be significant. It could be important, for example, to secure financial support for one or more graduate students per year. Clarify your current and eventual levels of access to the following: office space, computer, E-mail, printing, fax, phone, copying, travel funds, research assistance, funds for library acquisitions, support for speakers and workshops. Are internal research funds available? Are stipends, reduced teaching loads, or other forms of support available for journal editorships or other professional responsibilities?

What kinds of service responsibilities are routine or otherwise expected? What committee responsibilities might you have? Will you have to oversee recruitment or do other heavy administrative work? What will be your responsibilities for graduate students? Will you be charged with expanding a program? The more senior your position, the more likely you will be expected to expand the recruitment of students and junior faculty. Does visibility through gender, race, or ethnicity lead to higher committee loads or multiple department obligations? If so, consider requesting extra compensation.

If a local goal is to build autonomy for a new program, you might examine the statuses of other freestanding units around campus. To whom would this program be responsible? What are the possible future sources of support? Who defines the unit's boundaries, especially regarding budgets, students, faculty, affiliated faculty, secretarial staff, space, library acquisitions, copying, phone, and mail? Gearing up a new program can be exceptionally difficult. Are other departments or programs friendly to the idea? You might consider contacting the highest-ranking person and asking her or his advice on how to proceed.

There was widespread consensus among our group that details of all employment agreements should be put in writing. Even this is no guarantee, but it is does provide a good source of shared memory. Finally, it is worth building career narratives here as well, both for yourself and for the unit(s) to which you will be committing time and effort.

Finding Money

One of the most surprising discoveries in our corridor talk at the SAR seminar was the diversity of our funding sources. Interestingly, only one of us had ever received support from the National Science Foundation's anthropology program, which might be considered the natural source for such funding. Most of us have turned instead to a variety of other sources for financial support, sometimes through creative social hacking; that is, by breaking into arenas to which we would not normally have had access. For example, members of our group have received funding from Aerojet Nuclear Corporation, American Council of Learned Societies, American Institute of Physics, Danforth, FIPSE, Fulbright, Hastings Institute, Institute for Advanced Study, the Japanese government, Luce Foundation, Murdock Foundation, National Endowment for the Humanities, National Science Foundation Ethics and Values Studies, Rockefeller Gender Research, Smithsonian, Spencer Foundation, Wenner-Gren Foundation for Anthropological Research, and grants from individual universities for pilot projects, travel, language study, and sabbaticals. Some of us have even worked for the people we studied, as lab technicians, tour guides, software developers, counseling assistants, historians, or co-authors. Aside from the residual professional taboo against being paid by one's informants, each situation posed a unique set of analytic and ethical issues. We found it reassuring to learn that others had encountered similar opportunities and had to cope with similar feelings of ambivalence, ambiguity, and even embarrassment.

Grant writing remains one of the trickiest areas of academic life. Decision-making processes are often opaque and can vary from review to review. Also, grant proposals constitute a literary genre that is underanalyzed, poorly understood except by good grant writers and reviewers, rarely a component of formal education, and often considered by their authors a huge waste of time.

Submitting a grant proposal is like buying a lottery ticket: it generates a nice fantasy, although one with a low probability of coming true. At the same time the odds are much better, the money will be given out to someone, and you will not get the grant if you do not apply for it. We offer some suggestions here based on our own experiences of grant writing and reviewing.

Some departments offer classes designed to help students find funding sources and write grants. Others leave it to the students to realize that they are supposed to make a career via grants and then to figure out which agencies to apply to and how to craft grants. To identify possible grant opportunities, seek assistance from friends and colleagues and from the office of sponsored programs or its equivalent at your college or university. On-line services such as SUNY SPIN (State University of New York Sponsored Programs Information Network) offer a huge database of grant programs that is updated daily.

In writing a journal article, you fit the audience to the work by choosing an appropriate journal. Writing a grant proposal is like trying to publish an article in the wrong journal: The audience is both fixed but largely unknown, and the author must adapt the writing accordingly. One of the most effective ways to assist grant writing is to make successful proposals and comments available to others. Surprisingly often, even a student's own mentors and advisors do not do this. We encourage grant writers to share their successful proposals with colleagues and students. Most large departments have senior faculty with experience serving on grant review committees; they are often in the best position to explain what makes one grant worthier than another.

Strategies for Grant Proposals

Begin working on a proposal at least six months prior to submission. The proposal should show evidence of strong preparation and a project already under way; it should convey the impression that the project fits an established trajectory, has already yielded interesting results, and simply needs support to come to completion.

Cultivate relationships with program officers. Write or call to request formal guidelines and any available reports on recent grants. Interview the program secretary about recent trends in the program and new initiatives that might be under way. Consider contacting previous grant recipients to learn how the process worked for them. If possible, obtain copies of successful proposals.

After reviewing guidelines and reports in detail, contact the program officer by formal letter or E-mail, briefly outlining your proposed research and indicating that you will follow up with a telephone call. Resist the urge to call right away. Program officers usually prefer to deal with written proposals. Initiating a relationship through letter or E-mail indicates your serious commitment to the process and gives the officer some control over how you fit into his or her work day. When you do make voice contact, elaborate your research objectives and plans as specifically as possible and mention the importance of your project.

The review process measures your proposal against both agency priorities and widely accepted standards of scholarship. Hence, most review mechanisms include representatives of both the organization and the field of research. The program officer at a private foundation may have great latitude in determining the appropriateness and value of a given proposal. At the National Endowment for the Humanities, in contrast, program officers must adhere closely to the recommendations of outside peer reviewers. National Science Foundation program officers are free to depart from peer review recommendations but must provide written justification for doing so.

With the NEH, NIH, or NSF, it may be worthwhile to map the agency's internal organization so as to understand your audience and its priorities. For example, the STS effort at NSF includes two main programs: Studies in Science, Technology, and Society (SSTS) and Societal Dimensions of Engineering, Science, and Technology (SDEST). SSTS was formerly the program in history and philosophy of science and is still dominated by history proposals. SDEST proposals, by contrast, should be more problem-oriented, advancing understanding of some important social, ethical, or value issue regarding contemporary science and technology.

Since the early 1990s virtually all new funds at NSF have gone to foundation-wide "initiatives" that are built around specific classes of societal problems. The division that proposes and gets approval for a new initiative serves as its home and receives a permanent allocation to its base budget, which makes new initiatives especially attractive to existing programs. Two initiatives relevant to our interests include Human Capital and Quality Studies. Contact the NSF directly to learn about new developments, by E-mail at nsf.gov and by the World Wide Web at URL: http://www.nsf.gov.

The most popular funding award is the individual research grant, which can vary significantly. The SSTS program at NSF, for example, includes Scholars Awards, which provide release-time and summer support for tightly organized, well-developed projects that only need time to complete. A Professional Development Fellowship helps a researcher cross disciplinary boundaries by initiating training in another field. Awards for dissertation research usually involve less money but more relaxed criteria. Grants for conferences and workshops are the most difficult to get and may depend upon including a book as an outcome in your proposal. If you want to go after a lot of money, consider building a collaborative research project that includes two or three principal investigators as well as training for graduate and undergraduate students.

The scholarly review process for most programs is likely to include both "specialist" and "panel" reviews. A specialist review is an ad hoc report written by an individual chosen by the program officer; it includes both a narrative evaluation and an overall grade or score, e.g., Excellent, Very Good, Good, Fair, Poor. A review panel consists of eight to twelve participants who meet for a day or two to evaluate all the proposals in a given pool, which can range from fewer than fifty to more than 120. At least two or three panelists provide written reviews and lead the discussion. Panel members who don't write reviews of a given proposal are expected to have read enough to participate in discussions.

Panel procedures vary from program to program. One NSF panel, for example, first ranks each proposal as a "1" (clearly fundable); "2" (fundable in an ideal world with unlimited funds, but not in this world); "3" (revise and resubmit, thus making a 3 better than a 2); or "4" (not fundable under any circumstances). At the end of the meeting, each panelist divides all the 1s into top, middle, and bottom, and the program officer calculates a numerical score for each proposal. In this program, approximately 15 to 20 percent of submitted proposals receive funding.

One important lesson in all this is that readers must work very fast. Write your proposal in such a way that someone can read and understand it in fifteen minutes. Include section headings to indicate major transitions, and write clean, clear paragraphs.

Writing a proposal can be an intensely alienating experience, for it often involves describing what you want to do in ways and language that others might want to hear. Furthermore, the process is rarely a linear one, tending to come together in sections that must be not only internally coherent but also connected systematically to other sections. A good proposal is highly self-referential, clear, and easy to read. What follows is a brief account of the links between how a finished NSF proposal might look and how you might get there:

How It Looks	*What You Do*
Project Summary	Objectives
Objectives	Budget
Background/Significance	Methods and Procedures
Research Design/Hypotheses	Expected Results
Methods and Procedures	Schedule
Schedule	Background
Expected results	Appendices
Bibliography	Research Design/Hypotheses
Biographical sketch/CV	Bibliography
Budget	Biographical Sketch/CV
Appendices	Project Summary

Objectives

Begin by evaluating the types of research you plan to do and what the likely products will be. For example, do you plan to do an ethnographic study and write a book? Might the research include other methodological strategies, such as archival work or a questionnaire survey, and might the products include a series of articles, new course development, or annotated data made available on the World Wide Web? Each of these commitments has direct implications for every major section of your proposal.

Budget

Although the most difficult to conceptualize and write, the budget is the most important section to you, the PI, since the whole purpose of the proposal is to get enough money to do your work. Given the type of research project you plan to do, about how much money do you think you will need? What is the total time period you foresee for the project? Do you want to pay yourself a salary, whether for support as an individual scholar, released time during an academic year, or summer support? Do you want to pay for research assistance? How much traveling do you intend to do, both to conduct the research and to report its results at professional meetings? What other expenses might you incur, such as costs for transcribing tapes, copying, telephone, supplies, publication, etc.? Include significant salary support, especially if you are adding 50 or 60 percent for overhead charges to your college or university.

If you fear your requested budget will be insufficient but you cannot justify asking for more, consider seeking some return of "overhead" funds to the project. Find out what percentages will be allocated to your department, your college or division, and your university or college as a whole, and ask each to return, say, a quarter or a third of the overhead funds it would receive so that you could support a particular need that would otherwise go unfunded.

Methods and Procedures

Think of this task as a detailed history or narrative of the future. Specify in as much detail as possible exactly what you plan to do with the grant money. A well-written methods and procedures section reaches out to virtually every other section of the proposal. It is linked tightly to the budget, every item of which must be justified in terms of methods and procedures. A well-written section is also linked to project objectives and proposed results, for reviewers will ask if the specific plans fit the objectives and will lead to the expected results. In other words, this section holds everything else together.

Expected Results

A good presentation of expected results can reassure reviewers that giving you all this money really will lead to something significant, so consider carrying your history of the future beyond research tasks and into writing and dissemination. If, for example, you intend to produce a book, provide a detailed outline, taking care to demonstrate that the frame of the chapters fits the frame of the research.

Schedule

Can you really complete the project in the time allotted? If not, rewrite the methods and procedures and recalculate the budget. Provide a detailed schedule of the proposed research even if you are unsure how long various steps will take. Consider describing the schedule in terms of overlapping phases. A key here is to demonstrate that you understand the different steps you will have to take and how these relate to one another.

Background

A main purpose of this section is to show the significance of the problem addressed by the project. Your probable inclination is to begin with this section, but too much work here raises the possibility that the rest of the proposal is irrelevant or poorly related to the background. Reviewers might view the proposal as a research article with a request for money tacked on at the end. If you put off writing this section until you have mapped the project in some detail, it will be more deeply connected to the section on methods and procedures and, hence, to the body of the research.

Appendices

Limits on proposal length generally do not extend to appendices. Can an appendix do any work that would otherwise be done in the proposal, thus freeing space for other things? For example, literature reviews often consume many pages. Have you completed or published a paper that already does some of this work, which you can attach at the end and cite in the proposal? How about a paper that displays the interpretive frame you plan to use in the proposed research?

Research Design (Descriptive Hypotheses?)

This section prepares the reviewer to understand and assess the detailed discussion of methods and procedures that follows. A key goal is to derive, describe, and justify your approach to the research problem by reviewing other approaches that have been or could be used. You will need to demonstrate an accomplished level of conceptual sophistication with a minimum of jargon.

Bibliography

Reviewers tend to look for their own research traditions in a bibliography and may take a PI to task if the proposal does not acknowledge a full range of perspectives. The bibliography usually is not included in the limits on page length. Make sure, however, that it focuses on the proposed project and does not appear to have been downloaded from that massive database you have been accumulating for years.

Biographical Sketch/CV

Remember that reviewers are evaluating both the proposal and the capabilities of the PI. Many reviewers begin by reading the CV in order to understand the career trajectory in which the project is supposed to fit. Be sure to account for every year of study/research/work, showing how active you have been. You want readers to have confidence that the project will indeed yield the results it promises.

Project Summary

This section will be read more closely than anything else in the proposal. It should provide an overview of the entire project rather than simply a restatement of project objectives with an additional sentence or two. Therefore it should be written last. For harried reviewers, a good summary can be a crucial resource for pointing out the strengths of a project in the midst of a debate. Be sure it includes material from every major section, especially Methods and Procedures and Expected Results.

Resources for the Anthropology of Science, Technology, and Medicine

To help readers identify departments, institutions, professional associations, publishers, and others with an interest in the anthropology of science, technology, and medicine, we offer the following listings as places to begin your search.

Departments of Anthropology or Anthropology and Sociology (English-speaking) with an Interest in the Study of Science, Technology, and Medicine

Arizona State University, Australian National University, Brandeis University, Bucknell University, University of California (UC) Berkeley, UCLA, UC Santa Cruz, Central Michigan University, University of Connecticut, University of Florida, Grinnell College, University of Houston–Clear Lake, Johns Hopkins University, University of Illinois, University of Kansas, Lewis and Clark College, University of Manchester, University of Massachusetts, Massachusetts Institute of Technology (MIT), Montclair State College, New School for Social Research, New York University, Oakland University, University of Oregon, University of Pennsylvania, Princeton University, Rice University, San Jose State University, Smith College, University of South Carolina, Stanford University, Tel Aviv University, Trent University, University of Virginia, Wayne State University.

Interdisciplinary Programs in the United States

(Note: At the graduate level, STS usually stands for Science and Technology Studies; in undergraduate programs, STS usually means Science, Technology, and Society.)

Ball State (Center for Communication and Information Sciences); UC Berkeley (Anthropology; Medical Anthropology); UCLA (Center for Cultural Studies of Science, Technology, and Medicine); UC Santa Cruz (History of Consciousness; Anthropology); UC San Diego (STS Program; Communications); UC San Francisco (Social and Behavioral Sciences); Carnegie Mellon (Technology and Public Policy); Cornell (STS Department; Program on Technology and Work); Delaware (Center for Energy and Urban Policy Research); Evergreen College (Feminist Studies; Environmental Studies); Georgia Tech (History, Technology, and Society; Literature, Communication, and Cultural Studies); Illinois (Sociology); Illinois Institute of Technology (Humanities); Indiana (Sociology); Lehigh (STS); Michigan Tech (Program in STS); MIT (STS Program); New Jersey Institute of Technology (Center for Technology Studies); New School for Social Research (Program in Feminist Research); Ohio State (Comparative Studies); Pennsylvania (History and Sociology of Science and Technology); Penn State (STS Program); Rensselaer (STS Department); Stanford (Center for Biomedical Ethics); State University of New York–Utica Institute of Technology (Technology Policy Center); Texas–Austin (Department of Radio, TV, and Film); Virginia (Division of Humanities, School of Engineering); Virginia Tech (STS Program; Center for Science and Technology Studies).

Interdisciplinary Programs Outside the United States

Australia (Australian National; Sydney; Wollongong); Canada (McGill; Quebec–Montreal); Britain (Aston; Bath; Brunel; Edinburgh; Keele; Lancaster; Manchester; Open University; Sussex); France (Ecole des Mines); Germany (Berlin; Bielefeld); Netherlands (STS Ph.D. program, Amsterdam; Limburg; Groningen; Twente); Free University of Amsterdam; Technical University of Delft); New Zealand (Women's Studies, Otago); Norway (Bergen; Oslo; Trondheim); Sweden (Goteburg; Linkoping).

Professional Societies

American Anthropological Association (Interest Group in Anthropology of Science, Technology, and Computing; Society for the Anthropology of Work; Society for Applied Anthropology; Society for Cultural Anthropology; Society for Medical Anthropology); European Association of Studies of Science and Technology; History of Science Society; Philosophy of Science Association; Society for Literature and Science; Society for the History of Technology; International Society for the History, Philosophy, and Social Studies of Biology; Society for the Social Studies of Science.

Electronic Contacts

CASTAC-L Listserv for announcements for the community of anthropologists of science, technology, and computing. Send E-mail to listserv@mitvma.mit.edu with the message: Subscribe CASTAC-L your name.

Journals

American Anthropologist (flagship journal of the American Anthropological Association; shows increasing interest in science due to widespread interest in the status of anthropological knowledge; submissions must cross subfields); *American Ethnologist* (journal of the American Ethnological Society; good place to publish ethnographic case studies); *Anthropological Quarterly* (good place to break in with work that makes both topical and conceptual contributions); *Configurations* (relatively new journal of the Society for Literature in Science; excellent location for work in cultural studies; long-term role of ethnographic work not yet clear); *Cultural Anthropology* (journal of the Society for Cultural Anthropology; strong interest in all work exploring science, technology, and medicine as cultural sites of meaning production); *Culture, Medicine, and Psychiatry* (actively encourages humanities/social science and science researchers to write together); *differences* (feminist and cultural studies); *Feminist Studies* (interdisciplinary feminist journal); *Human Organization* (journal of the Society for Applied Anthropology); *ISIS* (journal of the History of Science Society; favors archive-based historical research with extensive citation); *Knowledge and Society* (annual journal generally focused on issues in the sociology of science and technology but open to other approaches as well); *Medical Anthropological Quarterly* (journal of the Society for Medical Anthropology); *Perspectives on Science: Historical, Philosophical, Social* (good place to publish pieces that engage questions in history and philosophy of science and/or sociology of science or scientific knowledge); *Science as Culture* (good place to publish critical accounts and attempts at formulating critical participation); *Science in Context* (interdisciplinary journal focusing on the social studies of science); *Science, Technology, and Human Values* (journal of 4S, Society for the Social Studies of Science); *Social Studies of Science* (heartland of constructivist science and technology studies; focuses on the sociology of scientific knowledge and technology but open to other perspectives; emphasis on analytic accounts with clear conceptual framework and data analysis); *Signs* (interdisciplinary feminist journal); *Social Science and Medicine*; *Technology in Society* (interdisciplinary journal that tends to be more interested in impacts of technology on society, less on the construction of particular technologies); *Technology Review* (MIT magazine focusing on policy issues concerning contemporary technologies in society; papers must be highly accessible with clear policy implications); *Technology Studies* (relatively new interdisciplinary journal exploring all areas of technological development and use).

Academic Presses
California; Cambridge; Chicago; Columbia; Duke; Edinburgh; Harvard; Indiana; Minnesota; MIT; Oxford; Princeton; Rutgers; SAR Press; Stanford; Temple; Wisconsin; Yale.

Commercial Presses
Beacon, Free Association Books (often with Routledge); Routledge; SAGE; St. Martin's; Verso.

REFERENCES

ACT UP/New York Women & Aids Book Group and Marion Banzhaf
1990   *Women, AIDS, and activism*. Boston: South End Press.

Ademuwagun, Z. A.
1979   *African therapeutic systems*. Los Angeles: Crossroads Press.

Alpers, Joseph
1994   Putting genes on a chip. *Science* 264:1400.

Althusser, Louis
1971   Ideology and ideological state apparatuses. In *Lenin and philosophy and other essays*, L. Althusser, ed. New York: Monthly Review Press.

Anderson, Christopher
1993   Researchers win decision on knockout mouse pricing. *Science* 260:23–24.

Andreasen, Nancy C.
1992   Neuroradiology and neuropsychiatry: A new alliance. *American Journal of Neuroradiology* 13:841–43.

Andreasen, Nancy C., ed.
1989   *Brain imaging: Applications in psychiatry*. Washington, DC: American Psychiatric Press.

Andrews, Edmund L.
1993   US resumes granting patents on genetically altered animals. *Los Angeles Times* 3 February:A1, C5.

Appadurai, Arjun
1991   Global ethnoscapes: Notes and queries for a transnational anthropology. In *Recapturing anthropology: Working in the present*, Richard G. Fox, ed. Santa Fe: School of American Research Press.

Aristotle
1941   Nicomachean ethics. In *The basic works of Aristotle*, Richard McKeon, ed. New York: Random House.

1973   *Introduction to Aristotle*, 2d ed. Edited by Richard McKeon. Chicago: University of Chicago Press.

Ashmore, Malcolm
1989   *The reflexive thesis*. Chicago: University of Chicago Press.

Barinaga, Marcia
1994   Knockout mice: Round two. *Science* 265:26–28.

Barnes, Barry
1977   *Interests and the growth of knowledge*. London: Routledge.

1981   On the "hows" and "whys" of cultural change. *Social Studies of Science* 11: 481–98.

Barnes, Barry, and R. G. A. Dolby
1970   The scientific ethos: A deviant viewpoint. *Archives of European Sociology* II: 3–25.

Barnes, Barry, and Donald MacKenzie
1979   On the role of interests in scientific change. In *On the margins of science*. Sociological Review Monograph no. 27, Roy Wallis, ed. Keele, UK: University of Keele.

Barnes, Barry, and Steven Shapin, eds.
1979   *Natural order*. Beverly Hills: Sage.

Barthes, Roland
1972   *Mythologies*. New York: Hill and Wang.

1975   *Roland Barthes par Roland Barthes*. Paris: Editions du Seuil.

1977   *Image, music, text*. New York: Hill and Wang.

1981   *Camera lucida: Reflections on photography*. New York: Hill and Wang.

1982   *Empire of signs*. New York: Hill and Wang.

Barzun, Jacques
1964   *Science: The glorious entertainment*. New York: Harper & Row.

Bayer, Ronald
1981   *Homosexuality and American psychiatry: The politics of diagnosis*. New York: Basic Books.

Beardsley, Tim
1994   Big-time biology. *Scientific American* 271(5):90–97.

Beauvoir, Simone de
1989   *The second sex*. 1949. Reprint. New York: Knopf.

Becker, William, and Michael Flower
1993   Science in the liberal arts curriculum. National Science Foundation grant proposal.

Begley, Sharon
1991   Thinking looks like this: PET scans show the brain recalling and cogitating. *Newsweek* 118(22):67(1).

Bereano, Philip
1995   Broad coalition challenges patents on life. Blue Mountain Declaration. Coalition press release, 6 June.

Berger, Bennett M.
1994   Taking arms. *Science* 264:985–89.

Bernal, J. D.
1939   *The social function of science*. London: Routledge.

Bhabha, Homi
1994   *The location of culture*. New York: Routledge.

Bijker, Wiebe
1993   Do not despair: There is life after constructivism. *Science, Technology, and Human Values* 18(1):113–38.

Bijker, Wiebe, Thomas Hughes, and Trevor Pinch, eds.
1987   *The social construction of technological systems*. Cambridge: MIT Press.

Black, Randall
1989   What's on your client's mind? PET scans of brain can let you look. *UCI Journal* Jan.–Feb.

Black, Rita Beck
1993   Psychosocial issues in reproductive genetic testing and pregnancy loss. *Fetal Diagnosis and Therapy* 8:164–73.

Bleier, Ruth
1986   *Feminist approaches to science*. The ATHENE series. New York: Pergamon Press.

Bloor, David
1982   Durkheim and Mauss revisited: Classification and the sociology of knowledge. *Studies in the History and Philosophy of Science* 13(4):267–97.

1991   *Knowledge and social imagery*. 2d ed. Chicago: University of Chicago Press.

Blumenberg, Hans
1983   *The legitimacy of the modern age*. Translated by Robert M. Wallace. Cambridge: MIT Press.

Boltanski, Luc
1991   *De la justification: Les economies de la grandeur*. Paris: Editions Gallimard.

Bordo, Susan
1990   Feminism, postmodernism and gender-skepticism. In *Feminism/postmodernism*, Linda Nicholson, ed. New York: Routledge.

Botelho, Antonio
1993   Cultural contagion: The spread of science in the Third World. *Science, Technology, and Human Values* 18(3):389–94.

Bourdieu, Pierre
1982   *LeHon sur la leHon*. Paris: Les Editions de Minuit.

1988   *Homo academicus*. Translated by Peter Collier. Cambridge: Polity Press in association with Basil Blackwell.

1993   *La misère du monde*. Paris: Editions du Seuil.

Bourdieu, Pierre, and Loic Wacquant
1992   *An introduction to reflexive sociology*. Chicago: University of Chicago Press.

Bowen, J. Ray
1988   The engineering student pipeline. *Engineering Education* 78(8):732–35.

Braverman, Harry
1975   *Labor and monopoly capital.* New York: Monthly Review Press.

Breggin, Peter Roger, and Ginger Ross Breggin
1994   *Talking back to Prozac: What doctors won't tell you about today's controversial drug.* New York: St. Martin's Press.

Brown, JoAnne
1986   Professional language: Words that succeed. *Radical History Review* 34:33–52.

Brown, Wendy
1995   *States of injury, power and freedom in late modernity.* Princeton: Princeton University Press.

Browner, Carole, and Nancy Anne Press
1995   The normalization of prenatal diagnostic screening. In *Conceiving the new world order: The global politics of reproduction,* Faye D. Ginsburg and Rayna Rapp, eds. Berkeley: University of California Press.

Bruner, Edward
1990   Tenure narratives. *Anthropology Newsletter* 48.

Bruno, Giuliana
1992   Spectatorial embodiments: Anatomies of the visible and the female bodyscape. *Camera Obscura* 28:239–62.

Bugos, Glenn E.
1992   Intellectual property protection in the American chicken-breeding industry. *Business History Review* 66:127–68.

1994   Making biodiversity public and private: Three eras of American biodiversity institutions. Unpublished manuscript in progress for Arnold Thackray, ed., *Private Science.*

Bugos, Glenn E., and Daniel J. Kevles
1992   Plants as intellectual property: American practice, law, and policy in world context. *Osiris* 7 (2nd series):75–104.

Bullard, Robert
1990   *Dumping in Dixie: Race, class, and environmental equity.* Boulder, CO: Westview Press.

Business–Higher Education Forum
1983   *America's competitive challenge: The need for a national response.* Washington, DC: Business–Higher Education Forum.

Bynum, Caroline Walker
1989   The female body and religious practice in the later Middle Ages. In *Fragments for a history of the human body,* part I, Ramona Naddaff, Michel Feher, and Nadia Tazi, eds. New York: Zone Books.

1991   *Fragmentation and redemption: Essays on gender and the human body in medieval religion.* New York: Zone Books.

Callon, Michel
1980   The state and technical innovation: A case study of the electrical vehicle in

France. *Research Policy* 9:358–76. (Reprinted in Bijker, Hughes, and Pinch 1987.)

1986　Some elements of a sociology of translation: Domestication of the scallops and the fishermen of St. Brieuc's Bay. In *Power, action and belief: A new sociology of knowledge?* Sociological Review Monograph no. 32, John Law, ed. Keele, UK: University of Keele.

Callon, Michel, and John Law
1982　On interests and their transformation: Enrollment and counterenrollment. *Social Studies of Science* 12:615–25.

Cambrosio, Alberto, and Peter Keating
1988　Going monoclonal: Art, science, and magic in the day-to-day use of hybridoma technology. *Social Problems* 35(3):244–60.

1992　Between fact and technique: The beginnings of hybridoma technology. *Journal of the History of Biology* 25(2):175–230.

Campbell, Neil A.
1993　*Biology.* 3d ed. Redwood City, CA: Benjamin Cummings.

Canguilhem, Georges
1989　*The normal and the pathological.* Translated by Carolyn R. Fawcett and Robert S. Cohen. New York: Zone Books.

Carrithers, Michael, Steven Collins, and Steven Lukes
1985　*The category of the person: Anthropology, philosophy, history.* Cambridge: Cambridge University Press.

Carson, Rachel
1987　*Silent spring.* 2d ed. Boston: Houghton Mifflin.

Cartwright, Lisa
1992　"Experiments of destruction": Cinematic inscriptions of physiology. *Representations* 40 (Fall):129–52.

1995　*Screening the body: Tracing medicine's visual culture.* Minneapolis: University of Minnesota Press.

Case, Sue-Ellen
1991　Tracking the vampire. *differences* 3(2):1–20.

Casper, Monica J.
1993　Deconstructing the fetus: Towards a sociology of fetal practices in science and medicine. Master's thesis, Dept. of Social and Behavioral Sciences, University of California, San Francisco.

Castoriadis, Cornelius
1984　The imaginary institution of society. In *The structural allegory: Reconstructive encounters with the new French thought,* John Fekete, ed. Minneapolis: University of Minnesota Press.

Chaiklin, Seth, and Jean Lave
1993　*Understanding practice: Perspectives on activity and context.* Cambridge: Cambridge University Press.

Cheng, John
1993　Asians, aliens, and science fiction in America, 1926–1945. Unpublished ms., University of California, Berkeley.

Chow, Rey
1993   *Writing diaspora.* Bloomington: Indiana University Press.

Christie, John R. R.
1993   A tragedy for cyborgs. *Configurations* 1:171–96.

Chubin, Daryl, and Sal Restivo
1983   The "mooting" of science studies: Research programmes and science policy. In *Science observed,* Karin Knorr-Cetina and Michael Mulkay, eds. London: Sage.

Claeson, Bjorn, E. Martin, W. Richardson, M. Schoch-Spana, and K.-S. Taussig
1996   Scientific literacy, what it is, why it's important, and why scientists think we don't have it: The case of immunology and the immune system. In *Naked science: Anthropological inquiry into boundaries, power, and knowledge,* Laura Nader, ed. New York: Routledge.

Clarke, Adele, and Joan Fujimura, eds.
1992   *The right tools for the job: At work in twentieth-century life sciences.* Princeton: Princeton University Press.

Clarke, Adele, and Teresa Montini
1993   The many faces of RU486: Tales of situated knowledges and technological contestations. *Science, Technology, and Human Values* 18(1):42–78.

Clifford, James, and George Marcus, eds.
1986   *Writing culture.* Berkeley: University of California Press.

Clynes, Manfred E., and Nathan S. Kline
1960   Cyborgs and space. *Astronautics* 26–27 (September):75–76.

Collier, Jane F., and Carol Delaney
1992   Reply to Cris Shore, virgin births and sterile debates. *Current Anthropology* 33(3):302–3.

Collins, Harry
1975   The seven sexes: A study in the sociology of a phenomenon or the replication of experiment in physics. *Sociology* 9:205–24.

1981   Understanding science. *Fundamenta Scientiae* 1:367–80.

1983   An empirical relativist programme in the sociology of scientific knowledge. In *Science observed,* Karin Knorr-Cetina and Michael Mulkay, eds. Beverly Hills: Sage.

1985   *Changing order: Replication and induction in scientific practice.* Beverly Hills: Sage.

1987   Expert-systems and the science of knowledge. In *The social construction of technological systems: New directions in the sociology and history of technology,* Wiebe Bijker, Thomas P. Hughes, and Trevor Pinch, eds. Cambridge: MIT Press.

1990   *Artificial experts: Social knowledge and the intelligent machine.* Cambridge: MIT Press.

1994   Scene from afar. *Social Studies of Science* 24(2):369–89.

Collins, Harry, and Trevor Pinch
1982   *Frames of meaning.* London: Routledge.

Collins, Randall, and Sal Restivo
1983   Robber barons and politicians in mathematics: A conflict model of science. *Canadian Journal of Sociology* 8(2):199–227.

Cone, Marla
1993   The mouse wars turn furious. *Los Angeles Times* 9 May: A1, A16–17.

Cooter, Roger
1984   *The cultural meaning of popular science: Phrenology and the organization of consent in nineteenth-century Britain.* Cambridge History of Medicine. Cambridge and New York: Cambridge University Press.

Corea, Gena
1985   *The mother machine: Reproductive technologies from artificial inseminations to artificial wombs.* New York: Harper and Row.

Cowan, Ruth Schwartz
1983   *More work for mother.* New York: Basic Books.

1992   Genetic technology and reproductive choice: An ethics for autonomy. In *The code of codes: Scientific and social issues in the human genome project*, Daniel J. Kevles and Leroy Hood, eds. Cambridge: Harvard University Press.

1993   Aspects of the history of prenatal testing. *Fetal Diagnosis and Therapy* 8:10–17.

Cozzens, Susan, and Thomas Gieryn, eds.
1990   *Theories of science in society.* Bloomington: University of Indiana Press.

Crary, Jonathan, and Sanford Kwinter
1992   *Incorporations.* New York: Zone.

Cruse, Julius M., and Robert E. Lewis, Jr.
1993   Transgenes today. *Transgene* 1(1):1–2.

Csordas, Thomas J.
1990   Embodiment as a paradigm for anthropology. *Ethos* 18(1):5–47.

Cuticchia, A. Jamie, Michael A. Chipperfield, Christopher J. Porter, William Kearns, and Peter L. Pearson
1993   Managing all those bytes: The human genome project. *Science* 262:47–49.

Davis, Jessica
1993   Reproductive genetic testing: What is the state of the art? *Fetal Diagnosis and Therapy* 8:28–37.

Davis-Floyd, Robbie
1992   *Birth as an American rite of passage.* Comparative studies of health systems and medical care, no. 35. Berkeley: University of California Press.

Dawkins, Richard
1982   *The extended phenotype: The gene as a unit of selection.* London: Oxford University Press.

de Lauretis, Teresa
1984   *Alice doesn't: Feminism, semiotics, cinema.* Bloomington: Indiana University Press.

de Lauretis, Teresa, and Stephen Heath, eds.
1980   *The cinematic apparatus.* New York: St. Martin's Press.

Deleuze, Gilles, and Felix Guattari
1987   *A thousand plateaus: Capitalism and schizophrenia.* Translated and foreword by Brian Massumi. Minneapolis: University of Minnesota Press.

Derrida, Jacques
1970   Structure, sign, and play in the discourse of the human sciences. In *The languages of criticism and the sciences of man,* Richard Macksey and Eugenio Donato, eds. Baltimore: Johns Hopkins University Press.

1976   *Of grammatology.* Translated by Gayatri Chakravorty Spivak. Baltimore: Johns Hopkins University Press.

Desjarlais, Robert R., ed.
1995   *World mental health: Problems and priorities in low-income countries.* New York: Oxford University Press.

Dewey, John
1916   Introduction. In *Essays in experimental logic.* Reprint, New York: Dover Books, 1953.

Dickson, David
1984   *The new politics of science.* New York: Pantheon.

Dietz, H. C., G. R. Cutting, R. E. Pyeritz, C. L. Maslen, L. Y. Sakai, G. M. Corson, E. G. Puffenberger, A. Hamosh, E. J. Nanthkumar, S. M. Curristin, G. Stetten, D. A. Meyers, and C. A. Francomano
1991   Defects in the fibrillin gene cause the Marfan syndrome. *Nature* 353:337–39.

DiMascio, A., M. M. Weissman, B. A. Prujoff, C. Neu, M. Zwilling, and G. L. Klerman
1979   Differential symptom reduction by drugs and psychotherapy in acute depression. *Archives in General Psychiatry* 36:1450–56.

Diprose, Rosalyn, and Robyn Ferrell, eds.
1991   *Cartographies: Poststructuralism and the mapping of bodies and spaces.* North Sydney, Australia: Allen & Unwin.

Douglas, Mary
1966   *Purity and danger: An analysis of the concepts of pollution and taboo.* London: Ark.

Douglas, Mary, and Aaron Wildavsky
1982   *Risk and culture: An essay on the selection of technological and environmental dangers.* Berkeley: University of California Press.

Downey, Gary Lee
1992   Agency in CAD/CAM technology. *Anthropology Today* 8(5) October:6–10.

Downey, Gary Lee, Arthur Donovan, and Timothy J. Elliott
1989   The invisible engineer: How engineering ceased to be a problem in science and technology studies. *Knowledge and Society* 8:189–216.

Downey, Gary Lee, and Juan C. Lucena
1994   Engineering studies. In *Handbook of science, technology, and society,* Sheila Jasanoff, Gerry Markle, and James Petersen, eds. Newbury Park, CA: Sage.

Downey, Gary Lee, Joseph Dumit, and Sarah Williams
1995   Granting membership to the cyborg image. In *The cyborg handbook,* C. H. Gray, H. Figueroa-Sarriera, and S. Mentor, eds. New York: Routledge.

Downey, Gary Lee, and Juan D. Rogers
1995   On the politics of theorizing in a postmodern academy. *American Anthropologist* 97(2):269–81.

Dreyfus, Hubert, and Paul Rabinow
1993   Is a science of meaningful action possible? In *Bourdieu: Critical perspectives*, Craig Calhoun, ed. Chicago: University of Chicago Press.

Dubinskas, Frank, ed.
1988   *Making time: Ethnographic studies of high-technology organizations.* Philadelphia: Temple University Press.

Dubois, W. E. B.
1906   *Health and physique of the Negro American.* Atlanta: Atlanta University Press.

Duden, Barbara
1991   *The woman beneath the skin.* Cambridge: Harvard University Press.

1993   *Disembodying women: Perspectives on pregnancy and the unborn.* Cambridge: Harvard University Press.

Dumit, Joseph
1995   Mindful images: PET scans and personhood in biomedical America. Ph.D. diss., History of Consciousness, University of California–Santa Cruz.

During, Simon
1989   Waiting for the post: Some relations between modernity, colonization, and writing. *Ariel* 20:31–61.

Durkheim, Emile
1965   *The elementary forms of religious life.* New York: The Free Press.

Dyson, Freeman J.
1988   *Infinite in all directions: Gifford lectures given at Aberdeen, Scotland, April–November 1985.* New York: Harper & Row.

Ehrenreich, Barbara, and Deirdre English
1973   *Witches, midwives, and nurses: A history of women healers.* 2d ed. Old Westbury, NY: Feminist Press.

Eisenhart, M.
1994   Women scientists and the norm of gender neutrality at work. *Journal of Women and Minorities in Science and Engineering* 1(3):193–207.

Eisenhart, M., E. Finkel, and S. Marion
1996   Creating conditions for scientific literacy. *American Educational Research Journal* 33(2):261–95.

Elfenbein, Debra
1995   *Living with Prozac and other selective serotonin reuptake inhibitors (SSRIs): Personal accounts of life on antidepressants.* San Francisco: Harper San Francisco.

Elkin, I., M. B. Parloff, S. W. Hadley, and J. H. Autry.
1985   The NIMH treatment of depression collaborative research program: Background and research plan. *Archives in General Psychiatry* 42:305–16.

Ellul, Jacques
1964   *The technology society.* New York: Knopf.

Escobar, Arturo
1994   Welcome to cyberia: Notes on the anthropology of cyberculture. *Current Anthropology* 35(3):211–31.

1995   *Encountering development*. Princeton: Princeton University Press.

Evans-Pritchard, E. E.
1976   *Witchcraft, oracles, and magic among the Azande*. London: Clarendon Press.

1978   *The Nuer: A description of the modes of livelihood and political institutions of a Nilotic people*. New York: Oxford University Press.

Fabian, Johannes
1983   *Time and the other: How anthropology makes its object*. New York: Columbia University Press .

Fabrega, Horacio
1989   An ethnomedical perspective of Anglo-American psychiatry. *American Journal of Psychiatry* 146: 588–96.

Farmer, Paul
1992   *AIDS and accusation: Haiti and the geography of blame*. Berkeley: University of California Press.

Farquhar, Judith
1992   Eating Chinese medicine. Paper presented at the American Anthropological Association annual meetings, 16–20 November, San Francisco.

Fausto-Sterling, Ann
1985   *Myths of gender: Biological theories about men and women*. New York: Basic Books.

Feldman, H. W., and P. Biernacki
1988   The ethnography of needle sharing among intravenous drug users and implications for public policies and intervention strategies. *NIDA Research Monograph* 80(34):28–39.

Feynman, Richard
1985   *"Surely you're joking, Mr. Feynman!": Adventures of a curious character as told to Ralph Leighton and edited by Edward Hutchings*. New York: W. W. Norton.

1988   *What do YOU care what other people think?: Further adventures of a curious character as told to Ralph Leighton*. New York: W. W. Norton.

Firth, H. V.
1991   Severe limb abnormalities after chorionic villus sampling. *Lancet* 337: 762–63.

Fischer, Michael M. J.
1995   Eye(I)ing the sciences and their signifiers (language, trope, autobiographers): InterViewing for a cultural studies of science and technology. In *Technoscientific imaginaries: Conversations, profiles, and memoirs*, George E. Marcus, ed. Chicago: University of Chicago Press.

Fischer, Roland
1990   Why the mind is not in the head but in the society's connectionist network. *Diogenes* 151:1–28.

Fish, Stanley Eugene
1980   *Is there a text in this class? The authority of interpretive communities.* Cambridge: Harvard University Press.

Fisher, Lawrence M.
1994   Gene project is already big business. *San Jose Mercury News* 30:9A.

Fleck, James
1994   Knowing engineers: Response. *Social Studies of Science* 24:105–13.

Fleck, Ludwik
1979   *Genesis and development of a scientific fact.* 1935. Reprint. Chicago: University of Chicago Press.

Flower, Michael
1994   A native speaks for himself: Reflections on technoscientific literacy. Paper read at the American Anthropological Association annual meetings, Washington, DC, 30 November 3 December.

n.d.   Technoscientific liberty. Unpublished paper. University Honors Program, Portland State University, Portland, OR.

Forsythe, Diana
1992   Using ethnography to build a working system: Rethinking basic design assumptions. In *Proceedings of the 16th Syposium on Computer Applications in Medical Care (SCAMC92)*, M. E. Frisse, ed. New York: McGraw-Hill.

1993a   Engineering knowledge: The construction of knowledge in artificial intelligence. *Social Studies of Science* 23:445–77.

1993b   The construction of "work" in artificial intelligence. *Science, Technology and Human Values* 18(4):460–79.

1994   STS (re)constructs anthropology: Reply. *Social Studies of Science* 24:113–23.

Foucault, Michel
1970   *The order of things: An archaeology of the human sciences.* New York: Vintage.

1979   *Discipline and punish: The birth of the prison.* New York: Vintage.

1983   On the genealogy of ethics. In *Michel Foucault: Beyond structuralism and hermeneutics*, Paul Rabinow and Hubert Dreyfus, eds. Chicago: University of Chicago Press.

1984   What is enlightenment? In *The Foucault reader*, Paul Rabinow, ed. New York: Pantheon Books.

1990   *The uses of pleasure: The history of sexuality*, vol. II. New York: Pantheon Books.

Foucault, Michel, Luther H. Martin, Huck Gutman, and Patrick H. Hutton, eds.
1988   *Technologies of the self: A seminar with Michel Foucault.* Amherst: University of Massachusetts Press.

Fox, Michael
1983   Genetic engineering: Human and environmental concerns. Unpublished briefing paper.

1992   *Superpigs and wondercorn: The brave new world of biotechnology and where it all may lead.* New York: Lyons & Burford.

Fox, Richard Gabriel
1991   *Recapturing anthropology: Working in the present.* School of American Research Advanced Seminar Series. Santa Fe, NM: School of American Research Press.

Frackowiak, Richard S. J.
1986   An introduction to positron tomography and its application to clinical investigation. In *New brain imaging techniques and psychopharmacology,* Michael R. Trimble, ed. Oxford: Oxford University Press.

Franklin, Sarah
1993   Life itself. Paper presented at the Center for Cultural Values, Lancaster University, Lancaster, England, 9 June.

Franklin, Sarah, Celia Lury, and Jackie Stacey, eds.
n.d.   *Second nature: Body techniques in global culture.* Forthcoming.

Friedkin, William, director
1987   *Rampage.* Miramax. Motion picture.

Fujimura, Joan
1988   The molecular biological bandwagon in cancer research: Where social worlds meet. *Social Problems* 35:261–83.

1992   Crafting science: Standardized packages, boundary objects, and "Translation." In *Science as practice and culture,* Andrew Pickering, ed. Chicago: University of Chicago Press.

Fuller, Steve
1993   *Philosophy of science and its discontents.* 2d ed. New York: Guilford Press.

Fyfe, Gordon, and John Law
1988   On the invisibility of the visual: Editors' introduction. In *Picturing power: Visual depiction and social relations,* Gordon Fyfe and John Law, eds., pp. 1–14. London: Routledge.

Gaines, Atwood D.
1985   The once- and the twice-born: Self and practice among psychiatrists and Christian psychiatrists. In *Physicians of western medicine: Anthropological approaches to theory and practice,* Robert A. Hahn and Atwood D. Gaines, eds. Dordrecht, Netherlands: D. Reidel Publishing Company.

Galison, Peter L.
1997   The trading zone: Coordinating action and belief. In *Image and logic: A material culture of microphysics,* Peter L. Galison, ed. Chicago: University of Chicago Press.

Gasner, Cheryll
1993   The joining circles. *American Journal of Medical Genetics* 47:136–42.

Geertz, Clifford
1973   *The interpretation of cultures: Selected essays.* New York: Basic Books.

Gelder, Ken
1994   *Reading the vampire.* New York: Routledge.

Geller, Jay
1992   Blood sin: Syphilis and the construction of Jewish identity. *Faultline* 1:21–48.

Gene Exchange, The
1991   *The Gene Exchange* 2(4) (December).

Gerson, E., and S. Star
1986   Analyzing due process in the workplace. *ACM Transactions on Office Information Systems* 4:257–70.

Gieryn, Thomas F.
1983   Boundary-work and the demarcation of science from non-science: Strains and interests in professional ideologies of scientists. *American Sociological Review* 48:781–95.

Gilbert, Scott F.
1994   *Developmental biology.* 4th ed. Sunderland, MA: Sinauer.

Gilman, Sander L.
1988   *Disease and representation: Images of illness from madness to AIDS.* Ithaca: Cornell University Press.

Gleick, James
1992   *Genius: The life and science of Richard Feynman,* New York: Pantheon Books.

Gomez, Jewelle
1991   *The Gilda stories.* Ithaca: Firebrand Books.

Good, Byron J.
1994   *Medicine, rationality and experience: an anthropological perspective.* Lewis Henry Morgan Lecture series. Cambridge: Cambridge University Press.

Good, Mary-Jo Del Vecchio
n.d.   *The quest for competence in American medicine.* Berkeley: University of California Press. Forthcoming.

Good, Mary-Jo, P. Brodwin, Byron J. Good, and Arthur Kleinman, eds.
1992   *Pain as human experience: An anthropological perspective.* Berkeley: University of California Press.

Good, Mary-Jo Del Vecchio, Tseunetsugu Munakata, Yasuki Kobayashi, Cheryl Mattingly, and Byron J. Good
1994   Oncology and narrative time. *Social Science and Medicine* 38(6):855–62.

Gould, Stephen Jay
1981   *The mismeasure of man.* New York: W. W. Norton.

Gray, Chris Hables
1991   Computers as weapons and metaphors: The US military 1940–90 and postmodern war. Ph.D diss., History of Consciousness, University of California–Santa Cruz.

Greenblatt, Stephen Jay
1980   *Renaissance self-fashioning: From More to Shakespeare.* Chicago: University of Chicago Press.

Griesemer, J. R., and W. C. Wimsatt
1989   Picturing Weismannism: A case study of conceptual evolution. In *What the philosophy of biology is: Essays dedicated to David Hull,* Michael Ruse, ed. Dordrecht, Netherlands: Kluwer Academic Publishers.

Gross, Paul R., and Norman Levitt
1994   *Higher superstition: The academic left and its quarrels with science.* Baltimore: Johns Hopkins University Press.

Grosveld, F., and G. Kollias, eds.
1992   *Transgenic animals.* San Diego: Academic Press.

Grund, Jean-Paul C., Charles D. Kaplan, and Nico F. P. Adriaana
1991   Featuring high-risk behavior: Needle sharing in the Netherlands: An ethno-
        graphic analysis. *American Journal of Public Health* 81(12):1602–7.

Gusterson, Hugh
1992   The rituals of science: Comments on Abir-Am (with response). *Social Episte-
        mology* 6(4):373–87.

Hacker, Sally
1989   *Pleasure, power, and technology: Some tales of gender, engineering, and the coop-
        erative workplace.* Boston: Unwin Hyman.

Hakken, David
1993   Computing and social change: New technology and workplace transforma-
        tion, 1980–1990. *Annual Review of Anthropology* 22:107–12.

Hakken, David, with Barbara Andrews
1993   *Computing myths, class realities: An ethnography of technology and working
        people in Sheffield, England.* Boulder, CO: Westview Press.

Hamilton, Joan O'C.
1994   Biotech: An industry crowded with players faces an ugly reckoning. *Business
        Week* (26 September):84–90.

Handler, Richard
1995   *Schneider on Schneider: The conversion of the Jews and other anthropological sto-
        ries.* Durham, NC: Duke University Press.

Haraway, Donna J.
1985   Manifesto for cyborgs: Science, technology, and socialist feminism in the late
        twentieth century. *Socialist Review* 80:65–108.

1988   Situated knowledges: The science question in feminism as a site of discourse
        on the privilege of partial perspective. *Feminist Studies* 14(3):575–99.

1989   *Primate visions: Gender, race, and nature in the world of modern science.* New
        York: Routledge.

1991a   *Simians, cyborgs, and women: The reinvention of nature.* New York: Routledge.

1991b   A cyborg manifesto: Science, technology, and socialist-feminism in the late
        twentieth century. In *Simians, cyborgs, and women: The reinvention of nature.*
        New York: Routledge.

1992   The promises of monsters: A regenerative politics for inappropriate/d others.
        In *Cultural studies*, L. Grossberg, C. Nelson, P. Treichler, eds. New York: Rout-
        ledge.

1994   A game of cat's cradle: Science studies, feminist theory, cultural studies.
        Configurations 2(1):59–72.

1997   *Modest_Witness@Second_Millennium.FemaleMan©_Meets_OncoMouse™*: Fem-
        inism and technoscience. New York: Routledge.

Harding, Sandra
1991   *Whose science? Whose knowledge? Thinking from women's lives.* Ithaca: Cornell
        University Press.

1992   After the neutrality ideal: Science, politics, and "strong objectivity." *Social Research* 59(3):567–87.

Harding, Sandra, ed.
1993   *The "racial" economy of science: Toward a democratic future.* Bloomington: Indiana University Press.

Harding, Susan
1987   Convicted by the holy spirit: The rhetoric of fundamental Baptist conversion. *American Ethnologist* 14(1):167–81.

Harrington, Anne, ed.
1992   *So human a brain: Knowledge and values in the neurosciences.* Boston: Dibner Institute/Birkhauser.

Harrison, Faye Venetia, ed.
1991   *Decolonizing anthropology: Moving further toward an anthropology for liberation.* Washington, DC: Association of Black Anthropologists, American Anthropological Association.

Hartouni, Valerie
1991   Containing women: Reproductive discourse in the 1980s. In *Technoculture*, Constance Penley and Andrew Ross, eds. Minneapolis: University of Minnesota Press.

1992   Fetal exposures: Abortion politics and the optics of allusion. *Camera Obscura* 29 (Fall):130–49.

Harvey, David
1989   *The condition of postmodernity: An enquiry into the origins of cultural change.* Oxford: Basil Blackwell.

Hawking, Stephen
1988   *A brief history of time: From the big bang to black holes.* New York: Bantam Books.

Heath, Deborah
1992   The view from the bench: Prosthetics and performance in molecular biotechnology. Paper presented at the American Anthropological Association meetings, San Francisco, December.

Heath, Deborah, and Paul Rabinow, eds.
1993   Bio-politics: The anthropology of the new genetics and immunology. *Medical Anthropology Quarterly* 17(1) (special issue on Culture, Medicine and Psychiatry). Dordrecht, Netherlands: Kluwer Academic Publishers.

Heim, Pat
1993   *Hardball for women.* New York: Plume.

Hess, David
1991   *Spirits and scientists.* University Park: Pennsylvania State University Press.

1992   The new ethnography and the anthropology of science and technology. In *Knowledge and society.* Vol. 9, *The anthropology of science and technology*, David Hess and Linda L. Layne, eds. Greenwich, CT: JAI Press.

1993   *Science in the new age.* Madison: University of Wisconsin Press.

1995   *Science and technology in a multicultural world.* New York: Columbia University Press.

1997a   *Can bacteria cause cancer?* New York: New York University Press.

1997b   *Science studies: An advanced introduction.* New York: New York University Press.

Hess, David J., and Linda Layne, eds.
1992   *Knowledge and society.* Vol. 9, *The anthropology of science and technology.* Greenwich, CT: JAI Press.

Hessen, Boris
1971   *The social and economic roots of Newton's* Principia. 1931. Reprint. New York: Howard Fertig.

Hightower, Jim
1973   *Hard tomatoes, hard times: A report of the Agribusiness Accountability Project on the failure of America's land grant college complex.* Cambridge, MA: Schenkman.

Hilgartner, Stephen
1994   Genome informatics: New communication regimes for biology. Paper read at the Society for the Social Studies of Science meetings, New Orleans, October.

Hobbelink, Henk
1991   *Biotechnology and the future of world agriculture.* London: Zed Books.

Hood, Leroy, Irving Weissman, William B. Wood, and John H. Wilson
1984   *Immunology.* 2d ed. Menlo Park, CA: Benjamin/Cummings Publishing Co.

Hoover, Cary, Alta Campbell, and Patrick J. Spain, eds.
1991   *Hoover's handbook of American business, 1992.* Austin: Reference Press.

Horkheimer, Max, and Theodor Adorno
1972   *Dialectic of enlightenment.* New York: Continuum.

Hounshell, David, and John Kenly Smith, Jr.
1988   *Science and corporate strategy: Du Pont R&D, 1902–1980.* New York: Cambridge University Press.

Hull, David
1988   *Science as a process.* Chicago: University of Chicago Press.

Irigaray, Luce
1985   *Speculum of the other woman.* Translated by Gillian C. Gill. Ithaca: Cornell University Press.

Jackson, Jean
1994   Chronic pain and the tension between the body as subject and object. In *Embodiment and experience: The existential ground of culture and self,* Thomas J. Csordas, ed. Cambridge: Cambridge University Press.

Jameson, Frederic
1984   Postmodernism, or the cultural logic of late capitalism. *New Left Review* 146:53–92.

Jasanoff, Sheila, Gerry Markle, James Peterson, and Trevor Pinch, eds.
1994   *Handbook of science and technology.* Beverly Hills: Sage.

Johnson, Jory
1988   Martha Schwartz's "splice garden": A warning to a brave new world. *Landscape Architecture* (July/August).

Jones, James H.
1981   *Bad blood: The Tuskegee syphilis experiment.* New York: The Free Press.

Jordanova, Ludmilla
1989   *Sexual visions: Images of gender in science and medicine between the eighteenth and twentieth centuries.* Madison: University of Wisconsin Press.

Juma, Calestous
1989   *The gene hunters: Biotechnology and the scramble for seeds.* Princeton: Princeton University Press.

Kane, S.
1990   AIDS, addiction and condom use: Sources of sexual risk for heterosexual women. *Journal of Sex Research* 27(3):427–44.

1991   HIV, heroin and heterosexual relations. *Social Science and Medicine* 32(9): 1037–50.

Kane, S., and Theresa Mason
1992   "IV drug users" and "sex partners": The limits of epidemiological categories and the ethnography of risk. In *The time of AIDS: Social analysis, theory, and method*, Gilbert Herdt and Shirley Lindenbaum, eds. Newbury Park, CA: Sage.

Kay, Lily E.
1993   *The molecular vision of life: Caltech, the Rockefeller Foundation, and the rise of the new biology.* New York: Oxford University Press.

Keller, Evelyn Fox
1983   *A feeling for the organism: The life and work of Barbara McClintock.* San Francisco: W. H. Freeman.

1985   *Reflections on gender and science.* New Haven: Yale University Press.

1992   *Secrets of life, secrets of death: Essays on language, gender and science.* New York: Routledge.

Kevles, Daniel J.
1977   *The physicists: The history of a scientific community in modern America.* New York: Knopf.

1985   *In the name of eugenics: Genetics and the uses of human heredity.* New York: Knopf.

Kittay, Jeffrey, and Judith Shulevitz
1993   Ivory bowers. In *Lingua franca* 4:48–53.

Kittler, Friedrich A.
1985   *Discourse networks 1800/1900.* Translated by Michael Metteer and Chris Cullens. Stanford: Stanford University Press.

Kleinman, Arthur
1986   *Social origins of distress and disease: Depression, neurasthenia, and pain in modern China.* New Haven: Yale University Press.

Kleinman, Arthur, and Byron Good
1985   *Culture and depression: Studies in the anthropology and cross-cultural psychiatry of affect and disorder.* Berkeley: University of California Press.

Knorr-Cetina, Karin
1981   *The manufacture of knowledge: An essay on the constructivist and contextual nature of science.* New York: Pergamon Press.

1992   The couch, the cathedral, and the laboratory: On the relationship between experiment and laboratory in science. In *Science as practice and culture,* Andrew Pickering, ed. Chicago: University of Chicago Press.

Knorr-Cetina, Karin, and Michael Mulkay
1983   Introduction: Emerging principles in social studies of science. In *Science observed,* Karin Knorr-Cetina and Michael Mulkay, eds. Beverly Hills: Sage.

Köhler, Georges, and Cesar Milstein
1975   Continuous cultures of fused cells secreting antibody of predefined specificity. *Nature* 256(5517):497–97 (August 7).

Kolker, Aliza, and Meredith Burke
1994   *Prenatal testing: A sociological perspective.* Westport, CT: Bergin & Garvey (Greenwood Publishing Group).

Konner, Melvin
1983   *The tangled wing: Biological constraints on the human spirit.* New York: Harper & Row.

1991   Human nature and culture: Biology and the residue of uniqueness. In *The boundaries of humanity: Humans, animals, machines,* James J. Sheehan and Morton Sosna, eds. Berkeley: University of California Press.

Kosman, L. A.
1980   Being properly affected: Virtues and feelings in Aristotle's *Ethics.* In *Essays on Aristotle's Ethics,* Amelie Rorty, ed. Berkeley: University of California Press.

Kramer, Peter D.
1993   *Listening to Prozac: A psychiatrist explores antidepressant drugs and the remaking of the self.* New York: Viking.

Krimsky, Sheldon
1991   *Biotechnics and society: The rise of industrial genetics.* New York: Praeger.

Kuhar, Michael J.
1989   Perspectives. In *Brain imaging: Techniques and applications,* N. A. Sharif and M. E. Lewis, eds. New York: Halsted Press.

Kuhn, Thomas
1970   *The structure of scientific revolutions.* 2d ed. Chicago: University of Chicago Press.

Laing, R. D.
1967   *The politics of experience* and *The bird of paradise.* Harmondsworth, UK: Penguin.

Latour, Bruno
1983   Give me a laboratory and I will raise the world. In *Science Observed,* Karin Knorr-Cetina and Michael Mulkay, eds. London: Sage.

1986   Visualization and cognition: Thinking with eyes and hands. *Knowledge and Society: Studies in the Sociology of Culture Past and Present* 6:1–40.

1987   *Science in action: How to follow scientists and engineers through society.* Cambridge: Harvard University Press.

1988   *The pasteurization of France*. Cambridge: Harvard University Press.

1990a   Postmodern? No, simply Amodern! Steps toward an anthropology of science. *Studies in the History and Philosophy of Science* 21(1):145–71.

1990b   Drawing things together. In *Representation in scientific practice*, Michael Lynch and Steve Woolgar, eds. Cambridge: MIT Press.

1992   Where are the missing masses? The sociology of a few mundane artifacts. In *Shaping technology/building society*, Wiebe Bijker and John Law, eds. Cambridge: MIT Press.

1993   *We have never been modern*. Translated by Catherine Porter. Cambridge: Harvard University Press.

Latour, Bruno, and Steve Woolgar
1986   *Laboratory life: The social construction of scientific facts*. 2d ed. Princeton: Princeton University Press.

Laudan, Larry
1990   *Science and relativism*. Chicago: University of Chicago Press.

Laughlin, Kim
1995   Rehabilitating science, imagining "Bhopal." In *Late editions 2: Technoscientific imaginaries*, George Marcus, ed. Chicago: University of Chicago Press.

Lave, Jean
1990   Views of the classroom: Implications for math and science learning research. In *Toward a scientific practice of science education*, M. Gardner, ed. Hillsdale, NJ: L. Erlbaum Association.

Lave, Jean, and Etienne Wenger
1991   *Situated learning: Legitimate peripheral participation*. Cambridge: Cambridge University Press.

Lawrence, Errol
1982   In the abundance of water the fool is thirsty: Sociology and black "pathology." In *The empire strikes back: Race and racism in 70s Britain*, Centre for Contemporary Cultural Studies, ed. London: Hutchinson.

Layne, Linda
1992   Of fetuses and angels: Fragmentation and integration in narratives of pregnancy loss. In *Knowledge and society*. Vol. 9, *The Anthropology of Science and Technology*, David Hess and Linda Layne, eds. Greenwich, CT: JAI Press.

LeFevre, Michael L., Raymond P. Bain, Bernard Ewigman, Frederick D. Frigoletto, James P. Crane, and Donald McNellis
1993   A randomized trial of prenatal ultrasonographic screening: Impact on maternal management and outcome. *American Journal of Obstetrics and Gynecology* 169(3):483–89.

LeVay, S. A.
1993   *The sexual brain*. Cambridge: MIT Press.

Lévi-Strauss, Claude
1963   *Totemism*. Translated by Rodney Needham. Boston: Beacon Press.

1966   *The savage mind*. Chicago: University of Chicago Press.

1985   *The view from afar.* Translated by Joachim Neugroschel and Phoebe Hoss. New York: Basic Books.

Lewontin, Richard, Steven Rose, and Leon Kamin
1984   *Biology, ideology, and human nature.* New York: Pantheon.

Long, Elizabeth
1985   *The American dream and the popular novel.* Boston: Routledge & Kegan Paul.

Longino, Helen E.
1990   *Science as social knowledge: Values and objectivity in scientific knowledge.* Princeton: Princeton University Press.

1994   In search of feminist epistemologies. *Monist* 77(4):472–85.

Lovell, A. M., and N. Scheper-Hughes
1986   Deinstitutionalization and psychiatric expertise: Reflections on dangerousness, deviancy, and madness (Italy and the United States). *International Journal of Law and Psychiatry* 9(3):361–81.

Lucena, Juan C.
1996   Making scientists and engineers for America: From sputnik to global competition. Ph.D. diss., Virginia Polytechnic Institute and State University.

Lynch, Michael
1985   *Art and artifact in the laboratory.* London: Routledge.

1992   Extending Wittgenstein: The pivotal move from epistemology to the sociology of science. In *Science as practice and culture,* Andrew Pickering, ed. Chicago: University of Chicago Press.

Lynch, Michael, and Steve Woolgar
1985   Introduction: Sociological orientations to representational practice in science. In *Representation in scientific practice,* Michael Lynch and Steve Woolgar, eds. Cambridge: MIT Press.

Lynch, Michael, and Steve Woolgar, eds.
1990   *Representation in scientific practice.* Cambridge: MIT Press.

Lyotard, Jean Francois
1984   *The postmodern condition: A report on knowledge.* Vol. 10, *Theory and history of literature.* Minneapolis: University of Minnesota Press.

MacIntyre, Alisdair
1981   *After virtue.* South Bend, IN: University of Notre Dame Press.

MacKenzie, Donald
1981   Interests, positivism, and history. *Social Studies of Science* 11:498–501.

1983   *Statistics in Britain.* Edinburgh: University of Edinburgh Press.

1984   Reply to Yearley. *Studies in the History and Philosophy of Science* 15(3):251–59.

Mackenzie, Michael, Peter Keating, and Alberto Cambrosio
1990   Patents and free scientific information in biotechnology: Making monoclonal antibodies proprietary. *Science, Technology and Human Values* 15(1): 65–83.

Mannheim, Karl
1966   *Ideology and utopia: An introduction to the sociology of knowledge.* 1936. Reprint, New York: Harcourt, Brace, and World.

Manning, Peter, and Horatio Fabrega
1973   The experience of self and body: Health and illness in the Chiapas highlands. In *Phenomenological sociology*, George Psathas, ed. New York: Wiley.

Marcus, George
1995   *Technoscientific imaginaries: Conversations, profiles, and memoirs.* Chicago: University of Chicago Press.

Marcus, George, and Michael Fischer
1986   *Anthropology as cultural critique.* Chicago: University of Chicago Press.

Margulis, Lynn, and Dorion Sagan
1995   *What is life?* New York: Simon and Schuster.

Markus, Gyorgy
1987   Why is there no hermeneutics of the natural sciences? Some preliminary theses. *Science in Context* 1.

Martin, Brian
1993   The critique of science becomes academic. *Science, Technology, and Human Values* 18(2):247–59.

n.d.   Sticking a needle into science: vaccines and the origins of AIDS. *Social Studies of Science* 26(2). In press.

Martin, Brian, C. M. A. Baker, C. Manwell, and C. Pugh
1986   *Intellectual suppression.* North Ryde, New South Wales: Angus and Robertson.

Martin, Brian, Evelleen Richards, and Pam Scott
1991   Who's a captive? Who's a victim? Response to Collins's method talk. *Science, Technology, and Human Values* 16(2):252–55.

Martin, Emily
1987   *The woman in the body: A cultural analysis of reproduction.* Boston: Beacon Press.

1992   The end of the body? *American Ethnologist* 19(1):121–40.

1994   *Flexible bodies: Tracking immunity in American culture from the days of polio to the age of AIDS.* Boston: Beacon Press.

Maslen, C. L., G. M. Corson, B. K. Maddox, R. W. Glanville, and L. Y. Sakai
1991   Partial sequencing of a candidate gene for the Marfan syndrome. *Nature* 353:334–37.

Mattingly, Cheryl
1994   The concept of therapeutic "emplotment." *Social Science and Medicine* 38(6): 811–22.

Mauss, Marcel
1985   A category of the human mind: The notion of person; the notion of self. In *The Category of the person*, Michael Carrithers, Steven Collins, and Steven Lukes, eds. Translated by W. D. Halls (1938). Cambridge: Cambridge University Press.

Mayberg, Helen S.
1992   Functional brain scans as evidence in criminal court: An argument for caution. *The Journal of Nuclear Medicine* 33(6):18N–19N, 25N.

McDonough, Peggy
1990   Congenital disability and medical research: The development of amniocentesis. *Women and Health* 16(3/4):137–53.

McLoughlin, David
1994   Immigration out of control. In *North and South*, pp. 44–55.

Merchant, Carolyn
1980   *The death of nature: Women, ecology, and the scientific revolution*. San Francisco: Harper & Row.

Merleau-Ponty, Maurice
1945   *Phenomenologie de la perception*. Paris: Gallimard.

1964   *Signs*. Translated by Richard C. McCleary. Evanston: Northwestern University Press.

Merton, Robert
1957   Priorities in scientific discovery: A chapter in the sociology of science. *American Sociological Review* 22(6):635–59.

1973   *The sociology of science*. Chicago: University of Chicago Press.

Mestrovic, Stjepan G., and John A. Cook
1986   The dangerous standard: What is it and how is it used? *International Journal of Law and Psychiatry* 8:443–69.

Michael, Mike, and Lynda Birke
1994   Accounting for animal experiments: Identity and disreputable others. *Science, Technology, and Human Values* 19(2):189–204.

Milunsky, Aubrey
1993   Commercialization of clinical genetic laboratory services: In whose best interest? *Obstetrics and Gynecology* 81(4):627–29.

Mitchell, Lisa
1993a   Making babies: Routine ultrasound imaging and the cultural construction of the fetus in Montreal, Canada. Ph.D. diss., Case Western Reserve University.

1993b   "Showing her the baby": Clinical interpretations of ultrasound fetal images for women. Unpublished ms.

1994   The routinization of the other: Ultrasound, women, and the fetus. In *Misconceptions: The social construction of choice and the new reproductive and genetic technologies*, Gwynne Basen, Margrit Eichler, and Abby Lippman, eds. Hull, Quebec: Voyageur Press.

Monahan, John
1988   Risk assessment of violence among the mentally disordered: Generating useful knowledge. *International Journal of Law and Psychiatry* 11:249–57.

Monahan, John, and Saleem A. Shah
1989   Dangerousness and commitment of the mentally disordered in the United States. *Schizophrenia Bulletin* 15(4):541–53.

Moore, Henrietta
1988   *Feminism and anthropology*. Minneapolis: University of Minnesota Press.

Morris, William, ed.
1979   *The American Heritage dictionary of the English language*. Boston: Houghton Mifflin Co.

Morrow, John F., Stanley N. Cohen, Annie C. Y. Chang, Herbert Boyer, Howard M. Goodman, and Robert Helling
1974   Replication and transcription of eucaryotic DNA in Escherichia coli. *Proceedings of the National Academy of Sciences, USA* 71:1743–47.

Moskowitz, Milton, Michael Katz, and Robert Levering
1980   *Everyone's business, an almanac: The irreverent guide to corporate America.* New York: Harper & Row.

Mulkay, Michael
1976   Norms and ideology in science. *Social Science Information* 15:637–56.

Mulkay, Michael, Jonathan Potter, and Steven Yearley
1983   Why an analysis of scientific discourse is needed. In *Science observed*, Karin Knorr-Cetina and Michael Mulkay, eds. Beverly Hills: Sage.

Mulvey, E. P., and C. W. Lidz
1984   Clinical considerations in the prediction of dangerousness in mental patients. *Clinical Psychology Review* 4:379–401.

Mumford, Lewis
1964   *Technics and civilization.* 1934. Reprint. New York: Harcourt, Brace, and World.

NIH (National Institutes of Health)
1979   Antenatal diagnosis: Report of a consensus development conference. Sponsored by the National Institute of Child Health and Human Development, 5–7 March. NIH publication no. 79–1973. Bethesda, MD: National Institutes of Health.

1984   Consensus development conference statement. *Diagnostic Ultrasound Imaging in Pregnancy* 5(1).

NSB (National Science Board)
1988   *The role of the National Science Foundation in economic competitiveness.* Washington, DC: US National Science Foundation.

1993   *Science and engineering indicators.* Washington, DC: GPO.

Nagle, Thomas
1986   *The view from nowhere.* Oxford: Oxford University Press.

National Academy of Engineering
1986   *A technology agenda to meet the competitive challenge.* Washington, DC: National Academy of Sciences Press.

National Academy of Sciences
1965   *Basic research and national goals.* A report to the Committee on Science and Astronautics, US House of Representatives. Washington, DC: GPO.

National Science Foundation
1990   The future supply of natural scientists and engineers. In *The state of academic science and engineering*, pp. 189–232. Washington, DC: US National Science Foundation.

Natowicz, Marvin
1994   Commercialization of genetic services: Preliminary results from a survey. Unpublished ms.

Nelkin, Dorothy, ed.
1992   *Controversy: Politics of technical decisions.* 3ʳᵈ ed. Sage Focus editions, vol. 8. Newbury Park, CA: Sage.

Nespor, Jan
1994   *Knowledge in motion: Space, time, and curriculum in undergraduate physics and management.* London: Falmer.

Nilsson, Lennart
1977   *A child is born.* New York: Dell.

Noble, David F.
1977   *America by design: Science, technology, and the rise of corporate capitalism.* New York: Oxford University Press.

1984   *Forces of production.* New York: Knopf.

1992   *A world without women: The Christian clerical culture of western science.* New York: Knopf.

Norden, Michael J.
1995   *Beyond Prozac: Brain-toxic lifestyles, natural antidotes & new generation antidepressants.* New York: HarperCollins.

Nussbaum, Martha
1993   Non-relative values: An Aristotelian view. In *The quality of life,* Martha Nussbaum and Amartya Sen, eds. Oxford: Clarendon Paperbacks.

Nyce, J. M.
1996   Belief, community, and technology on the internet. *Practicing Anthropology* 18(2):33–35.

Nyce, J. M., and Jonas Loewgren
1995   Toward foundational analysis in human-computer interaction. In *The social and interactional dimensions of human-computer interfaces,* P. J. Thomas, ed. Cambridge: Cambridge University Press.

OTA (Office of Technology Assessment)
1989   *New developments in biotechnology: Patenting life.* Washington, DC: GPO.

1992   *The biology of mental disorders.* New Developments in Neuroscience series. Washington, DC: US Congress.

Oakley, Ann
1984   *The captured womb.* Oxford: Basil Blackwell.

1993   A history lesson: Ultrasound in obstetrics. In *Essays on women, medicine and health,* Ann Oakley, ed. Edinburgh: Edinburgh University Press.

O'Neill, Patrick Geoffrey
1984   *Essential kanji: 2,000 basic Japanese characters systematically arranged for learning and reference.* New York: Weatherhill.

Ong, Aihwa
1987   *Spirits of resistance and capitalist discipline: Factory women in Malaysia.* SUNY Series in the Anthropology of Work. Albany: State University of New York Press.

1991   The gender and labor politics of postmodernity. *Annual Review of Anthropology* 20:279–309.

Orr, Julian E.
1990   *Talking about machines: An ethnography of a modern job*. Ithaca: ILR Press.

Ortner, Sherry
1973   On key symbols. *American Anthropologist* 75:1338–46.

1984   Theory in anthropology since the sixties. *Comparative Studies in Society and History* 26(1).

Osaka University of Foreign Studies
1969   *The first step to kanji, part one*. Osaka: Osaka University of Foreign Studies.

Oudshoorn, Nellie
1994   *Beyond the natural body*. London: Routledge.

Pardes, Herbert, and Harold Alan Pincus
1985   Neuroscience and psychiatry: An overview. In *The integration of neuroscience and psychiatry*, Herbert Pardes and Harold Alan Pincus, eds. Washington, DC: American Psychiatric Press.

Pechesky, Rosalind
1987   Fetal images: The power of visual cultures in the politics of reproduction. *Feminist Studies* 13(2):206–46.

Penley, Constance, and Andrew Ross, eds.
1991   *Technoculture*. Minneapolis: University of Minnesota Press.

Pfaffenberger, Bryan
1988   The social meaning of the personal computer; Or, why the personal computer revolution was no revolution. *Anthropological Quarterly* 61(1):39–47.

1990   *Democratizing information: Online databases and the rise of end-user searching*. Boston: G. K. Hall.

1992   Social anthropology of technology. *Annual Review of Anthropology* 21:491–516.

Pickering, Andrew, ed.
1992   *Science as practice and culture*. Chicago: University of Chicago Press.

Piercy, Marge
1976   *Woman on the edge of time*. New York: Fawcett Crest.

Pollock, Nathan L.
1990   Accounting for predictions of dangerous. *International Journal of Law and Psychiatry* 13:207–15.

Post, Robert M., and James C. Ballenger, eds.
1984   *Neurobiology of mood disorders*. Baltimore: Williams & Wilkins.

Potter, Elizabeth
n.d.   Making gender/making science: Gender ideology and Boyle's experimental philosophy. In *Making a difference: Feminist critiques in the natural sciences*, Bonnie Spanier, ed. Bloomington: Indiana Univerity Press. Forthcoming.

Prakash, Gyan
1992   Science "gone native" in colonial India. In *Representations* 40:153–78.

President's Commission on Industrial Competitiveness
1985   *Global competition: The new reality*. Washington, DC: GPO.

Press, Nancy Anne, and Carole Browner
1993   "Collective Fictions": Similarities in reasons for accepting maternal serum alpha-fetoprotein screening among women of diverse ethnic and social class backgrounds. *Fetal Diagnosis and Therapy* 8(supplement 1):97–106.

Pyeritz, Reed, and Julia Conant
1989   *The Marfan syndrome.* Port Washington, NY: National Marfan Foundation.

Rabinow, Paul
1986   Representations are social facts. In *Writing culture: The poetics and politics of ethnography*, James Clifford and George Marcus, eds. Berkeley: University of California Press.

1992   Artificiality and enlightenment: From sociobiology to biosociality. In *Incoporations*, Jonathan Crary and Sanford Kwinter, eds. New York: Zone Books (distributed by MIT Press).

Radway, Janice A.
1984   *Reading the romance: women, patriarchy, and popular literature.* Chapel Hill: University of North Carolina Press.

Rapp, Rayna
1990   Constructing amniocentesis: Maternal and medical discourses. In *Uncertain terms: Negotiating gender in American culture*, Faye Ginsburg and Anna Lowenhaupt Tsing, eds. Boston: Beacon.

1993   Accounting for amniocentesis. In *Knowledge, power, and practice: The anthropology of medicine and everyday life*, Shirley Lindenbaum and Margaret Lock, eds. Berkeley: University of California Press.

1998   *Moral pioneers: Fetuses, families and amniocentesis.* New York: Routledge.

Reder, Karen
n.d.   Making mice: Clarence Little, the Jackson Laboratory, and the standardization of mus faeroensis for research. Ph.D. diss., Department of the History and Philosophy of Science, Indiana University, in progress.

Reeve, C. D. E.
1992   *Practices of reason: Aristotle's* Nicomachean Ethics. Oxford: Clarendon Press.

Reich, Robert
1983   *The next American frontier.* Baltimore: Johns Hopkins University Press.

Restivo, Sal
1983   The myth of the Kuhnian revolution. In *Sociological theory 1983*, Randall Collins, ed. San Francisco: Jossey-Bass.

1988   Modern science as a social problem. *Social Problems* 35(3):206–25.

Restivo, Sal, and Julia Loughlin
1987   Critical sociology of science and scientific validity. *Knowledge: Creation, Diffusion, Utilization* 8(3):486–508.

Rhodes, Lorna Amarasingham
1991   *Emptying beds: The work of an emergency psychiatric unit.* Comparative studies of health systems and medical care, vol. 27. Berkeley: University of California Press.

Rifkin, Jeremy
1984a *Algeny: A new word, a new world.* New York: Penguin Books.

1984b Letter to William Gartland. *Federal Register* 49(September 30):37016.

Rojas-Burke, J.
1993 PET scans advance as tool in insanity defense. *Journal of Nuclear Medicine* 34(1):N13–14.

Romanucci-Ross, Lola, Daniel E. Moerman, and Laurence R. Tancredi, eds.
1991 *The anthropology of medicine: From culture to method.* New York: Bergin & Garvey.

Rose, Hilary
1983 Hand, brain and heart: Toward a feminist epistemology for the natural sciences. *Signs* 9(1):73–96.

1994 *Love, power and knowledge: Toward a feminist transformation of the sciences.* Bloomington: Indiana University Press

Rose, Hilary, and Steven Rose, eds.
1976a *The political economy of science: Ideology of/in the natural sciences.* London: Macmillan.

1976b *The radicalisation of science: Ideology of/in the natural sciences.* London: Macmillan.

Rothman, Barbara Katz
1986 *The tentative pregnancy: Prenatal diagnosis and the future of motherhood.* New York: W. W. Norton.

1989 *Recreating motherhood: Ideology and technology in a patriarchal society.* New York: W. W. Norton.

Rouse, Joseph
1993 What are cultural studies of scientific knowledge? *Configurations* 1(1):1–22.

1996 *Engaging science: How to understand its practices philosophically.* Ithaca: Cornell University Press.

1997 New philosophies of science in North America—twenty years later. *Zeitschrift für allgemeine Wissenschaftstheorie* 28. In press.

Ruby, Jay, ed.
1982 *A crack in the mirror.* Philadelphia: University of Pennsylvania.

Russ, Joanna
1975 *The female man.* New York: Bantam Books.

Safire, William
1993 Only the factoids. *New York Times Magazine,* 5 December, p. 32.

Sahlins, Marshall
1976a *Culture and practical reason.* Chicago: University of Chicago Press.

1976b *The use and abuse of biology: An anthropological critique of sociobiology.* Ann Arbor: University of Michigan Press.

Said, Edward W.
1989 Representing the colonized: Anthropology's interlocutors. *Critical Inquiry* 15(2):205–25.

Sakai, Lynn Y., Douglas R. Keene, and Eva Engvall
1986   Fibrillin, a new 350-kD glycoprotein, is a component of extracellular micro-
        fibrils. *Journal of Cell Biology* 103(6):2499–2509.

Saunders, Barry Ferguson
1989   Inspection and decryption: Edgar Poe's "The Gold Bug" and the diagnostic
        gaze. M.A. thesis, University of North Carolina, Chapel Hill.

Scheper-Hughes, Nancy, and Margaret Lock
1993   The mindful body: A prolegomenon to future work in medical anthropology.
        Medical Anthropology Quarterly 1:1–36.

Schiebinger, Londa L.
1989   *The mind has no sex? Women in the origins of modern science.* Cambridge: Har-
        vard University Press.

Schrage, Michael
1993   Biomedical researchers scurry to make genetically altered mice. *San Jose
        Mercury News* 8 February:3D8.

1994   NIH boss seeks to squeeze more science out of stagnant budget. *San Jose
        Mercury News* 16 May:3D.

Schumacher, E. F.
1973   *Small is beautiful.* Harmondsworth, UK: Penguin.

Schutz, Alfred
1971   *The stranger: An essay in social psychology.* The Hague: Martinus Nijhoff.

Schwartz, Douglas
1992   President's message: Celebrating twenty-five years of creative scholarship. In
        Ideas in anthropology: 1992 annual report of the School of American Research.
        Santa Fe: School of American Research.

ScienceScope
1994a   Egypt to build science city in desert. *Science* 263:1551.

1994b   A Maryland motif for biotech ventures. *Science* 264:1071.

Sclove, Richard
1995   *Democracy and technology.* New York: Guilford Press.

Scott, Pam, Evelleen Richards, and Brian Martin
1990   Captives of controversy: The myth of the neutral social researcher in con-
        temporary scientific controversies. *Science, Technology, and Human Values*
        15(4):474–94.

Seymour, E., and N. Hewitt
1994   *Talking about leaving: Factors contributing to the high attrition rates among sci-
        ence, mathematics, and engineering undergraduate majors.* Boulder, CO: Bureau
        of Sociological Research.

Shapin, Steven
1979a   Homo phrenologicus: Anthropological perspectives on an historical problem.
        In *Natural order: Historical studies of scientific culture,* Barry Barnes and Steven
        Shapin, eds. Beverly Hills: Sage.

1979b   The politics of observation: Cerebral anatomy and social interests in the
        Edinburgh phrenology disputes. In *On the margins of science: The social con-
        struction of rejected knowledge,* Roy Wallis, ed. Keele, UK: University of Keele.

Shapin, Steven, and Simon Schaffer
1985   *Leviathan and the air pump: Hobbes, Boyle, and the experimental life.* Princeton: Princeton University Press.

Sheehan, James J., and Morton Sosna
1991   *The boundaries of humanity: Humans, animals, machines.* Berkeley: University of California Press.

Shiva, Vandana
1993   *Monocultures of the mind: Perspectives on biodiversity and biotechnology.* London: Zed Books.

Shriners Hospitals for Crippled Children
1992   *Annual report, Medical Research Programs.* Tampa: SHCC.

Sibthorpe, Beverly
1992   The social construction of sexual relationships as a determinant of HIV risk perception and condom use among injection drug users. *Medical Anthropology Quarterly* 6(3):255–70.

Skomal, Susan
1994   Lessons for the field—ethics in fieldwork. In *Anthropology Newsletter* 35:1, 4.

Smith, Brian Cantwell
1994   Coming apart at the seams: The role of computation in a successor metaphysics. Paper read at Intersection of the Real and the Virtual conference, Stanford University, June.

Spanier, Bonnie
1991   Gender and ideology in science: A study of molecular biology. *NWSA Journal* 3(2):167–98.

Spivak, Gayatri Chakravorty, and Sarah Harasym
1990   *The post-colonial critic: Interviews, strategies, dialogues.* New York: Routledge.

Stabile, Carol
1992   Shooting the mother: Fetal photography and the politics of disappearance. *Camera Obscura* 28:179–206.

Stacey, Judith
1988   Can there be a feminist ethnography? In *Women's Studies International Forum* 11:21–27.

1990   *Brave new families: Stories of domestic upheaval in late twentieth century America.* New York: Basic Books.

1994   Imagining feminist ethnography: A response to Elizabeth Wheatley. In *Women's Studies International Forum* 17:417–19.

Stafford, Barbara Maria
1991   *Body criticism: Imaging the unseen in enlightenment art and medicine.* Cambridge: MIT Press.

Star, Susan Leigh
1989   *Regions of the mind: Brain research and the quest for scientific certainty.* Stanford: Stanford University Press.

1991   Power, technology and the phenomenology of conventions: On being allergic to onions. In *A sociology of monsters: Power, technology and the modern world.* Sociological Review Monograph no. 38, John Law, ed. Oxford: Basil Blackwell.

1992   The skin, the skull, and the self: Toward a sociology of the brain. In *So human a brain: Knowledge and values in the neurosciences*, Anne Harrington, ed. Boston: Birkhauser.

1994   Misplaced concretism and concrete situations: Feminism, method and information technology. In *Gender-Nature-Culture*, working paper, Odense University, Denmark.

Statistics New Zealand
1994   *New Zealand official yearbook*. Auckland: Statistics New Zealand.

Stepan, Nancy Leys
1986   Race and gender: The role of analogy in science. *Isis* 77:261–77.

Stepan, Nancy Leys, and Sander L. Gilman
1993   Appropriating the idioms of science: The rejection of scientific racism. In *"Racial" economy of science*, Sandra Harding, ed. Bloomington: Indiana University Press.

Stipp, David
1992   The insanity defense in violent-crime cases gets high-tech help. In *Wall Street Journal* 4 March:A11.

Stone, Allucquere Roseanne
1992   Virtual systems. In *Incorporations*, Jonathan Crary and Sanford Kwinter, eds. New York: Zone.

Strathern, Marilyn
1988   *The gender of the gift: Problems with women and problems with society in Melanesia*. Studies in Melanesian Anthropology, vol. 6. Berkeley: University of California Press.

1992   *After nature: English kinship in the late twentieth century*. The Lewis Henry Morgan Lectures series, 1989. Cambridge: Cambridge University Press.

Stratmeyer, Melvin E., and Christopher L. Christman
1983   Biological effects of ultrasound. *Woman and Health* 8:65–81.

Suchman, Lucy
1987   *Plans and situated actions: The problem of human-machine communication*. Cambridge: Cambridge University Press.

1992   Technologies of accountability: On lizards and airplanes. In *Technology in working order: Studies of work, interaction, and technology*, G. Button, ed. London: Routledge.

1994   Working relations of technology production and use. *Computer Supported Cooperative Work* 2:21–39.

Szasz, Thomas Stephen
1970   *The manufacture of madness: A comparative study of the Inquisition and the mental health movement*. New York: Harper & Row.

Tapper, Marion
1993   *Ressentiment and power, some reflections on feminist practice*. London: Routledge.

Task Force on Science Policy, Committee on Science and Technology, US House of Representatives
1985   *Scientists and engineers: Supply and demand.* Vol. 9, Science Policy Study. Washington, DC: GPO.

Taussig, Michael
1992   *The nervous system.* New York: Routledge.

1993   *Mimesis and alterity: A particular history of the senses.* New York: Routledge.

Taylor, Janelle
1992   The public fetus and the family car: From abortion politics to a Volvo advertisement. *Public Culture* 4(2):67–80.

1993   Envisioning kinship: Fetal imagery and relatedness. Presentation for an invited symposium, "Reproducing Reproduction," American Anthropological Association annual meetings, Washington, DC.

1994   Agents of change: Women, midwives, doctors, and the routinization of ultrasound. Paper presented at the Society for the Social Studies of Science annual meetings, New Orleans, October.

1996   Image of contradiction: Obstetrical ultrasound in American culture. In *Reproducing reproduction*, Sarah Franklin and Helena Ragone, eds. College Park: University of Pennsylvania Press.

Teitelman, Robert
1994   *Profits of science: The American marriage of business and technology.* New York: Basic Books.

Terry, Jennifer
1989   The body invaded: Medical surveillance of women as reproducers. *Socialist Review* 19(July–Sept.):13–31.

Tonso, K. L.
1996a   The impact of cultural norms on women. *Journal of Engineering Education* 85(3):217–25.

1996b   Student learning and gender issues. *Journal of Engineering Education* 85(2): 143–50.

Toumey, Christopher
1994   *God's own scientists: Creationists in a secular world.* New Brunswick: Rutgers University Press.

Traweek, Sharon
1988   *Beamtimes and lifetimes: The world of high energy physicists.* Cambridge: Harvard University Press.

1992   Border crossings: Narrative strategies in science studies and among physicists in Tsukuba Science City, Japan. In *Science as practice and culture*, Andrew Pickering, ed. Chicago: University of Chicago Press.

1993   An introduction to cultural, gender, and social studies of sciences and technologies. *Culture, Medicine and Psychiatry* (special issue: *Biopolitics: The anthropology of the new genetics and immunology)* 17:3–25.

1995a   Bachigai in Ibaraki: Tsukuba Science City, Japan. In *Technoscientific Imaginaries* (Late Editions, vol. II), George Marcus, ed. Chicago: University of Chicago Press.

1995b   Bodies of evidence: Law and order, sexy machines, and the erotics of field-work among physicists. In *Choreographing History*, Susan Foster, ed. Bloomington: Indiana University Press.

1996a   Kokusaika, gaiatsu, and bachigai. In *Naked science: Anthropological inquiry into boundaries, power, and knowledge*, Laura Nader, ed. New York: Routledge.

1996b   Unity, dyads, triads, quads, and complexity: Cultural choreographies of science. In *Social Text* (special issue, Stanley Aronowitz and Andrew Ross, eds.) 46–47(spring–summer):129–39.

1996c   When Eliza Doolittle studies 'enry 'iggins. In *Technoscience, power, and cyberculture: Implications and strategies*, Stanley Aronowitz, Barbara Marinsons, Michael Menser, and Jennifer Rich, eds. New York: Routledge.

n.d.   Warning signs: Acting on images. In *Revisioning women, health, and healing: Feminist, cultural, and technoscience perspectives*, Adele Clarke and Virginia Olesen, eds. New York: Routledge. Forthcoming.

Treichler, Paula A.
1987   AIDS, homophobia and biomedical discourse: An epidemic of signification. *Cultural Studies* 1(3):263–305.

1991   How to have theory in an epidemic: The evolution of AIDS treatment activism. In *Technoculture: Cultural politics*, vol. 3, Constance Penley and Andrew Ross, eds. Minneapolis: University of Minnesota Press.

Treichler, Paula A., and Lisa Cartwright
1992   Introduction: Imaging technologies, inscribing science. *Camera Obscura* 28:5–20.

Tsing, Anna Lowenhaupt
1993   *In the realm of the diamond queen: Marginality in an out-of-the-way place*. Princeton: Princeton University Press.

Turkle, Sherry
1984   *The second self: Computers and the human spirit*. New York: Simon and Schuster.

Turner, Victor
1983   Body, brain, and culture. *Zygon* 18(3):221–45.

US Congress Joint Committee on Atomic Energy
1957   *Development of scientific, engineering, and other professional manpower*. Washington, DC: GPO.

US Congress Joint Economic Committee
1987   *Investment in research and development*. Washington, DC: GPO.

Ulmer, Gregory
1985   *Applied grammatology*. Baltimore: Johns Hopkins University Press.

Villeneuve, Claude, Catherine Larouche, Abby Lippman, and Myriam Marrache
1988   Psychological aspects of ultrasound imaging during pregnancy. *Canadian Journal of Psychiatry* 33(August):530–35.

Visvanathan, Shiv
1991   Mrs. Bruntland's disenchanted cosmos. *Alternatives* 16(3):377–84.

Vlahof, David, Alvaro Munoz, James C. Anthony, Sylvia Cohn, David D. Celentano, and Kenrad E. Nelson
1990   Association of drug injection patterns with antibody to human immuno-deficiency virus type 1 among intravenous drug users in Baltimore, Maryland. *American Journal of Epidemiology* 132(5):847–56.

Vlahof, David, J. C. Anthony, A. Munoz, J. Margolick, K. E. Nelson, D. D. Celentano, L. Solomon, and B. Frank Polk
1991   The ALIVE study, a longitudinal study of HIV-1 infection in intravenous drug users: Description of methods and characteristics of participants. In *Longitudinal studies of HIV infection in intravenous drug users: Methodological issues in natural history research.* Research monograph no. 109, Peter Hartsock and Sander G. Genser, eds. Rockville, MD: US Department of Health and Human Services.

Wagner, Henry N., Jr.
1986   Images of the brain: Past as prologue. *Journal of Nuclear Medicine* 27(12): 1929–37.

Wagner, Henry N., Jr., and Linda E. Ketchum
1989   *Living with radiation: The risk, the promise.* Baltimore: Johns Hopkins University Press.

Wajcman, Judith, ed.
1991   *Feminism confronts technology.* University Park: Pennsylvania State University Press.

Waldby, Catherine
1996   *AIDS and the body politic: Biomedicine and sexual difference.* London: Routledge.

Weber, Max
1946   Religious rejections of the world. In *From Max Weber*, H. Gerth and C. Wright Mills, eds. New York: Oxford University Press.

Webster, Andrew
1991   *Science, technology, and society.* New Brunswick: Rutgers University Press.

White, Tobin Frye
1996   How to solve a physics problem: Negotiation of knowledge and identity in introductory university physics. M.S. thesis, Virginia Polytechnic Institute and State University.

Williams, Sarah
1987   An "archae-logy" of Turkana beads. In *The archaeology of contextual meanings*, Ian Hodder, ed. Cambridge: Cambridge University Press.

Winner, Langdon
1986   *The whale and the reactor.* Chicago: University of Chicago Press.

1993   Upon opening the black box and finding it empty: Social constructivism and the philosophy of technology. *Science, Technology, and Human Values* 18(3): 362–78.

Wolf, Ron
1994   Biotech boom. *San Jose Mercury News* 14 March:1D, 9D.

Wood, William P.
1985   *Rampage.* New York: St. Martin's Press.

Woodman, William F., Mack C. Shelly II, and Brian J. Reichel
1989   *Biotechnology and the research enterprise: A guide to the literature.* Ames: Iowa State University Press.

Woolgar, Steve
1981a   Critique and criticism: Two readings of ethnomethodology. *Social Studies of Science* 11:504–14.

1981b   Interests and explanation in the social study of science. *Social Studies of Science* 11: 365–94.

1982   Laboratory studies: A comment on the state of the art. *Social Studies of Science* 12:481–98.

1983   Irony in the social study of science. In *Science observed*, Karin Knorr-Cetina and Michael Mulkay, eds. Beverly Hills: Sage.

1988a   Reflexivity is the ethnographer in the text. In *Knowledge and reflexivity*, Steve Woolgar, ed. Beverly Hills: Sage.

1988b   *Science: The very idea.* London: Tavistock.

1991a   The turn to technology in social studies of science. *Science, Technology, and Human Values* 16(1):20–50.

1991b   What is "anthropological" about the anthropology of science? *Current Anthropology* 32(1):79–81.

1992   Some remarks about positionism: A reply to Collins and Yearley. In *Science as practice and culture*, Andrew Pickering, ed. Chicago: University of Chicago Press.

Woolgar, Steve, and Malcolm Ashmore
1988   Introduction to the reflexive project. In *Knowledge and reflexivity*, Steve Woolgar, ed. Beverly Hills: Sage.

Worthington, Richard
1993   Theme section: Science and technology as a global system. *Science, Technology, and Human Values* 18(2):176–230.

Wright, Susan
1986   Recombinant DNA technology and its social transformation. *Osiris* 2 (second series):303–60.

1994   *Molecular politics: Developing American and British regulatory policy for genetic engineering, 1972–1982.* Chicago and London: University of Chicago Press.

Wurtzel, Elizabeth
1994   *Prozac nation: Young and depressed in America.* Boston: Houghton Mifflin.

Yanagisako, Sylvia, and Carol Delaney, eds.
1995   *Naturalizing power: Essays in feminist cultural analysis.* New York: Routledge.

Yates, Frances
1974   *The art of memory.* Chicago: University of Chicago Press.

Yearley, Stephen
1982   The relationship between epistemological and sociological cognitive interests: Some ambiguities underlying the use of interest theory in the study of scientific knowledge. *Studies in the History and Philosophy of Science* 13(4): 353–88.

Young, Robert

1972   The anthropology of science. *New Humanist* 88(3):102–5.

1977   Science is social relations. *Radical Science Journal* 5:65–129.

Yoxen, Edward

1981   Life as a productive force: Capitalizing the science and technology of molecular biology. In *Science, technology and the labour process*, L. Levidow and B. Young, eds. London: CSE Books.

1984   *The gene business.* New York: Harper & Row.

1989   Seeing with sound: A study of the development of medical images. In *The social construction of technological systems*, Wiebe E. Bijker, Thomas P. Hughes, and Trevor Pinch, eds. Cambridge: MIT Press.

Zuckerman, Harriet

1989   The other Merton thesis. *Science in Context* 3(1):239–67.

INDEX

Participants in the School of American Research advanced seminar "Cyborg Anthropology." Santa Fe, October 1993. From left: Joseph Dumit, Sarah Williams, David Hess, Deborah Heath, Sharon Traweek, Gary Downey, Donna Haraway, Rayna Rapp, Emily Martin, Paul Rabinow.